The Anterior Pituitary

ULTRASTRUCTURE IN BIOLOGICAL SYSTEMS

Edited by

Albert J. Dalton

National Cancer Institute
National Institutes of Health
Bethesda, Maryland

Françoise Haguenau

Laboratoire de Médicine Expérimentale
Collège de France
Paris, France

VOLUME 1 / Tumors Induced by Viruses: Ultrastructure Studies, 1962

VOLUME 2 / Ultrastructure of the Kidney, 1967

VOLUME 3 / The Nucleus, 1968

VOLUME 4 / The Membranes, 1968

VOLUME 5 / Ultrastructure of Animal Viruses and Bacteriophages:
An Atlas, 1973

VOLUME 6 / C. E. Challice and S. Virágh, eds. Ultrastructure of the
Mammalian Heart, 1973

VOLUME 7 / A. Tixier-Vidal and Marilyn G. Farquhar, eds.
The Anterior Pituitary, 1975

The Anterior Pituitary

EDITED BY

A. TIXIER-VIDAL

Laboratoire de Biologie Moleculaire
Collège de France
Paris, France

AND

MARILYN G. FARQUHAR

Section of Cell Biology
Yale University
School of Medicine
New Haven, Connecticut
and
The Rockefeller University
New York, New York

ACADEMIC PRESS   *New York San Francisco* **London**   *1975*

A Subsidiary of Harcourt Brace Jovanovich, Publishers

ACADEMIC PRESS, INC.
111 Fifth Avenue, New York, New York 10003

United Kingdom Edition published by
ACADEMIC PRESS, INC. (LONDON) LTD.
24/28 Oval Road, London NW1

Library of Congress Cataloging in Publication Data

Tixier-Vidal, A
    The anterior pituitary.

    (Ultrastructure in biological systems series)
    Includes bibliographies.
        1.   Adenohypophysis.   2.   Ultrastructure (Biology)
I.   Farquhar, Marilyn G., joint author.   II.   Title.
III.   Series.   [DNLM: 1.   Pituitary gland–Anatomy and
histology.   2.   Pituitary hormones, Anterior.   WL UL755
v. 7 / WK510 A627]
QL868.T57   612'.492   74-18464
ISBN 0–12–692050–8

Contents

Introduction

MARC HERLANT

Mechanisms Involved in the Release of Adenohypophysial Hormones

JACOB KRAICER

Identification of Anterior Pituitary Cells by Immunoelectron Microscopy

PAUL K. NAKANE

Localization of Hormones in the Pituitary:
Receptor Sites for Hormones from
Hypophysial Target Glands and the Brain

WALTER E. STUMPF, MADHABANANDA SAR, AND DONALD A. KEEFER

Structure and Function of the Anterior
Pituitary and Dispersed Pituitary Cells.
In Vitro Studies

MARILYN G. FARQUHAR, EHUD H. SKUTELSKY, AND COLIN R. HOPKINS

Separation of Organelles and Cells from the Mammalian
Adenohypophysis

W. C. HYMER

Ultrastructure of Anterior Pituitary
Cells in Culture

A. TIXIER-VIDAL

Ultrastructure of Pituitary Tumor Cells:
A Critical Study

L. OLIVIER, E. VILA-PORCILE, O. RACADOT, F. PEILLON, AND J. RACADOT

List of Contributors

Numbers in parentheses indicate the pages on which the authors' contributions begin.

MARILYN G. FARQUHAR* (83), Section of Cell Biology, Yale University School of Medicine, New Haven, Connecticut and The Rockefeller University, New York, New York

MARC HERLANT (1), Laboratoire d'Histologie, Faculté de Médecine, Université Libre de Bruxelles, Bruxelles, Belgique

COLIN R. HOPKINS† (83), The Rockefeller University, New York, New York

W. C. HYMER (137), Department of Biology, The Pennsylvania State University, University Park, Pennsylvania

DONALD A. KEEFER (63), Laboratories for Reproductive Biology, Departments of Anatomy and Pharmacology, University of North Carolina, Chapel Hill, North Carolina

JACOB KRAICER (21), Department of Physiology, Queen's University at Kingston, Kingston, Ontario, Canada

PAUL K. NAKANE (45), Department of Pathology, University of Colorado School of Medicine, Denver, Colorado

L. OLIVIER (231), Laboratoire d'Histologie-Embryologie, Faculté de Médecine Pitié-Salpêtrière, Paris, France

* Present address: Section of Cell Biology, Yale University School of Medicine, New Haven, Connecticut.

† Present address: Department of Histology and Cell Biology (Medical), The University of Liverpool, Liverpool, England.

F. PEILLON (231), Laboratoire d'Histologie-Embryologie, Faculté de Médecine Pitié-Salpêtrière, Paris, France

J. RACADOT (231), Laboratoire d'Histologie-Embryologie, Faculté de Médecine Pitié-Salpêtrière, Paris, France

O. RACADOT (231), Laboratoire d'Histologie-Embryologie, Faculté de Médecine Pitié-Salpêtrière, Paris, France

MADHABANANDA SAR (63), Laboratories for Reproductive Biology, Departments of Anatomy and Pharmacology, University of North Carolina, Chapel Hill, North Carolina

EHUD H. SKUTELSKY* (83), Section of Cell Biology, Yale University School of Medicine, New Haven, Connecticut, and The Rockefeller University, New York, New York

WALTER E. STUMPF (63), Laboratories for Reproductive Biology, Departments of Anatomy and Pharmacology, University of North Carolina, Chapel Hill, North Carolina

A. TIXIER-VIDAL (181), Laboratoire de Biologie Moleculaire, Collège de France, Paris, France

E. VILA-PORCILE (231), Laboratoire d'Histologie-Embryologie, Faculté de Médecine, Pitié-Salpêtrière, Paris, France

* Present address: Section of Biological Ultrastructure, The Weizman Institute of Science, Rehovot, Israel.

Preface

The aim of this volume is to integrate ultrastructure in the context of present knowledge on pituitary cell biology. In the introductory chapter, Professor Herlant, the well-known pioneer of modern pituitary histophysiology, presents the status of the field. The following chapter deals with a physiological approach—the mechanisms involved in the release of adenohypophysial hormones. Two chapters are concerned with a morphological approach to fundamental aspects of pituitary cell biology—cellular and subcellular localization of anterior pituitary hormones by immunoelectron microscopy; localization of possible receptor sites for hormonal messengers on anterior pituitary cells. Three chapters are then devoted to *in vitro* systems which have undergone an important development in recent years—ultrastructure and function of dispersed anterior pituitary cells; separation of organelles and cells; ultrastructure and function of cultured anterior pituitary cells. The volume concludes with a chapter on the ultrastructure of pituitary tumors which we believed would add interesting and pertinent information to a book devoted to various approaches to the study of normal anterior pituitary cells.

A. Tixier-Vidal

The Anterior Pituitary

INTRODUCTION

Marc Herlant

LABORATOIRE D'HISTOLOGIE, FACULTÉ DE MÉDECINE
UNIVERSITÉ LIBRE DE BRUXELLES, BELGIQUE

Over the last few years, the use of the electron microscope applied to the study of the pituitary has so increased our knowledge both as regards its morphology and the functioning of its glandular cells that the authors of this volume have found it timely to strike a global balance of the multiform research done on the ultrastructural analysis of the adenohypophysis.

A study of this sort is primarily intended for all endocrinologists whether they are originally histologists, pathologists, or biochemists. For it is owing to its close association with biochemistry and physiology that electron microscopy has been found to be so particularly useful in the analysis of the functioning of the pituitary cells, and, at the present moment, an initial study of the ultrastructure of these cells cannot be dispensed with if one wishes to achieve a proper understanding of their physiology.

Indeed more than for any other gland the multiplicity of the hormonal activities of the adenohypophysis requires a correct identification of the cells producing each of these hormones. It seems logical to study the localization of a functional activity before analyzing it in itself. And any morphological research on adenohypophysis begins with an identification problem, whether light or electron microscopy is used. As far as the multiplicity of cellular forms is concerned it is only fair to acknowledge that we owe to light microscopy the greater part of what we have learned. Classic staining methods had revealed the existence of acidophil, basophil, and chromophobe cells. Histochemical studies then showed the glycoprotein nature of the basophil cells, and a combination of histochemical reactions with usual staining methods gave evidence that the glycoprotein cells as well as the acidophils consisted of distinct classes. Moreover, the functional significance of each one of these categories of cells had been elucidated a long time before the undertaking of ultrastructural studies.

Evidently the techniques used in light microscopy are far from perfect, and identification problems can still raise serious difficulties. In order to find glycoprotein-containing cells on a pituitary slide, a simple PAS reaction may be amply sufficient. But at present there are no standardized techniques at hand enabling us, in a consistent manner, to distinguish among the diverse categories of glycoprotein cells; and the division of acidophilic cells into distinct classes, which rests entirely on an arbitrary differentiation between closely related acidic stains, is even more uncertain.

Furthermore, the light microscope does not always distinctly reveal the granulations to which pituitary cells owe their tinctorial affinities. If the acidophil granulations are often resistant to fixatives, this is not so for the glycoprotein granulations of the basophilic cells. These granulations are delicate and burst easily under the effect of the fixatives, so that their contents generally appear as a magma of dusty or filiform particles. Nevertheless it has been known for a long time that the basophil cells could contain corpuscles of variable sizes and irregular shapes. However, before the electron microscope came into use and proved that those were lysosomes, these organelles were mistaken for specific granulations.

There is one last, but not least fault to be found with the light microscope as regards identification of cellular shapes: It has often led to error in the identification of the chromophobe cells. Now if one puts any faith in numbers, the chromophobe cells represent more than half the cells in the anterior hypophysis. We know finally that most of them do have a specific hormonal activity and that in fact they correspond to corticotropic cells but in photon microscopy it is extremely difficult to make the distinction between them and undifferentiated cells or chromophil cells that are emptied of their granulations, or follicular cells.

The imperfect techniques used in light microscopy are responsible for having cast doubts on functional localizations as well as on the nature of the cells where they reside. To a certain degree these variations account for the diffidence with which many biochemists look on histology.

I. The Identification of the Pituitary Cells with Electron Microscopy

It is easy to understand that many histologists, hampered by the imprecisions of light microscopy, have turned to the electron microscope in attempting to solve the problem that is the object of unceasing debate, i.e., the identification of pituitary cells.

However, one should constantly bear in mind this essential truth, to wit, that barring a few and very limited circumstances, electron microscopy very rarely helps in this identification. More often, ultrastructural examinations bear only on one cell, or a fraction of a cell, and its nature is more difficult to determine than in reading a histology slide.

A. Value of Classic Criteria: Size of Secretory Granules, Structure of Cellular Organelles

Electronic microscopy apparently provided two seemingly indisputable criteria: the size of the secretory granules which seemed to differ according to the nature of each cell; the structure of the cellular organelles—particularly that of the ergastoplasm—which seemed also to have some specific characteristics.

However, a serious analysis of these criteria shows that they are hardly safer than histochemical reactions or tinctorial affinities. One thing is certain: When a preparation of anterior hypophysis is examined at small magnifications the immediate impression is that the cells differ from one another in the size of their granulations; some bear large secretory granules, others much finer ones. After having resorted to measurements it was found that each cellular shape existing in a pituitary elaborated granulations of a consistent size.

In trying to identify pituitary cells by measuring their secretory granules, one is impeded by a first difficulty. It is hardly necessary to use the vernier in order to establish the existence of cells with large granulations or those with very fine ones; but, between these two extremes, one often finds cells whose granules are very similar in size. In that case, it is very often impossible to conclude that they are indeed different cell types.

There is an even more serious objection to this criterion, i.e., numerous observations prove that cells belonging to the same category contain granules which are larger when at rest than during their phase of secretory activity.

Prolactin cells are a particularly striking example of this. Following the now old observations made by Farquhar and Rinehart in 1954, it was admitted that prolactin cells were easily distinguishable due to the size of their granulations, the diameter of which may reach 600 nm, exceeding therefore that of the secretory granules of all other categories of cells, whose greater diameters hardly exceed 350 nm. But such granulations in fact are characteristic of prolactin cells at rest. Their size is not at all the same when one observes prolactin cells during secretion. In the lactating female for example, the prolactin cells are the seat of an intense elaboration of secretory granules whose diameter hardly exceeds 250 nm. The examination of prolactin cells *in vitro* is more demonstrative of this diminution in the size of secretory granules in relation with the cell secretory activity. Within the first days in culture, the prolactin cells contain two distinct classes of secretory granules, either large or small. The first group corresponds to those contained in the cells before culture. The second group was formed when the cells were grown *in vitro*. Indeed, under these conditions, the cells are free from the inhibition normally exerted by the hypothalamus and their secretory activity is increased (Pasteels, 1973).

It must be borne in mind that pituitary cells are finally subjected to the control of hypothalamic mediators and that they enter upon a secreting phase either under the influence of a mediator which stimulates their activity or following the neutralization of an inhibiting mediator. Some observations, still quite fragmentary, seem to show that variations in the size of granules contained in any one cell reflect the action of these mediators. Examination of LH gonadotropic cells, subjected *in vitro* to the action of the LRH mediator which stimulates their secretory activity (Tixier-Vidal *et al.,* 1971; Gourdji *et al.,* 1972), again reveals the presence of granules of unequal dimensions. The same phenomenon also characterizes the LH cells of moles in sexual repose to which the LRH mediator has been administered (Pieters and Herlant, 1972).

While these observations place restrictions on a prospect which seemed so attractive, they do not imply, however, that one should from now on renounce the use of granule size to characterize the pituitary cells. While its value is not as absolute as was thought before, this criterion still is of interest because even if the diameter of secretory granules does oscillate with the functional state of the cell, its variations are contained within certain limits, themselves maintaining a relative specificity. It will never be possible, for example, to mistake somatotropic cells, with invariably large granules, for corticotropic cells, which, whatever their functional state, are always characterized by very much smaller secretory granules. The measurement of secretory granules remains just as valid a criterion for the distinction of TSH cells from FSH cells because the former have granules of much smaller diameter then the latter.

For some time one was also under the delusion that variations in electron density of granules could lead to the identification of cells, in particular, glycoprotein cells, the secretory granules of which seemed much less dense than those of serous cells. These variations in density are especially apparent when one fixes first with osmium tetroxide. Under such conditions, the FSH, TSH, and MSH cells are easily detected and distinguished from the other cell types owing to the much lower density of their granulations caused by a swelling up and often a rupture of these granules. But in fact these are artifacts produced by a defective fixation; with the use of a double fixation (glutaraldehyde, osmium-tetroxide), the granules of the glycoprotein cells appear as dense as those of other cellular types.

Let us now examine whether the ultrastructure of cellular organelles can yield valid criteria of identification. Here again examination of a pituitary preparation at low magnification in the electron microscope is sufficient to convince one that the appearance of the ergastoplasm, the Golgi apparatus, and the mitochondria may vary from cell to cell. May we feel authorized to speak of a specialized structure really characteristic of cellular types? Are we not faced with simple functional modifications of these organelles? We will endeavor to answer these questions.

To begin with, as far as the ergastoplasm is concerned, it is certain that its structure may vary with the chemical nature of the hormone elaborated by the cell. Somatotropic cells and prolactin cells are easy to recognize because of their ergastoplasm, characterized by flat saccules, richly lined with ribosomes, and arranged in parallel rows. This is not surprising. All glandular cells which secrete proteins of a high molecular weight have a similar ergastoplasm (Haguenau, 1958). This similarity in the appearance of the ergastoplasm in somatotropic and in prolactin cells implies however that one cannot distinguish these cells on this basis. Similarly in some species, MSH cells, the ergastoplasm of which is also constituted by flattened saccules lined with numerous ribosomes (Tixier-Vidal and Picart, 1967), could be mistaken for somatotropic or prolactin cells.

On the other hand, it is certain that the structure of the ergastoplasm facilitates the identification of FSH and TSH cells. Indeed, like many glycoprotein-secreting cells, the ergastoplasm of these two types of pituitary cells have vacuoles apparently isolated from one another and only poorly studded with ribosomes. However, as an identification characteristic, the ergastoplasm may be very misleading when there is a defect in fixation; a defective fixation indeed can induce a vacuolization phenomenon of all the pituitary cells at the level of the ergastoplasm.

Variations in mitochondrial structure have also been called on as criteria of cellular identification. It was observed that the mitochondria of FSH cells have many more packed cristae than those of LH cells; and it has also been noted that the mitochondrial cristae of the latter showed an irregular disposition. These criteria have no real specificity. Moreover, mitochondria are highly susceptible to artifacts so that their structure often varies not with the nature of cells, but with their fixation.

As for variations of structure in the Golgi apparatus, they are of great interest for the detection of secretory activity.

B. The Identification of Chromophobe Cells

The role of electron microscopy in the identification of chromophobe cells has been decisive in histology as well as in pathology. As soon as pituitary material was examined in the electron microscope, it was revealed that cells seemingly chromophobe could nevertheless contain granulations (Haguenau and Bernhard, 1955). Although this observation was made on pathological material, i.e., on cells from a chromophobe adenoma, the discovery was far reaching and it was soon observed that the same was true of chromophobe cells in normal pituitary.

There are indeed very few cells in which the electron microscope does not detect some granules, but frequently they are so far apart that they evidently cannot be seen by light microscopy. Evidence was thus brought forth that the so-called chromophobe cells might well correspond either to cells actually differentiating or to cells emptying themselves of their granulations in the course of a particularly intense secretion.

Analysis of chromophobe cells by means of electron microscopy led to an even more important discovery: the existence of a new line of functional cells. It revealed that the greater number of these chromophobe cells constituted in fact a homogeneous class of elements, characterized by the extreme fineness of their granulations. Measurements confirmed that these secretory granules had a diameter not exceeding 150 nm. This is why they cannot be seen with the conventional microscope unless they are massed inside the cells which electron microscopy proved to be rare; it showed, in fact, that more often these granulations are widely dispersed and that they tend to assemble only along the plasma membrane. It is easy therefore to understand why it was an achievement to make the distinction between them and other cells of chromophobe appearance. The discovery of these cells is fundamental, because it is through the exploration of their function that they were proved to be the ACTH-secreting cells (Herlant and Klastersky, 1963a,b; Siperstein and Allison, 1965; Kurosomi and Kobayashi, 1966; Herlant, 1970; Siperstein and Miller, 1970; Pelletier and Racadot, 1971). In pathology, on the other hand, the use of the electron microscope in the study of pituitary adenomas has brought about the elucidation of apparently unexplainable facts: Often adenomas which have engendered an acromegaly, a galactorrhea, or an hypercorticism all seem to be constituted of chromophobe cells only. Their ultrastructural analysis allows one to reveal their true nature (Porcile *et al.,* 1964; Racadot *et al.,* 1964, 1971; Olivier *et al.,* 1965).

On the other hand, the interstitial cell of the adenohypophysis, whose existence has been postulated for a long time, had received scarcely any attention for the simple reason that, in light microscopy, they are hardly distinguishable from chromophobe cells. Electron microscopy has confirmed the existence of follicular cells which sometimes enclose the glandular cords within a large network (Salazar, 1968; Bergland and Torack, 1969; Vila-Porcile *et al.,* 1971a,b). It revealed also the existence of stellate cells that according to these authors belong to the same category.

C. The Use of Specific Techniques for the Identification of Pituitary Cells in the Electron Microscope

1. *The Contribution of Histochemistry*

We hardly need to stress the decisive part played by the PAS reaction in the study of the pituitary with the light microscope. Great hopes were thus raised by the adaptation to electron microscopy of histochemical techniques permitting the detection of polysaccharides. Undoubtedly the proposed techniques (Rambourg, 1967; Thiery, 1967) allowed more precise localization of glycoprotein secretory granules in pituitaries of widely different origins (Follenius and Doerr-Schott, 1966; Rambourg and Racadot, 1968; Tixier-Vidal and Picart, 1970, 1971; Dubois, 1971). There is certainly no greater specificity in these techniques than in that of the PAS in light microscopy, but they do yield an element of confirmation when there is a suspicion of the glycoprotein nature of a cell. The results must, however, be interpreted with the greatest care when they alone represent the element of identification.

2. *The Contribution of Immunocytochemistry*

At first glance, there seems to be no better way than immunology to offer total reliability in the identification of pituitary cells. Accordingly the introduction of immunocytochemistry in electron microscopy awoke the highest interest (Nakane, 1970, 1971). However, whether with light or electron microscopy, there is much to be done before immunocytological techniques may be considered infallible. One of the objections is that hormones with a similar chemical nature may have in common the same antigenic groups so that antisera prepared against one determined hormone might react with other cells than those which naturally secrete that hormone. It has not been possible up to now to overcome this obstacle, particularly for the identification of the two types of gonadotropic cells. Whatever the hormone used in the preparation of the antisera, whatever the origin of the pituitary on which they are tested, they react indifferently with both categories of gonadotropic cells, whether in light (Leleux and Robyn, 1970, 1971; Tougard *et al.,* 1971) or in electron microscopy (Nakane, 1970). There are similar problems in the detection of corticotropic cells with immunological techniques. ACTH is chemically closely related to both MSH (α,β). It is therefore understandable that an antiserum, prepared even from synthetic ACTH, reacts indifferently with corticotropic cells and with those cells responsible for the secretion of the MSH(β) (Herlant *et al.*, 1972).

The introduction of immunocytological techniques in electron microscopy is in an initial stage. The hopes it has raised will certainly be confirmed in the future (see chapter by Nakane).

The above critical analysis concerning the identification of pituitary cells with the electron microscope should not give the impression that no cellular type may be defined with a total certainty. This would not be exact and we are very often able to pronounce unhesitatingly on the nature of a cellular type. It seemed necessary, however, to stress the relative value of all identification criteria.

True, certain pituitary cells are more easily identified on an histological slide than from an electron micrograph. When in doubt, we can always resort to the semi-fine slides method. This practice can be particularly useful in the identification of glycoprotein cells. Though too rarely put to use, the practice of semi-fine sections has brought about brilliantly successful results (Doerr-Schott, 1965).

II. The Contribution of Electron Microscopy to the Functional Exploration of the Anterior Pituitary

It is in the dynamic analysis of pituitary cells that electron microscopy has most certainly achieved its greatest goals.

Light microscopy obviously could yield but elementary information about the functioning of these cells. It had proved, however, that pituitary hyperactivity is usually accompanied by morphological modifications which can be frequently localized and assigned to one class of cells. These modifications generally consist of hypertrophy and sometimes in apparent multiplication and degranulation phenomena. Histological slides had also revealed that the states of cellular hyperfunction were accompanied by an hypertrophy of the Golgi apparatus and of the nucleolus; sometimes also, in particular at the level of basophil cells, by a vacuolar modification. Again it is from the study of histological slides that the ergastoplasm was discovered, and it is precisely in pituitary cells that its first description was given in detail (Celestino da Costa, 1923).

It was later proved that the basophilia observed in certain classes of pituitary cells were not to be ascribed to the nature of their granulations, but rather to the endowment in ribonucleic acid of their cytoplasm (Desclin, 1940; Herlant, 1943). By that date, a number of authors had already established that the amount of ribonucleic acid in the cells could vary with their functional state and they suspected the part played by this component in the elaboration of the secretion material.

These observations are still wholly valid and it is on the basis of such criteria that a definition of the functional state of each cellular type of the anterior lobe of the hypophysis could be reached. But it is also in fact the limit of the aid provided by the light microscope in the analysis of function. Many unanswered questions remained for which answers could be found by means of resolution that the electron microscope only was able to provide.

A. ORIGIN AND FATE OF SECRETORY GRANULES

At once, the electron microscope was to reveal the manner in which specific granulations are formed and was to show the secretion process itself. Whereas even with the most sophisticated techniques, the light microscope only gave an imperfect image of the Golgi apparatus, but the electron microscope revealed its ultrastructure formed of saccules and vacuolae. It suggested also that in all glandular cells elaborating their secretion material as granulations, the granules are formed at the level of the Golgi apparatus cisternae. The phenomenon is particularly well illustrated in the glandular cells of the hypophysis and has been known for a long time (Rinehart and Farquhar, 1953; Farquhar and Wellings, 1957). Indeed the multiplication of the Golgi vacuoles and the appearance therein of forming granules are considered at present as the classic criteria for secretion activity in all pituitary cells. It is equally well known that when secretory granules spread into the hyaloplasm, they remain encapsulated. The membrane which holds them is in fact of Golgi origin.

Electron microscopy also demonstrated that the process of granular secretion corresponds to the extrusion of the contents of the granulations after fusion of their own membrane with the cell membrane (Farquhar, 1961; Sano, 1962). This is considered a fundamental phenomenon at least for serous cells.

Fixation artifacts had led to the belief in a distinct mode of secretion at the level of glycoprotein cells (Herlant, 1963), for when those cells are improperly fixed, their granulations burst easily, and apparently they set free their contents into the hyaloplasm. Such pictures of the bursting open of secretory granules are much rarer when the glycoprotein cells have been subjected to correct fixation; but even in the case of an intense secretion activity, the excretion phenomena in these cells are not as obvious as at the level of serous cells. At most, one observes small indentations along the basal membrane of the cells. The occurrence of such indentations suggests that the contents of glycoprotein granules are dissolved as soon as free.

On the other hand, it can be now considered as established that secretory granules are set free into the perivascular spaces. We mainly owe to electron microscopy the discovery of these spaces which separated glandular cells from the sinusoid walls (Rinehart and Farquhar, 1953; Farquhar, 1961). Classic histology deluded us into believing that the cells and the vessels were in contact, and very naturally it was thought for a long time that the pituitary cells spilled their secretion products directly into the vessels. These perivascular spaces are connected into a wide lacunary system which penetrates among the glandular cords, a long distance away from the capillaries. Secretory granules are thus not necessarily set free close to the vessels. In these spaces it seems that the granules dissolve immediately, so that it is impossible to follow the secretion transit up to the capillary itself. It has been established that the endothelial cells are fenestrated and show pores which are bridged by thin diaphragms. This disposition, however, is common to all so-called visceral or fenestrated capillaries whatever their localization. One can suppose that the secretion filters through these diaphragms; but it cannot be ruled out that it may be previously absorbed by the endothelial cells, for numerous pinocytosis vesicles are often observed at their level.

From these morphological features, one may ask if the contents of the granules correspond to the hormones themselves? This probably is true for most pituitary cells. But one may question the fact for MSH cells that, although secreting small polypeptides, elaborate granules which are manifestly glycoprotein. As with neurosecretory granules the glycoprotein fraction of MSH granules corresponds perhaps to a carrier molecule bound to the hormone.

But the formation of secretory granules in cisternae of the Golgi apparatus, their spreading out in the hyaloplasm, and their extrusion into the perivascular spaces are only terminal stages. It is now well known that in protein-secreting cells this material is synthesized at the level of the ribosomes lining the ergastoplasmic saccules. It should be remembered that according to present concepts the ribosomes are RNA-loaded corpuscles which represent one of the essential links in the synthesis of proteins in animal cells. Their activation, however, implies the intervention of a messenger RNA which conveys to them the information emitted by the nuclear DNA. The action of the messenger RNA finds its morphological expression in the grouping of ribosomes into small clusters built of a definite number of elements held together by a mesh which, most probably, corresponds to the RNA messenger itself (Rich, 1963). These clusters are known as polysomes. It is admitted that in secreting glandular cells, the proteins are synthesized at the level of the ribosomes organized into polysomes lining the walls of the ergastoplasm. The elaborated material is then discharged into the cavities of the endoplasmic reticulum, the mechanisms of its passage through the membranes of the reticulum being unknown. It is then directed toward the Golgi apparatus where it is condensed into granules.

These now very classic notions may be applied to pituitary cells. They explain why, in light microscopy, certain categories of pituitary cells exhibit a more basophilic cytoplasm while secreting actively. Such a phenomenon is particularly characteristic of prolactin cells during lactation (Desclin, 1940, 1945). When studied in the electron microscope they show a hypertrophy of the ergastoplasm (Hedinger and Farquhar, 1957; Barnes, 1962; Hymer et al., 1961; Pasteels, 1963). These changes are so typical that they are now used as identification criteria for prolactin cells. Somatotropic cells show a similar disposition in their ergastoplasm and, apparently, the synthesis of their hormonal material proceeds in the same manner. The phenomena of secretion stimulation are, however, rarely as intense.

Many questions nevertheless have remained unanswered as to the part played by the ergastoplasm in the elaboration of pituitary hormones. Not all classes of cells possess ergastoplasms comparable to those of prolactin or somatotropic cells. In corticotropic cells or in LH cells, for example, the ergastoplasm is represented only by a reduced number of saccules or vacuoles; on the other hand, their ground cytoplasm is richly endowed with free ribosomes. In this respect corticotropic cells merit special attention. Their activation is undoubtedly accompanied by a rise in the number of free polysomes. This can easily be seen in the electron microscope but has further been confirmed by differential centrifugation which allows the separation of the polysomes from the free ribosomes (Kraicer, 1964). By using this technique, it seems well established that there is a parallelism between the quantities of polysomes in corticotropic cells and the rate of secretion of ACTH. On the other hand, the various stages of secretory activity of corticotropic cells are not accompanied by important modifications of the endoplasmic reticulum. Should the conclusion be that the synthesis of certain pituitary hormones can take place at the level of free polysomes? If this hypothesis were to be proved, it would imply that it is possible for hormones to be set free in the hyaloplasm.

The problem raised by the TSH and FSH glycoprotein cells is a reverse one. When active they display in the electron microscope a vacuolar dilatation of their endoplasmic reticulum, such dilatation may reach considerable proportions, but the polysomes do not appear perceptibly augmented.

As this rapid survey has shown, as soon as applied to the study of the adeno-hypophysis, the electron microscope has yielded information of high interest about two distinct phases in the elaboration of secretory material. It has brought indisputable proof that in all pituitary cells without exception, the secretion granules are formed in the cavities of the Golgi apparatus. It has also shown that the states of activation of pituitary cells are accompanied by modifications in the ergastoplasm which may take sundry appearances, but all suggest the participation either of the ribosomes or the endoplasmic reticulum in the elaboration of the secretion material.

However, in order to gain a complete view of the process several questions remain unanswered, which appeared beyond the possibilities of the electron microscope alone.

The transfer of the material elaborated in the cavities of the ergastoplasm and its arrival in the Golgi apparatus were only postulates. The condensation of the secretory material into electron dense corpuscles—so-called "intracisternal granules"—inside the cavities of the endoplasmic reticulum, as described in the cells of the exocrine pancreas (Palade, 1956; Caro and Palade, 1964), seems to be an exception at the level of pituitary cells. The phenomenon has been observed only in TSH cells of animals subjected to thyroidectomy (Farquhar, 1969). Similar corpuscles in the endoplasmic reticulum were also observed *in vitro* in cells of the intermediate lobe (Hopkins, 1972). Such observations are rare, however.

As regards the transit of the synthesized material from the ergastoplasm into the Golgi apparatus it has long been shown that the Golgi cisternae are in continuity with the ergastoplasmic cisternae (see review by Haguenau, 1958; Jamieson and Palade, 1967a,b). Several other authors have confirmed this observation such as Claude for the liver (1970), Flickinger (1969) for the epididymus cells, and Tixier-Vidal and Picart (1971) for the duck pituitary. The functional significance of such morphological continuity was shown through autoradiography and will be discussed in the following section.

B. The Contribution of "Autoradiography"

It is through the application of autoradiography to electron microscopy and the use of labeled amino acids that the mechanisms of protein synthesis in glandular cells were elucidated; and it is also owing to the use of this technique that it was possible to follow the transit of the elaborated material. This was accomplished with a propitious material—the cells of the exocrine pancreas (Caro, 1961; Caro and Palade, 1964; Jamieson and Palade, 1967a,b).

As far as the pituitary cells are concerned the present information concerns only the prolactin cells and the MSH cells in the duck (Tixier-Vidal *et al.,* 1965; Tixier-Vidal and Picart, 1967), the somatotropic cells in the rat (Racadot *et al.,* 1965), and the cells of the intermediary lobe in the *Xenopus* (Hopkins, 1972). Some observations were obtained in tissue cultures (Tixier-Vidal *et al.,* 1965; Hopkins, 1972), others *in vivo* (Racadot *et al.,* 1965). In spite of their small number the results confirm in a striking manner the conclusions reached for the exocrine pancreas. In all types of cells studied the labeled amino acid, whether leucine or glycine, is to be found first at the surface of the endoplasmic reticulum, its presence being most probably the sign of the synthetic processes taking place at the level of the ribosomes. Shortly after, the isotope is detected in the saccules of the endoplasmic reticulum and finally it is localized on the secretory granules during their formation in the Golgi apparatus. That a similar sequence of events occurs in the synthesis processes in MSH cells as well as in prolactin cells and in somatotropic cells is indeed of high interest. The hormones secreted by MSH cells are peptides of very small molecular weight. Those peptides are already present at the level of the ribosomes and are to be found again in the granules formed at the level of the Golgi apparatus (Hopkins, 1972). It remains to reconcile this with the fact that the granulations of MSH cells have the same histochemical behavior as glycoproteins.

The use of autoradiography in the functional study of the hypophysial cells also draws attention to a much neglected factor: It introduces the time notion in the analysis of secretion processes. We do now realize that these biosynthetic phenomena proceed in a well-determined chronological manner. Synthesis of the secretion material at the ribosomes is very rapid, its transfer up to the Golgi apparatus is very much slower. It is also obvious that the secretion granules may remain in the cell ground cytoplasm for a long period before their excretion.

C. The Role of Lysosomes in Pituitary Cells

Glandular cells obviously show some delaying stages in their secretory cycle. This general rule applies to the cells of the adenohypophysis. Through light microscopy, it had long been established that the stages of functional repose are generally manifested by granule retention. It is easy to understand that cells that have stopped secreting accumulate in their cytoplasm the material they have elaborated. Autoradiography also reveals that the elaboration processes and the excretion processes are not necessarily synchronous. Light microscopy, however, does not yield any information about the future of granulations that are being accumulated inside resting pituitary cells. Again the problem has been solved by the electron microscope. The pituitary cells rid themselves of the granulations elaborated in excess through their lysosomes. These organelles of Golgi origin constitute real digestive vacuolae filled with enzymes, particularly, hydrolases, the main role of which is to absorb and degrade the metabolites which might overcrowd the cells (de Duve, 1959, 1963a,b). These may be foreign bodies as well as constituents of the cell itself. The lysosomes absorb foreign bodies phagocytized by the cells as well as pinocytosis vesicles, altered mitochondria, and membranes from the endoplasmic reticulum, when the latter's activity has ceased.

Electron microscopy has revealed the rich content of lysosomes in pituitary cells especially "primary lysosomes," which means that they are still in the initial stage of their evolution cycle. They appear then as dense and homogeneous corpuscles. This discovery entailed as a first consequence the conviction that, in light microscopy, specific granulations had often been mistaken for lysosomes. This then was the explanation for the apparently paradoxical fact that in Batrachians, the gonadotropic cells may contain both basophilic and acidophilic secretion granules; the latter in fact correspond to lysosomes (Doerr-Schott, 1964). Also, the so-called specific granulations that served so long for the identification of LH cells in the human hypophysis are, in fact, lysosomes, particularly abundant in those cells (Herlant, 1965).

The digestive function of lysosomes can also exert itself at the expense of secretion granules accumulating in pituitary cells when they enter a period of rest. When lactation is interrupted, for example, the lysosomes of the prolactin cells engulf secretory granules that they progressively digest; they can also phagocytize ribosomes and fragments of endoplasmic reticulum membrane (Smith and Farquhar, 1966). The last phenomenon does indeed correspond to a cessation in the elaborating activity; the organelles which have ceased to function are immediately eliminated. The capture of the granulations by the lysosomes when the cells enter a resting phase has been confirmed, in particular at the level of somatotropic and thyrotropic cells (Farquhar, 1969). It is very easy to put TSH cells to rest: One simply injects thyroxine into animals who have undergone thyroidectomy. Under these conditions, one can observe a massive absorption of granulations in the lysosomes.

D. The Effect of Hypothalamic Mediators on the Ultrastructure of Pituitary Cells

It is common knowledge that each class of pituitary cells is subjected to the control of hypothalamic mediators that now tend to be considered as hormones. The mediators may either stimulate or diminish the cell secretion activity. Hypersecretion may result from the action of a stimulating mediator as well as from the blockage of an inhibiting factor. Conversely, a cessation of secretory activity may be due to the slowing down of the liberation of a stimulating mediator at the level of the hypothalamus or to the action of an inhibiting mediator.

These mediators are peptides with a small molecular weight. Some have been isolated, purified, and even synthesized. *In vivo* and *in vitro,* they have proved to be potent liberators of hormones. It was only logical that their action on the pituitary cells should be studied morphologically. First, the administration of nonpurified hypothalamic extracts was shown to induce a massive extrusion of secretory granules at the level of the somatotropic cells (De Virgilis *et al.,* 1968). This action seemed due to a specific growth hormone mediator. Indeed a similar effect was observed after the administration of this purified mediator *in vivo* (Couch *et al.,* 1969; Coates *et al.,* 1970). One must be cautious however in interpreting these results. It has not been formally proved that the phenomenon of granular extrusion can be attributed to the mediator only. Secretion of somatotropic cells like that of corticotropic cells can be stimulated by stress resulting from the intravascular injection. Gonadotropic cells which apparently do not react to stress are undoubtedly a more favorable material to demonstrate an exogenous action. One of their specific stimulating mediators, LRH, has been isolated in its purified state, and even synthesized (Schally *et al.,* 1971). According to some (Jutisz *et al.,* 1967), LRH induces a liberation of LH only, while according to others (Schally *et al.,* 1971), it could also stimulate FSH cells. The effects of this last mediator have been analyzed morphologically. To demonstrate, however, a specific action on the LH cells, one had to resort to certain experimental expedients. Indeed the first observations had not revealed granule extrusion at the level of LH cells after the administration of purified LRH in the rat (Rennels *et al.,* 1971). Such images nevertheless were observed either *in vitro* with purified LRH on pituitary from castrated ewes previously treated with progestagens (Tixier-Vidal *et al.,* 1971) or *in vivo* by intraarterial injection of LRH to female rats in permanent estrus (Shiino *et al.,* 1972). Under these conditions, extrusion images are frequent. They are localized in LH cells alone and are manifest only in the minutes following the mediator's injection. That is probably the reason why they had escaped earlier detection.

Later, during the hours following the administration of the mediators, one has invariably observed that reactive cells showed a marked hypertrophy of the Golgi apparatus; this hypertrophy being always accompanied by an intense neoformation of secretion granules. It thus seems that secretion is sufficient in itself to trigger hormonal biosynthesis.

As to the mode of action of the hypothalamic factors on pituitary cells (see chapter by Kraicer), it is subjected to the general rule of double mediation (Sutherland *et al.,* 1968; Sutherland, 1972). The hypothalamic mediators apparently act on the adenyl cyclases present in the membranes of pituitary cells, and, in turn, the latter converts the intracellular ATP into cyclic AMP, which would participate in the phenomena of granule extrusion. In this connection, it has been proved recently that cyclic AMP produces on the ultrastructure of somatotropic and prolactin cells the same effects as those which can be attributed to the mediators themselves. Thus, adjunction of cyclic AMP to an incubation medium in which pituitary fragments are floating, immediately induces granule extrusion and in the following hours Golgi apparatus hypertrophy with intense granule neoformation.

In the introduction to this volume we have found it necessary to trace at the outset a global picture of the essential acquisitions due to the electron microscope applied in conjunction with other techniques of the cell biologists to the study of the adenohypophysis. In the following chapters the problems here merely glossed over are analyzed in greater detail. The reader now will have undoubtedly realized to what extent electron microscopy has helped to enlarge our knowledge of the morphology of pituitary cells, of their function, and of the control mechanisms which govern them. These notions may no longer be ignored by those who, for whatever reason, have an interest in the pituitary gland. However electron microscopy has also shown the necessity to correlate the data with all those obtained from other methods, either biochemical or physiological. The present volume attempts such a confrontation.

REFERENCES

Barnes, B. G. (1962). Electron microscope studies on the secretory cytology of the mouse anterior pituitary. *Endocrinology* **71,** 618–628.

Bergland, R. M., and Torack, R. (1969). An ultrastructural study of follicular cells in the human anterior pituitary. *Amer. J. Pathol.* **57**, 273–297.

Caro, L. (1961). Electron microscopic radioautography of thin section. The Golgi zone as a site of protein concentration in pancreatic acinar cells. *J. Biophys. Biochem. Cytol.* **10**, 37–44.

Caro, L., and Palade, G. E. (1964). Protein synthesis, storage and discharge in the pancreatic exocrine cell. An autoradiographic study. *J. Cell Biol.* **20**, 473–495.

Celestino da Costa, A. (1923). Sur les aspects histologiques du fonctionnement de l'hypophyse. *C. R. Ass. Anat.* **18**, 108–117.

Claude, A. (1970). Growth and differentiation of cytoplasmic membranes in the course of lipoprotein granule synthesis in the hepatic cell. Elaboration of elements of the Golgi complex. *J. Cell Biol.* **47**, 745–766.

Coates, P. W., Ashby, E. A., Krulich, L., Dhariwal, A. P. S., and McCann, S. M. (1970). Morphologic alterations in somatotrophs of the rat adenohypophysis following administration of hypothalamic extracts. *Amer. J. Anat.* **128**, 389–412.

Couch, E. F., Arimura, A., Schally, A. V., Saito, M., and Swano, S. (1969). Electron microscope studies of somatotrophs of rat pituitary after injection of purified growth hormone releasing factor (GRF). *Endocrinology* **85**, 1084–1091.

de Duve, C. (1959). Lysosomes, a new group of cytoplasmic particles. *In* "Subcellular Particles" (T. Hayashi, ed.), pp. 128–159. Ronald Press, New York.

de Duve, C. (1963a). Lysosome concept. *Lysosomes, Ciba Found. Symp., 1963* pp. 1–35.

de Duve, C. (1963b). The lysosome. *In* "Living Cell" (D. Kennedy, ed.), pp. 62–70. Freeman, San Francisco (excerpted from *Scientific American*).

Desclin, L. (1940). Détection de substances pentose-nucléique dans les cellules du lobe antérieur de l'hypophyse du rat et du cobaye. *C. R. Soc. Biol.* **123**, 457–459.

Desclin, L. (1945). Contribution à l'étude de la structure et du fonctionnement de l'hypophyse pendant le post-partum. *Arch. Biol.* **56**, 261–323.

De Virgilis, G., Meldolesi, J., and Clementi, F. (1968). Ultrastructure of growth hormone producing cells of rat pituitary after injections of hypothalamic extract. *Endocrinology* **83**, 1278–1284.

Doerr-Schott, J. (1964). Localisation au microscope électronique de l'activité phosphatasique acide dans les cellules beta de l'hypophyse de la grenouille rousse. *Rana temporaria. C. R. Acad. Sci.* **258**, 1621–1623.

Doerr-Schott, J. (1965). Etude comparative de la cytologie et de l'ultrastructure de l'hypophyse distale de trois espèces d'Amphibiens anoures: Rana temporaria L. Bufo vulgaris Laur, Xenopus laevis D. *Gen. Comp. Endocrinol.* **5**, 631–653.

Dubois, P. (1971). Signification fonctionnelle d'une catégorie cellulaire de l'antéhypophyse foetale humaine. *C. R. Acad. Sci.* **273**, 880–882.

Farquhar, M. G. (1961). Origin and fate of secretory granules in cells of the anterior pituitary gland. *Trans. N. Y. Acad. Sci.* [2] **23**, 346–351.

Farquhar, M. G. (1969). Lysosome function in regulating secretion: Disposal of secretory granules in cells of the anterior pituitary gland. *In* "Lysosomes in Biology and Pathology" (J. T. Dingle and H. B. Fell, eds.), Vol. 2, pp. 462–482. North-Holland Publ., Amsterdam.

Farquhar, M. G., and Rinehart, J. F. (1954). Electron microscopic studies of the anterior pituitary gland of castrate rats. *Endocrinology* **54**, 516–541.

Farquhar, M. G., and Wellings, S. R. (1957). Electron microscopic evidence suggesting secretory granules formation within Golgi apparatus. *J. Biophys. Biochem. Cytol.* **3**, 319–322.

Flickinger, C. J. (1969). The pattern of growth of the Golgi complex during the foetal and postnatal development of the rat epididymis. *J. Ultrastruct. Res.* **27**, 344.

Follenius, E., and Doerr-Schott, J. (1966). Mise en évidence sélective au microscope électronique des cellules à sécrétion glycoprotidiques de l'hypophyse par la réaction à l'acide périodique-nitrate d'argent. *C. R. Acad. Sci.* **262**, 912–914.

Gourdji, D., Kerdelhue, B., and Tixier-Vidal, A. (1972). Ultrastructure d'un clone de cellules hypophysaires sécrétant de la prolactine (clone GH3). Modifications induites par l'hormone hypothalamique de libération de l'hormone thyréotrope (TRF). *C. R. Acad. Sci.* **274**, 437–440.

Haguenau, F. (1958). The ergastoplasm: Its history, ultrastructure and biochemistry. *Int. Rev. Cytol.* **7**, 425–483.

Haguenau, F., and Bernhard, W. (1955). L'appareil de Golgi dans les cellules normales et cancéreuses de vertébrés. Rappel historique et étude au microscope électronique. *Arch. Anat. Microsc. Morphol. Exp.* **44**, 27–55.

Hedinger, C. E., and Farquhar, M. G. (1957). Elektronenmikroskopische Untersuchungen von zwei Typen acidophiler Hypophysenvorderlappenzellen bei der Ratte. *Schweiz. Z. Pathol. Bakteriol.* **20**, 766–768.

Herlant, M. (1943). Recherches sur la localisation histologique des hormones gonadotropes femelles au niveau de l'hypophyse antérieure. *Arch. Biol.* **54**, 225–257.

Herlant, M. (1963). Apport de la microscopie électronique à l'étude du lobe antérieur de l'hypophyse. *In* "Cytologie de l'adénohypophyse" (J. Benoit and C. Da Lage, eds.), pp. 73–90. CNRS, Paris.

Herlant, M. (1965). Les cellules responsables de l'activité gonadotrope dans l'hypophyse humaine. *Rev. Eur. Endocrinol.* **2**, 113–134.

Herlant, M. (1970). Le siège de l'activité corticotrope dans le lobe antérieur de l'hypophyse. *Arch. Sci. Biol. (Belgrade)* **20**, 91–104.

Herlant, M., and Klastersky, J. (1963a). Etude au microscope électronique des cellules corticotropes de l'hypophyse. *C. R. Ass. Anat.* **118**, 720–729.

Herlant, M., and Klastersky, J. (1963b). Etude au microscope électronique des cellules corticotropes de l'hypophyse. *C. R. Acad. Sci.* **256**, 2709–2711.

Herlant, M., Ectors, F., and Dessy, C. (1972). Détection dans le lobe antérieur de l'hypophyse chez les mammifères de cellules sécrétrices d'intermédine au moyen d'un immunsérum anti-corticotropine. *C. R. Acad. Sci.* **274**, 1183–1186.

Hopkins, C. R. (1972). The biosynthesis, intracellular transport and packaging of melanocyte-stimulating peptides in the amphibian pars intermedia. *J. Cell Biol.* **53**, 642–653.

Hymer, W. C., McShan, W. H., and Christianssen, R. G. (1961). Electron microscopic studies of anterior pituitary glands from lactating and estrogen treated rats. *Endocrinology* **69**, 81–90.

Jamieson, J. D., and Palade, G. E. (1967a). Intracellular transport of secretory proteins in the pancreatic exocrine cell. I. Role of the peripheral elements of the Golgi complex. *J. Cell Biol.* **34**, 577–597.

Jamieson, J. D., and Palade, G. E. (1967b). II. Transport to condensing vacuoles and zymogen granules. *J. Cell Biol.* **34**, 597–618.

Jutisz, M., Berault, A., Novella, M. A., and Ribot, G. (1967). Etude de l'action du facteur hypothalamique LRF (LH releasing factor) chez le rat *in vivo* et *in vitro*. *Acta Endocrinol. (Copenhagen)* **55**, 481–496.

Kraicer, J. (1964). Pituitary polysomes and adrenocorticotropin secretion. *Biochim. Biophys. Acta* **87**, 703–706.

Kurosomi, K., and Kobayashi, Y. (1966). Corticotrophs in the anterior pituitary glands of normal and adrenalectomized rats as revealed by electron microscopy. *Endocrinology* **78**, 745–758.

Leleux, P., and Robyn, C. (1970). Etude en immunofluorescence d'une réaction immunologique croisée entre les gonadotropines humaines et le gonadotropines du rat. *Ann. Endocrinol.* **31**, 181–191.

Leleux, P., and Robyn, C. (1971). Immunohistochemistry of individual antehypophysial cells. *Karolinska Symp. Res. Methods Reprod. Endocrinol.* **3**, 168–189.

Nakane, P. K. (1970). Classification of anterior pituitary cell types with immuno-enzyme histochemistry. *J. Histochem. Cytochem.* **18**, 9–20.

Nakane, P. K. (1971). Application of peroxydase-labelled antibodies to the intracellular localization of hormones. *In vitro* methods in reproductive endocrinology. *Karolinska Symp. Res. Methods Reprod. Endocrinol.* **3**, 190–204.

Olivier, L., Porcile, E., de Brye, C., and Racadot, J. (1965). Etude de quelques adénomes hypophysaires chez l'homme en microscopie électronique. *C. R. Ass. Anat.* **127**, 1258–1265.

Palade, G. E. (1956). Intracisternal granules in the exocrine cells of the pancreas. *J. Biophys. Biochem. Cytol.* **2**, 417–422.

Pasteels, J. L. (1963). Recherches morphologiques et expérimentales sur la sécrétion de prolactine. *Arch. Biol.* **74**, 439–553.

Pasteels, J. L. (1973). Tissue culture of human hypophyses. Evidence of a specific prolactin in man. *Lactogenic Horm., Ciba Found. Symp.* pp. 269–286.

Pelletier, G., and Racadot, J. (1971). Identification des cellules hypophysaires sécrétant l'ACTH chez le rat. *Z. Zellforsch. Mikrosk. Anat.* **116**, 228–239.

Pieters, A., and Herlant, M. (1972). Modifications saisonnières des cellules à prolactine dans l'antéhypophyse de la Taupe mâle. *C. R. Acad. Sci.* **284**, 3002–3006.

Porcile, E., de Brye, C., and Racadot, J. (1964). Données ultrastructurales concernant une tumeur adenohypophysaire humaine. *J. Microsc. (Paris)* **3**, 49 (abstr.).

Racadot, J., Olivier, L., Porcile, E., de Brye, C., and Klotz, H. P. (1964). Adénome hypophysaire de type "mixte" avec symptomatologie acromégalique. II. Etude au microscope optique et électronique. *Ann. Endocrinol.* **25**, 503–507.

Racadot, J., Olivier, L., Porcile, E., and Droz, B. (1965). Appareil de Golgi et origine des grains de sécrétion chez le rat. Etude radioautographique en microscopie électronique après injection de leucine tritiée. *C. R. Acad. Sci.* **261**, 2972–2974.

Racadot, J., Vila-Porcile, E., Peillon, F., and Olivier, L. (1971). Adénomes hypophysaires à cellules à prolactine. Etude structurale et ultrastructurale. Corrélations anatomo-cliniques. *Ann. Endocrinol.* **32**, 298–305.

Rambourg, A. (1967). Détection des glycoprotéines en microscopie électronique: Coloration de la surface cellulaire et de l'appareil de Golgi par un mélange acide chromique-phosphotungstique. *C. R. Acad. Sci.* **265**, 1426–1428.

Rambourg, A., and Racadot, J. (1968). Identification en microscopie électronique de six types cellulaires dans l'antéhypophyse du rat à l'aide d'une technique de coloration par le mélange acide chromique-phosphotungstique. *C. R. Acad. Sci.* **266**, 153–155.

Rennels, E. G., Bogdanove, E. M., Arimura, A., Saito, M., and Schally, A. V. (1971). Ultrastructural observations of rat pituitary gonadotrophs following injection of purified porcine LH–RH. *Endocrinology* **88**, 1318–1326.

Rich, A. (1963). Polyribosomes. *In* "Living Cell" (D. Kennedy, ed.), pp. 166–169. Freeman, San Francisco (excerpted from *Scientific American*).

Rinehart, J. F., and Farquhar, M. G. (1953). Electron microscopic studies of the anterior pituitary gland. *J. Histochem. Cytochem.* **1**, 93–113.

Salazar, H. (1968). Ultrastructural evidence for the existence of a non-secretory sustentacular cell in the human adenohypophysis. *Anat. Rec.* **160**, 419–420.

Sano, M. (1962). Further studies on the theta cells of the mouse anterior pituitary as revealed by electron microscopy with special reference to the mode of secretion. *J. Cell Biol.* **15**, 85–97.

Schally, A. V., Arimura, A., Kastin, A. J., Matsuo, H., Baba, Y., Redding, T. W., Nair, M. G., and Debeljuk, L. (1971). Gonadotropin-releasing hormone: One polypeptide regulates secretion of luteinizing and follicle-stimulating hormones. *Science* **173**, 1036–1037.

Schally, A. V., Redding, T. W., Matsuo, H., and Arimura, A. (1972). Stimulation of FSH and LH release *in vitro* by natural and synthetic LH and FSH releasing hormones. *Endocrinology* **90**, 1561–1568.

Shiino, M., Arimura, A. V., and Rennels, E. G. (1972). Ultrastructural observations of granule extrusion from rat anterior pituitary cells after injection of LH-releasing hormone. *Z. Zellforsch. Mikrosk. Anat.* **128**, 152–161.

Siperstein, E. R., and Allison, V. F. (1965). Fine structure of the cells responsible for secretion of adrenocorticotrophin in the adrenalectomized rats. *Endocrinology* **76**, 70–79.

Siperstein, E. R., and Miller, J. K. (1970). Further cytophysiologic evidence for the identity of the cells that produce adrenocorticotropic hormone. *Endocrinology* **86**, 451–486.

Smith, R. E., and Farquhar, M. G. (1966). Lysosome function in the regulation of the secretory process in cells of the anterior pituitary gland. *J. Cell Biol.* **31**, 319–347.

Sutherland, E. W. (1972). Studies on the mechanism of hormone action. *Science* **177**, 401–408.

Sutherland, E. W., Robison, G. A., and Butcher, R. W. (1968). Some aspects of the biological role of adenosine 3′,5′ monophosphate (cyclic AMP). *Circulation* **37**, 279–306.

Thiery, J. P. (1967). Mise en évidence des polysaccharides sur coupes fines en microscopie électronique. *J. Microsc. (Paris)* **6**, 987–1018.

Tixier-Vidal, A., and Picart, R. (1967). Etude quantitative par radio-autographie au microscope électronique de l'utilisation de la DL-Leucine 3 H par les cellules de l'hypophyse du canard en culture organotypique. *J. Cell Biol.* **35**, 501–519.

Tixier-Vidal, A., and Picart, R. (1970). Localisation ultrastructurale des glycoprotéines acides et des structures osmiophiles dans la zone golgienne des cellules glycoprotidiques de l'adénohypophyse. *C. R. Acad. Sci.* **271**, 767–769.

Tixier-Vidal, A., and Picart, R. (1971). Electron microscopic localization of glycoproteins in pituitary cells of duck and quail. *J. Histochem. Cytochem.* **19**, 775–797.

Tixier-Vidal, A., Fiske, S., Picart, R., and Haguenau, F. (1965). Autoradiographie au microscope électronique de leucine tritiée par l'hypophyse du Canard en culture organotypique. *C. R. Acad. Sci.* **261**, 1133–1136.

Tixier-Vidal, A., Kerkelhue, B., Berault, A., and Jutisz, M. (1970). Cinétique de la sécrétion de l'hormone lutéinisante (LH) par l'antéhypophyse de rat en culture organotypique; influence d'une préparation purifiée du facteur hypothalamique de libération de LH (LRF). *C. R. Acad. Sci.* **271**, 523–526.

Tixier-Vidal, A., Kerdelhue, B., Berault, A., Picart, R., and Jutisz, M. (1971). Action *in vitro* du facteur hypothalamique de libération de l'hormone lutéinisante (LRF) sur l'antéhypophyse d'agnelle. Etude ultrastructurale des tissus incubés. *J. Comp. Endocrinol.* **17**, 33–59.

Tougard, C., Kerdelhue, B., Tixier-Vidal, A., and Jutisz, M. (1971). Localisation par cyto-immuno-enzymologie de la LH, de ses sous-unités alpha et beta et de la FSH dans l'adénohypophyse de la ratte castrée. *C. R. Acad. Sci.* **273**, 897–900.

Vila-Porcile, E., and Olivier, L. (1971). Les cellules folliculaires et stellaires du lobe antérieur de l'adénohypophyse de rat adulte. *C. R. Ass. Anat.* **152**, 812.

Vila-Porcile, E., Olivier, L., and Racadot, J. (1971). Cellules folliculaires du lobe antérieur de l'hypophyse humaine. *C. R. Ass. Anat.* **152**, 813.

MECHANISMS INVOLVED IN THE RELEASE OF ADENOHYPOPHYSIAL HORMONES

Jacob Kraicer

DEPARTMENT OF PHYSIOLOGY, QUEEN'S UNIVERSITY AT KINGSTON
KINGSTON, ONTARIO, CANADA

I. Introduction

Until about fifteen years ago, the major effort in studies of secretion of adenohypophysial hormones was in the direction of elucidating the way in which the brain controlled secretion from the gland. The landmark monograph of Harris (1955) signaled the onset of the modern era in our knowledge of the control of secretion of adenohypophysial hormones. It is now firmly established that secretion of adenohypophysial hormones is under the control of the central nervous system and that secretion is controlled by releasing and inhibiting hormones which enter the hypophysial–portal circulation to reach the cells of the adenohypophysis (Porter *et al.,* 1971). Major advances have been made in the last several years in

21

defining the chemical nature of several of these releasing and inhibiting hormones (Burgus and Guillemin, 1970).

The establishment of the mechanisms controlling secretion from the adeno-hypophysis has led logically to studies of the mechanisms by which the releasing and inhibiting hormones act to control secretion within the gland. Although a definitive description of the physicochemical events within adenohypophysial cells, beginning with the interaction of the releasing hormone with the plasma membrane and culminating in the release of hormone, is not yet at hand, certain models have been proposed which are under active study in a number of laboratories.

In discussing the mechanisms involved in the secretion of adenohypophysial hormones, one must, of necessity describe the way in which the hypothalamic releasing and inhibiting hormones interact with the cells of the adenohypophysis to regulate secretion. This review is restricted to the hormones of the pars distalis, thus excluding consideration of the neurohypophysial hormones and the pars inter-media hormone, MSH. Furthermore, there will be no discussion of the mechanisms by which the inhibiting hormones PIH and GIH act to suppress release of pro-lactin and growth hormone since there are as yet no substantial data on this sub-ject. Finally, it will be assumed that all of the releasing hormones act in a similar way to initiate secretion of their respective hormones. As far as we know, this is a valid assumption.

There have been a number of *in vivo* and *in vitro* studies demonstrating that the releasing hormones stimulate both synthesis and release of adenohypophysial hormones. The question immediately arises as to whether the primary action is on release, with synthesis being secondary to the release process, whether synthesis is primary, with release following as a consequence of new synthesis, or whether there is a primary action on both synthesis and release. Since release can still occur when new synthesis is blocked (McCann, 1971) there would appear to be a pri-mary action on the release process. A primary action on the synthesis of hormone as well cannot as yet be ruled out.

This review will concern itself primarily then, with the mechanisms governing the release of adenohypophysial hormones, with only passing reference to possible synthetic mechanisms.

II. The Release Process

A. Tentative Model

A tentative model can be proposed to describe the sequence of events initiated by an interaction of the releasing hormone with the plasma membrane of the adenohypophysial cell and culminating in the release of preformed hormone. This model is an extrapolation of the "stimulus-secretion" coupling hypothesis proposed by Douglas (1968) and co-workers.

The process begins with an interaction between the releasing hormone and the plasma membrane of its specific target cell, resulting in an alteration in the "chemical conformation" of the plasma membrane (Fig. 1). There is a concurrent change in the permeability characteristics of the plasma membrane to various ions, in particular to Ca^{2+}, with a resultant influx of Ca^{2+}, and perhaps other ions, into the cells. This redistribution of ions would also alter the electrical characteristics of the plasma membrane resulting in a depolarization, with a possible decrease in electrical resistance. The increase in intracellular Ca^{2+} (or redistribution of intracellular Ca^{2+}) would then lead to hormone release.

Lacy et al. (1968) have suggested that the granules of the beta cells of the islets of Langerhans may be interlinked with each other and with the plasma membrane by an internal cytoskeleton composed of microtubules and contractile filaments. They proposed that as a result of stimulation, contraction of this system would result in a fusion of the sacs enclosing the granules with the plasma membrane, causing the ejection of the hormone-containing granules into the extracellular space.

We suggested (Kraicer et al., 1969b) that this hypothesis, linked with that proposed by Douglas and co-workers, would lead to a unifying hypothesis that the release of hormone may not, in fact, be best described by the term "stimulus-secretion" coupling, but by "excitation-contraction" coupling, since the process of hormone release may be akin to that leading to muscle contraction. Rasmussen (1970) recently expanded this concept to speculate on the possible interaction among Ca^{2+}, cyclic AMP, and the contractile elements in the release process. The following discussion expands on the experimental data relevant to this model.

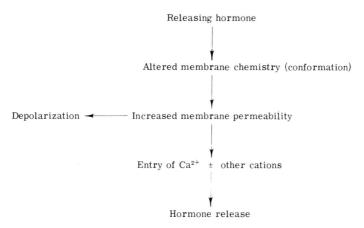

FIG. 1. Stimulus-secretion coupling hypothesis. Modified from Douglas (1968).

B. INTERACTION BETWEEN THE RELEASING HORMONE AND
     THE PLASMA MEMBRANE

The first step in the sequence of events leading to hormone release would be an interaction, or binding, between the releasing hormone and the plasma membrane of its target cell. Recent studies, using synthetic TRH have demonstrated a binding of TRH to plasma membrane fractions of adenohypophysial cells and have described some of the characteristics of this binding (Labrie *et al.,* 1972; Wilber and Seibel, 1973; Grant *et al.,* 1972). The data suggest, but do not establish unequivocally, that the binding described is restricted or specific to the plasma membranes of the thyrotrophs. Furthermore, these types of studies do not rule out, of course, a primary action within the cells as well.

C. ALTERATION IN THE PERMEABILITY OF THE
     PLASMA MEMBRANE TO IONS

Three secretagogues have been used most extensively to study the release process, (1) High [K$^+$], (2) crude extracts of hypothalamus-stalk-median eminence (HSME), and (3) purified, and more recently, synthetic hypothalamic releasing hormone preparations.

1. *High [K$^+$]*

The use of high [K$^+$] as a secretagogue warrants special comment. According to the Nernst relationship, if one assumes that K$^+$ is the most permeable ion species, then increasing the extracellular [K$^+$] of incubating tissue should decrease transmembrane potentials (TMP) (Ling and Gerard, 1950), and would, according to the Douglas hypothesis, stimulate hormone release. It is on this basis that high [K$^+$] has been used as a secretagogue. It has been repeatedly reported that incubating adenohypophyses in high [K$^+$] will increase the release of TSH, FSH, LH, ACTH, and GH, but will not alter prolactin release (Table I). Furthermore, the augmented release induced by high [K$^+$] is reversible (Samli and Geschwind, 1968; Kraicer *et al.,* 1969b; Wakabayashi *et al.,* 1969) and repeatable (Kraicer *et al.,* 1969b; Schofield and Cole, 1971). As well, the high [K$^+$]-induced release of TSH (Vale and Guillemin, 1967) and of ACTH (Kraicer *et al.,* 1969a,b) is inhibited by thyroxine and corticosterone, respectively.

However, the interpretation of these findings is clouded. TMP of adenohypophysial cells recorded both *in vitro* (Milligan and Kraicer, 1970; Martin *et al.,* 1973) and *in vivo* (York *et al.,* 1971, 1973) are low, averaging about 12–24 mV (inside negative). Incubating glands in media containing five times the normal [K$^+$] does lower TMP significantly, but the depolarization is only 4 mV (Martin *et al.,* 1973). If the Nernst potential for [K$^+$] were to hold for the cells of the adeno-

TABLE I

Hɪɢʜ [K⁺] Sᴛɪᴍᴜʟᴀᴛᴇs Rᴇʟᴇᴀsᴇ ᴏғ Aᴅᴇɴᴏʜʏᴘᴏᴘʜʏsɪᴀʟ Hᴏʀᴍᴏɴᴇs

Hormone	[K⁺] of incubation medium, mM	Release	Reference
TSH	25 or 50	↑[a]	Vale and Guillemin, 1967
FSH	over 20	↑	Jutisz and de la Llosa, 1968
	60	↑	Wakabayashi et al., 1969
	59	↑	Jutisz and de la Llosa, 1970
LH	59	↑, Rev.[b]	Samli and Geschwind, 1968
	60 or 90	↑, Rev.	Wakabayashi et al., 1969
ACTH	28	↑	Kraicer et al., 1969a
	25	↑, Rev. Rep.[c]	Kraicer et al., 1969b
GH	21	↑	MacLeod and Fontham, 1970
	32 or 54	↑	Parsons, 1970
	over 20	↑, Rev. Rep.	Schofield and Cole, 1971
Prolactin	21	No effect	MacLeod and Fontham, 1970
	54	No effect	Parsons, 1970

[a] ↑ = Stimulates.
[b] Rev. = reversible.
[c] Rep. = repeatable.

hypophysis, this would indicate an extremely low intracellular [K⁺] as compared with other tissues. This is not the case since the intracellular [K⁺] of the rat adenohypophysis is about 112 mM/kg wet weight (Kraicer and Milligan, 1973) (Table III). Thus if the cells were selectively permeable to K⁺, the TMP should be about 75 mV (inside negative) and a fivefold increase in [K⁺] would cause a depolarization of about 40 mV. The intracellular K⁺ does not appear to be bound, since uptake of ⁴²K proceeds rapidly with no indication of saturation (Kraicer and Milligan, 1973). Therefore the calculated concentration is close to the effective concentration.

Although an increase in extracellular [K⁺], *in vitro,* does decrease TMP, the absolute values of TMP observed, and the magnitude of the depolarization found are not those predicted from the Nernst relationship. Thus, the initial premise on which high [K⁺] was used as a secretagogue does not hold.

2. *Ca²⁺*

According to the stimulus-secretion coupling hypothesis, Ca²⁺ is the link or coupling between the stimulus and secretion. Two interrelated questions have been asked concerning the role of Ca²⁺ in the release process. The first is whether

TABLE II

THE RELEASE OF ADENOHYPOPHYSIAL HORMONES INDUCED BY HIGH $[K^+]$
AND BY RELEASING HORMONES IS Ca^{2+}-DEPENDENT[a]

Secretagogue	Hormone	Ca^{2+}-free condition	Release	Reference
A. High $[K^+]$				
	TSH	Repeated preinc. in Ca^{2+}-free; inc. in Ca^{2+}-free	↓, Rev.	Vale and Guillemin, 1967
	FSH	2 hr preinc. in Ca^{2+}-free; inc. in Ca^{2+}-free	↓	Wakabayashi et al., 1969
		45 min preinc. in Ca^{2+}-free + EDTA; inc. in Ca^{2+}-free + EDTA	↓, Rev.	Jutisz and de la Llosa, 1970
	LH	0.5–2 hr preinc. in Ca^{2+}-free; inc. in Ca^{2+}-free	↓	Wakabayashi et al., 1969
		1 hr preinc. in Ca^{2+}-free + EDTA; inc. in Ca^{2+}-free	↓, Rev.	Samli and Geschwind, 1968
	ACTH	2 quick rinses in Ca^{2+}-free; inc. in Ca^{2+}-free	↓, Rev., Rep.	Kraicer et al., 1969b
	GH	Direct transfer to Ca^{2+}-free inc.	↓	Parsons, 1970
		Wash in Ca^{2+}-free + EDTA; inc. in Ca^{2+}-free + EDTA	↓	Schofield and Cole, 1971
B. Releasing hormone preparation				
Purified ovine TRF	TSH	Repeated preinc. in Ca^{2+}-free; inc. in Ca^{2+}-free	↓, Rev.	Vale et al., 1967
Stalk-median eminence extract	TSH	1 hr preinc. in Ca^{2+}-free + EDTA; inc. in Ca^{2+}-free + EDTA	↓	Steiner et al., 1970
Purified ovine FRF	FSH	2 hr preinc. in Ca^{2+}-free; inc. in Ca^{2+}-free + EDTA	↓	Wakabayashi et al., 1969
Purified ovine FRF	FSH	45 min–2 hr preinc. in Ca^{2+}-free; inc. in Ca^{2+}-free	No effect	Jutisz and de la Llosa, 1970
Hypothalamic extract	LH	1 hr preinc. in Ca^{2+}-free + EDTA; inc. in Ca^{2+}-free	↓, Rev.	Samli and Geschwind, 1968
Purified ovine LRF	LH	2 hr preinc. in Ca^{2+}-free; inc. in Ca^{2+}-free + EDTA	↓	Wakabayashi et al., 1969

(*Continued*)

TABLE II (*Continued*)

Secretagogue	Hormone	Ca²⁺-free condition	Release	Reference
Rat hypothalamic extract or lysine vasopressin	ACTH	3 hr preinc. in Ca²⁺-free; inc. in Ca²⁺-free	↓	Zimmerman and Fleischer, 1970
Stalk-median eminence extract	GH	1 hr preinc. in Ca²⁺-free + EDTA; inc. in Ca²⁺-free + EDTA	↓	Steiner *et al.*, 1970

[a] Abbreviations and symbols used are ↓ = supresses; Rev. = reversible; Rep. = repeatable; inc. and preinc. = incubation and preincubation, respectively.

hormone release is Ca^{2+}-dependent, and the second is whether there is an influx of Ca^{2+} into the cells associated with the release process.

There have been a number of studies reporting that the augmented release of hormone induced by high $[K^+]$ and by releasing hormone preparations is suppressed when Ca^{2+} is removed from the incubation medium (Table II). However in most of the studies using Ca^{2+}-free media, very rigorous procedures were used to rid the system of Ca^{2+} (prolonged and repeated incubation in Ca^{2+}-free media or the use of Ca^{2+}-chelating agents). One cannot conclude from such studies that release of hormone was prevented solely by the removal of Ca^{2+} from the extracellular compartment, thus preventing influx into the cells. Such rigorous procedures could have prevented release, not only by preventing influx, but also by the removal of an essential intracellular Ca^{2+} compartment.

We attempted to define more closely the Ca^{2+} requirement for ACTH release (J. V. Milligan and J. Kraicer, unpublished data) using two types of Ca^{2+}-free washes; two quick rinses in Ca^{2+}-free medium followed by a 1-hr incubation in Ca^{2+}-free medium, and a more rigorous multiple Ca^{2+}-free wash procedure, with repeated prolonged incubations in Ca^{2+}-free media. Vasopressin, like elevated

TABLE III

INTRACELLULAR CONCENTRATIONS OF Na⁺, K⁺, AND Ca²⁺ IN THE RAT ADENOHYPOPHYSIS AFTER INCUBATION IN KREBS–RINGER BICARBONATE[a]

Ion	Intracellular concentration (mM/kg wet weight)	Measured media concentrations (mM/liter)	Theoretical values from formula (mM/liter)
Na⁺	15–34	132–151	143
K⁺	98–127	4.1–5.6	5.92
Ca²⁺	4–7	2.1–2.4	2.54

[a] From Kraicer and Milligan, 1973; J. V. Milligan and J. Kraicer, unpublished.

[K$^+$] (Kraicer *et al.*, 1969a,b) requires only a loosely bound Ca^{2+}-compartment in order to provoke ACTH release, since two quick rinses with Ca^{2+}-free medium abolishes the augmented release induced by these two secretagogues. On the other hand, the release of ACTH induced by a crude acid extract of rat hypothalamus-stalk-median eminence is unchanged by the quick rinse procedure and can only be reduced by prolonged and repeated washings in Ca^{2+}-free media. These findings would indicate that Ca^{2+} is essential for the release process, but that different pools may be involved, depending on the secretagogue, some secretagogues requiring a loosely bound or easily accessible Ca^{2+} pool, others a more tightly bound and less accessible pool.

Of more direct relevance are studies on the effect of secretagogues on the influx of Ca^{2+} into the cells of the adenohypophysis. Hormone release, *in vitro,* induced by high K$^+$ is associated with an influx of Ca^{2+} (Milligan and Kraicer, 1971; Schofield and Stead, 1971), while that induced by vasopressin and a crude acid extract of hypothalamus-stalk-median eminence is not (Milligan and Kraicer, 1971). On the other hand, the augmented release of GH induced by a purified GRH preparation is accompanied by an increased influx of Ca^{2+} (Milligan *et al.,* 1972). We have no explanation for the apparent dichotomy in the response to the hypothalamus-stalk-median eminence extract and to the GRH preparation.

There are as yet few published data on the actual content of calcium in the adenohypophysis and none on the intracellular localization of calcium. We have estimated the intracellular calcium concentration of the rat adenohypophysis to be about 3.75 mM/kg wet weight (Milligan and Kraicer, 1971), somewhat higher than the values reported for a variety of other soft tissues, but lower than that reported by Schofield (1971) in pituitary slices from heifers.

A pivotal role for Ca^{2+} in the release process would appear to hold. However, further studies are required to identify the intracellular Ca^{2+} compartments and to explore the intracellular events associated with the role of Ca^{2+} in the release process.

3. Na$^+$ and K$^+$

There has been some scattered evidence indicating that Na$^+$ and K$^+$ may be involved in the release process. Data demonstrating a Na$^+$ requirement for release has been inconsistent (Wakabayashi *et al.,* 1969; Parsons, 1970; Schofield and Cole, 1971). A role for K$^+$ in the normal release process (as opposed to the high [K$^+$]-induced release phenomenon) is unclear although Reinberg and Stolkowski (1953) had suggested that lowered intracellular [K$^+$] may lead to decreased hormone release, and elevated intracellular [K$^+$] to increased hormone release.

We have recently carried out a series of *in vitro* studies (Kraicer and Milligan, 1973) (1) to determine if changes in plasma membrane permeability or intracellular uptake of Na$^+$ or K$^+$ occur under conditions of augmented hormone release and if such changes could be related to the release mechanism, and (2) to see if the low TMP recorded both *in vivo* and *in vitro* (see below) might be explained in terms of preferential permeability to Na^+ and/or K^+. These studies demonstrated (a) that the cells of the adenohypophysis are essentially impermeable to Na^+ and reasonably permeable to K^+, and (b) that both the influx of Na^+ and K^+ and the intracellular content of Na^+ and K^+ are essentially unaltered by the augmented hormone release induced by a number of secretagogues, including dibutyryl cyclic AMP, theophylline, vasopressin, a crude acid extract of hypothalamus-stalk-median eminence, and a purified GRH preparation. Thus hormone release from the adenohypophysis is not associated with changes in Na^+ and K^+ uptake or permeability. Unlike Ca^{2+}, these two major cations appear not to be involved in the mechanisms of hormone release.

D. Electrical Characteristics of the Plasma Membrane and Hormone Release

Intracellular recording techniques have been used to characterize the electrical properties of the cells of the adenohypophysis both *in vivo* (York *et al.,* 1971, 1973) and *in vitro* (Milligan and Kraicer, 1970; Martin *et al.,* 1973). *In vivo* studies demonstrate TMP of about -20 mV, with membrane resistance of about 40 MΩ and membrane capacitance of about 1.2 pF. These membrane parameters appear to follow a normal distribution. *In vitro* TMP is about -12 mV. The early report of positive TMP *in vitro* (Milligan and Kraicer, 1970) could not be verified (Martin *et al.,* 1973).

Two questions of immediate relevance are (a) what is the ionic basis for the low TMP observed, and (b) are there changes in the electrical characteristics of the cells of the adenohypophysis with hormone release.

In vitro intracellular concentration of Na^+, K^+, and Ca^+ in the rat adenohypophysis are given in Table III. The low TMP observed both *in vitro* and *in vivo* cannot be simply described as following the Nernst relationship, with K^+ being the most permeable ion species as is the case in nerve and muscle. The cells of the adenohypophysis are reasonably permeable to K^+ and Ca^{2+} and essentially impermeable to Na^+ (Milligan and Kraicer, 1971; Milligan *et al.,* 1972; Kraicer and Milligan, 1973). If the internal free $[Ca^{2+}]$ is low, as in most cells (Borle, 1971), then the opposing concentration gradients of the most permeable ions, K^+ and Ca^{2+}, could account for the TMP observed in adenohypophysial cells.

Incubating glands in elevated $[K^+]$ increases hormone release from the adenohypophysis (see above). Associated with this augmented release is an increase in K^+ and Ca^{2+} permeability with no change in Na^+ permeability (Milligan and Kraicer, 1971; Kraicer and Milligan, 1973). Associated also with this release is a significant but small depolarization of adenohypophysial cells (Milligan and Kraicer, 1970; Martin *et al.,* 1973). A smaller, but still significant depolarization was produced by elevated $[K^+]$ in a Ca^{2+}-free medium, indicating, as Douglas *et al.* (1967) had found for chromaffin cells, that depolarization can still occur under conditions where augmented hormone release has been prevented. These data suggest that the depolarization produced by high $[K^+]$ is due to the increased Ca^{2+} permeability as well as the altered K^+ gradient. The depolarization due to elevated K^+ is smaller in Ca^{2+}-free media, since there would now be no increase in Ca^{2+} influx, just the altered $[K^+]$ gradient.

A small, but significant depolarization of adenohypophysial cells is also observed when Ca^{2+} alone is removed from the incubation medium containing the normal $[K^+]$ (Martin et al., 1973). This is associated with an increase in Na^+ permeability, but no change in K^+ permeability (Kraicer and Milligan, 1973) and a consistent increase in ACTH release (Kraicer and Milligan, 1969b). Under these conditions, the cells of the adenohypophysis appear to have a TMP related to $[K^+]$ and $[Na^+]$. The functional significance of this has yet to be explored.

While studies have yet to be reported on the effects of more "physiological" secretagogues on the electrical characteristics of the adenohypophysial cells in vitro, several in vivo studies using microelectrode techniques have been carried out. Following the chronic administration of propylthiouracil to rats, there is a massive hypertrophy and hyperplasia of the thyrotrophs of the adenohypophysis associated with the feedback-induced increase in secretion of TSH. The electrical characteristics of the cells of the adenohypophysis of such animals were compared with a group of controls (York et al., 1971). The "stimulus-secretion" coupling hypothesis would predict a depolarization of cells with a decrease in membrane resistance. However, this anticipated response was not seen. There was, in fact, a shift in the frequency distribution of TMP to a more negative mean value, with no change in membrane resistance. In a subsequent in vivo study (York et al., 1973), TMP and electrical resistance were measured following the intracarotid injection of crude hypothalamus-stalk-median eminence extract, rich in releasing hormone activity. The most consistent finding was not a depolarization and decrease in membrane resistance, but an increase in membrane resistance in the majority of cells with no change in TMP.

The results of a limited number of studies concerning the electrical characteristics of the cells of the adenohypophysis, and the changes observed with altered secretory activity, so far do not lead to a consistent pattern. The direct in vitro experiment to test the effect of a pure or synthetic releasing hormone directly on its target cell has not been reported.

E. The Role of Cyclic AMP in Hormone Release

1. Introduction

It is now firmly established that adenosine 3',5'-monophosphate (cyclic AMP) mediates the action of many protein hormones (for a detailed discussion, see Robison et al., 1971). According to currently accepted dogma, the protein hormone, or first messenger (in this case, the hypothalamic releasing hormone), first interacts with and binds to a specific receptor site on the plasma membrane. This receptor site is closely related to, or part of, an adenyl cyclase system. This interaction then leads to an increase in adenyl cyclase activity. Adenyl cyclase then catalyzes the conversion of ATP to cyclic AMP. Cyclic AMP is the "second messenger" which then effects action. Cyclic AMP is normally inactivated by a cyclic nucleotide phosphodiesterase that converts cyclic AMP to 5′-AMP. Methylxanthines such as aminophylline and theophylline inhibit this phosphodiesterase activity. Thus potentiation of a hormone by a methylxanthine is taken as good presumptive evidence that the hormone acts by stimulating adenyl cyclase activity.

Finally, cyclic AMP is relatively inert when applied to tissue *in vitro* or administered *in vivo,* or it may have a paradoxical effect. It is usual then, to administer derivatives of cyclic AMP that are extremely potent and have effects similar to that of the hormone which increases the endogenous level of cyclic AMP within the cell. The most commonly used is the acyl derivative, dibutyryl cyclic AMP. A caution has recently been raised, however, concerning the action of dibutyryl cyclic AMP. It may act, in part, by inhibiting phosphodiesterase activity, thus elevating endogenous cyclic AMP levels.

In order to conclude that cyclic AMP mediates the action of a given hormone on a given tissue, the following criteria must be met (Robison *et al.,* 1971): (1) the hormone will stimulate adenyl cyclase activity in broken cell preparations; (2) the hormone will increase intracellular cyclic AMP levels in intact cells and this increase in cyclic AMP should precede or at least not follow the physiological response; (3) exogenous cyclic AMP, or its derivatives should duplicate the physiological effect of the hormone; and (4) the effect of the hormone should be potentiated by administering the hormone together with a phosphodiesterase inhibitor. The hormone and the phosphodiesterase inhibitor should act synergistically.

As reviewed below, on the basis of these four criteria, cyclic AMP would indeed mediate the action of the releasing hormones.

2. Releasing Hormone Preparations Stimulate Adenyl Cyclase Activity in the Adenohypophysis

Crude extracts of hypothalamic tissue will increase adenohypophysial adenyl cyclase activity in cell-free preparations (Zor *et al.,* 1969, 1970; Steiner *et al.,* 1970) while control extracts will not. These same hypothalamic extracts had no effect on adenyl cyclase activity in a variety of other tissues. More recently, Poirier *et al.* (1973) reported that synthetic TRH would increase adenyl cyclase activity, and that this activity was localized to the plasma membrane fraction.

3. Releasing Hormone Preparations Increase Cyclic AMP Levels in the Adenohypophysis

It has been repeatedly observed that releasing hormone preparations, from crude hypothalamic extracts to synthetic TRH or LH–RH/FSH–RH, increase adenohypophysial cyclic AMP levels (Fleischer *et al.,* 1969; Zor *et al.,* 1969, 1970; Steiner *et al.,* 1970; Bowers, 1971; Peake *et al.,* 1972; Borgeat *et al.,* 1972). The crude releasing hormone preparations did not alter cyclic nucleotide phosphodiesterase or ATP'ase activity (Zor *et al.,* 1970; Borgeat *et al.,* 1972). An interesting recent finding is that a crude preparation of rat hypothalamic extract increased cyclic GMP levels as well (Peake *et al.,* 1972). It is relevant to note that hormone release induced by high [K⁺] is reported not to be associated with an increase in adenyl cyclase or cyclic AMP levels (Zor *et al.,* 1970).

4. *Cyclic AMP and Its Derivatives Stimulate Hormone Release from the Adenohypophysis*

A deluge of *in vitro* studies has demonstrated that cyclic AMP, or more commonly, its acyl derivatives, will increase release of TSH (Cehovic, 1969; Wilber *et al.,* 1969; Bowers, 1971), FSH (Jutisz and de la Llosa, 1969, 1970), LH (Ratner, 1970), ACTH (Fleischer *et al.,* 1969; Pelletier *et al.,* 1972), GH (MacLeod and Lehmeyer, 1970; Cehovic *et al.,* 1970; Ewart and Taylor, 1971; Hertelendy *et al.,* 1971a,b; Labrie *et al.,* 1971a; Hertelendy, 1971; Lemay and Labrie, 1972; Pelletier *et al.,* 1972; Peake *et al.,* 1972), and prolactin (Lemay and Labrie, 1972; Pelletier *et al.,* 1972). The augmented release of TSH is suppressed by thyroxine (Wilber *et al.,* 1969) and that of ACTH by dexamethasone (Fleischer *et al.,* 1969). Also, the response to cyclic AMP or its derivatives is potentiated by phosphodiesterase inhibitors (Cehovic, 1969; Cehovic *et al.,* 1970; Peake *et al.,* 1972).

Pelletier *et al.* (1972) have recently described the ultrastructural changes in the cells of the rat adenohypophysis, incubated with monobutyryl cyclic AMP. They report the ultrastructural changes in the somatotrophs, corticotrophs and prolactin cells to be quite similar to those found with the addition of releasing hormone preparations.

5. *Methylxanthenes Stimulate Hormone Release from the Adenohypophysis*

It has been repeatedly demonstrated, in *in vitro* studies, that the cyclic nucleotide phosphodiesterase inhibitors, theophylline and aminophylline, will increase adenohypophysial cyclic AMP levels (Fleischer *et al.,* 1969; Zor *et al.,* 1970; Steiner *et al.,* 1970; Hertelendy *et al.,* 1971a; Bowers, 1971; Peake *et al.,* 1972) and will stimulate the release of TSH (Cehovic, 1969; Wilber *et al.,* 1969; Steiner *et al.,* 1970), LH (Ratner, 1970), ACTH (Fleischer *et al.,* 1969), and GH (Schofield, 1967; Steiner *et al.,* 1970; MacLeod and Lehmeyer, 1970; Cehovic *et al.,* 1970; Hertelendy, 1971; Hertelendy *et al.,* 1971a,b; Ewart and Taylor, 1971; Peake *et al.,* 1972). Furthermore, theophylline or aminophylline (1) potentiates the increase in cyclic AMP induced by a crude hypothalamic extract (Zor *et al.,* 1970), (2) acts synergistically with vasopressin to increase ACTH secretion (Fleischer *et al.*, 1969), (3) potentiates the action of a purified FRF preparation to increase FSH release (Jutisz and de la Llosa, 1969), (4) potentiates the action of TRH to stimulate TSH release (Bowers, 1971), and (5) potentiates the action of a crude hypothalamic extract to stimulate GH release (Steiner *et al.*, 1970).

Clearly, the criteria outlined earlier have been met. Cyclic AMP does mediate the action of the hypothalamic releasing hormones.

6. *Mechanism of Action of Cyclic AMP*

The exact mechanism by which an increase in intracellular cyclic AMP activates the release process is not yet known. This is not too surprising when we recall that our knowledge concerning the physicochemical events associated with the extrusion of membrane-bound hormone-containing granules is still rudimentary. As well, there is as yet no unanimity of thought as to whether the process of exocytosis is the *only* mechanism of hormone release. However, if one assumes that granule extrusion is the basic mechanism for hormone release, then a tentative hypothesis can be developed.

A cyclic AMP-dependent protein kinase has been described in a number of tissues, which is activated allosterically by cyclic AMP (Robison *et al.*, 1971). These protein kinases catalyse the transfer of a terminal phosphate from ATP to another protein. It has been proposed that a large number of the effects induced by cyclic AMP may be mediated by the action of this protein kinase. Rasmussen (1970) has developed several models which link cyclic AMP-dependent protein kinase with the release process.

Lacy *et al.* (1968) suggested that the granules of the beta cells of the islets of Langerhans may be interlinked with each other and with the plasma membrane by an internal cytoskeleton composed of a contractile microtubule system. They proposed that as a result of stimulation, contraction of this system would result in a fusion of the sacs enclosing the granules with the plasma membrane, with the ejection of the granules into the extracellular space. This model was supported by the finding that colchicine, which interferes with the organization of gelated protoplasmic elements involved in cellular and intracellular movement, and more specifically, binds to microtubules, thereby interfering with their function (Malawista, 1965; Borisy and Taylor, 1967; Weisenberg *et al.*, 1968), blocks the increased secretion of insulin induced by elevated glucose. We subsequently suggested (Kraicer *et al.*, 1969b) that this hypothesis, linked with the "stimulus-secretion" coupling model might lead to a unifying hypothesis, that release of hormone from the adenohypophysis may, in fact, be akin to "excitation-contraction" coupling in muscle, where contraction is associated with activation of the actomyosin complex.

Rasmussen speculated (1970) that the interaction of the hormone with its receptor site on the plasma membrane would activate adenyl cyclase activity, which would lead to an increase in cyclic AMP levels. The increase in cyclic

AMP then would activate a protein kinase which would lead to the phosphorylation of one or more enzymic or contractile proteins within the cell, thus stimulating release via a process suggested by Lacy *et al.* (1968).

This model is currently being tested in the adenohypophysis. A cyclic AMP-dependent protein kinase has been found in bovine adenohypophysial tissue. It has been partially purified, and some of its properties determined (Labrie *et al.*, 1971b). Approximately 50% of the protein kinase activity was found free in the soluble fraction and the remaining activity was distributed widely amongst the various subcellular fractions (Lemaire *et al.*, 1971). Furthermore, a granule fraction obtained from bovine anterior pituitary tissue was phosphorylated by an endogenous protein kinase (Labrie *et al.*, 1971c). However, in the latter study no information was provided as to the actual component of the secretory granule fraction protein which was actually phosphorylated.

Is there evidence for a contractile process mediating hormone release? Several studies have concerned themselves with the effect on hormone release of blocking an alleged contractile system of microtubules and microfilaments. Kraicer and Milligan (1971) and Schofield and Cole (1971) have reported that colchicine blocks the augment release of ACTH and growth hormone, respectively, induced by high [K⁺] *in vitro*. However the augmented release of ACTH induced by a crude extract of rat hypothalamus-stalk-median eminence was not suppressed by colchicine (Kraicer and Milligan, 1971). In further studies Gauthier *et al.* (1972) have reported that deuterium oxide, a microtubule stabilizer, inhibited the release of growth hormone induced by high [K⁺], monobutyryl cyclic AMP, and a purified GRH. Vincristine, a mitotic spindle inhibitor, also inhibited the release of growth hormone induced by these secretagogues.

These data, although preliminary, indicate that the model proposed, that cyclic AMP may act via a protein kinase which then activates a contractile microtubule system involved in the active extrusion of secretory granules, certainly warrants further detailed investigation.

F. Interrelation of Cyclic AMP and Ca^{2+}

According to the "stimulus-secretion" coupling hypothesis of Douglas (1968) and collaborators, Ca^{2+} plays a pivotal role in coupling the stimulus to the resultant release of hormone. Since the cyclic AMP system surely plays a role in the release process, it is relevant to ask how cyclic AMP and Ca^{2+} are interrelated in the mechanisms governing hormone release. Rasmussen (1970) has suggested the following interrelations: (1) Ca^{2+} is required for the specific stimulus to activate adenyl cyclase; (2) both Ca^{2+} and cyclic AMP have independent effects in the cell and changes in their concentration occur simultaneously; or (3) the intracellular action of cyclic AMP requires the presence of Ca^{2+}.

It is most convenient to begin an attempt to unravel the interrelation between cyclic AMP at the beginning of the process. Hormone release induced by a number of secretagogues is Ca^{2+} dependent, since the removal of Ca^{2+} from the incubation medium prevents release (see above). There is as yet no consistent pattern as to whether release induced by "physiological" secretagogues is associated with an influx of Ca^{2+} into the cells (Milligan and Kraicer, 1971; Milligan et al., 1972). The increase in adenyl cyclase activity associated with the action of a crude hypothalamic extract was not prevented when Ca^{2+} was removed from the incubation medium and EDTA then added (Zor et al., 1970) indicating that Ca^{2+} may not be essential for adenyl cyclase activation. Since the increase in cyclic AMP activity induced by a crude hypothalamic extract (Zor et al., 1970; Steiner et al., 1970) is not inhibited in a Ca^{2+}-free environment, Ca^{2+} would not appear to be essential for cyclic AMP formation. However, definitive studies will require the use of purified or synthetic releasing hormone preparations.

The release of adenohypophysial hormones induced by cyclic AMP and its derivatives is Ca^{2+} dependent. Release will not occur when the glands are incubated in a Ca^{2+}-free environment produced by prolonged incubation in a Ca^{2+}-free medium with or without the addition of Ca^{2+}-chelating agents (Jutisz and de la Llosa, 1970; Zimmerman and Fleischer, 1970; Lemay and Labrie, 1972). Nor will cyclic nucleotide phosphodiesterase inhibitors stimulate hormone release under similar Ca^{2+}-free conditions (Zimmerman and Fleischer, 1970; Steiner et al., 1970; Ewart and Taylor, 1971). In these Ca^{2+}-free experiments, release could have been suppressed either by preventing an influx of Ca^{2+}, which might be essential for the action of cyclic AMP, or by the removal, or leaching out, of an intracellular Ca^{2+} pool necessary for the action of cyclic AMP. The former can be ruled out since there is no increased influx of Ca^{2+} associated with the augmented hormone release induced by either dibutyryl cyclic AMP or theophylline (Milligan and Kraicer, 1971). Thus it would appear that cyclic AMP requires Ca^{2+}, and that Ca^{2+} can be obtained from an intracellular, rather than an extracellular pool. In fact, the increase in Ca^{2+} influx seen with the augmented release of GH induced by a purified GRH preparation (Milligan et al., 1972) may not be a determinant in the release process per se, but may reflect mechanistically important changes in Ca^{2+} distribution within cell components, the latter being essential to the release process.

The most specific conclusion that can be made at present is that cyclic AMP, in its action, probably exerts a change in the distribution of Ca^{2+} within intracellular compartments and that this redistribution is an essential component in the release process. The altered intracellular Ca^{2+} distribution might then result in the activation of a protein kinase which then specifically activates, by phosphorylation, a protein (enzyme?) involved in the release process. This activated phosphorylated protein could be one element in the contractile cytoskeleton–vesicle complex which, when activated, leads to contraction, with subsequent extrusion of hormone-containing granules.

G. PROSTAGLANDINS

1. *Introduction*

The physiological role of the prostaglandins in body function is being investigated at an explosive rate (Weeks, 1972; Hinman, 1972; Horton, 1972). More relevant to this discussion is the accumulating evidence that the prostaglandins may be involved in the release of adenohypophysial hormones through an interrelation with cyclic AMP. This review will attempt only to assemble the scattered recent reports of *in vitro* studies implicating the prostaglandins in the mechanism of hormone release. A more definitive review must await the digestion and assimilation of a great number of reports appearing almost daily.

2. *Chemistry of Prostaglandins*

The prostaglandins are analogs of the parent compound prostanoic acid, a saturated twenty-carbon monocarboxylic acid containing a cyclopentane ring. The prostaglandins are divided into four series designated by the letters A, B, E, and F according to differences in the cyclopentane ring. All the primary prostaglandins are hydroxylated in the 15 position and contain a 13, 14-*trans* double bond. The degree of unsaturation of the side chains is indicated by a subscript numeral after the letter. Thus prostaglandins A_1, B_1, E_1, and F_1 have only the *trans* double bond. Prostaglandins A_2, B_2, E_2, and F_2 have in addition a *cis* double bond in the 5, 6-position, while prostaglandins A_3, B_3, E_3, and F_3 have an additional *cis* double bond in the 17, 18-position (Horton, 1972; Oesterling *et al.,* 1972).

3. *Prostaglandins and Hormone Release from the Adenohypophysis*

Let us begin with the hypothesis that the prostaglandins stimulate hormone release from the adenohypophysis by activating membrane-bound adenyl cyclase which in turn will increase cyclic AMP content. This hypothesis would require the following criteria to be met: (1) the adenohypophysis will contain prostaglandin synthetase and prostaglandins; (2) releasing hormone preparations will increase prostaglandin synthetase activity resulting in increased prostaglandin synthesis, which in turn will increase adenyl cyclase activity. The addition of prostaglandin synthetase inhibitors will prevent this increase in prostaglandin synthesis and the resultant increased adenyl cyclase activity and hormone release; (3) the administration of prostaglandins will increase adenyl cyclase activity, resulting in increased cyclic AMP content and hormone release. This effect will be blocked by prostaglandin antagonists; and (4) the effect of prostaglandins will be potentiated by the addition of cyclic AMP derivatives and/or phosphodiesterase inhibitors.

The following observations have been made: (1) the rat adenohypophysis does contain radioimmunoassayable prostaglandin (Orczyk and Behrman, 1972); (2) prostaglandins will increase adenyl cyclase activity (Zor *et al.,* 1969; MacLeod and Lehmeyer, 1970) and cyclic AMP levels (Zor *et al.,* 1969, 1970; Cooper *et al.,* 1972) and increase the release of growth hormone (MacLeod and Lehmeyer, 1970; Schofield, 1970; Hertelendy, 1971; Hertelendy *et al.,* 1971b; Cooper *et al.,* 1972), TSH (Vale *et al.,* 1971; Dupont and Chavancy, 1972, 1973), and ACTH (Vale *et al.,* 1971), with the increase in tissue cyclic AMP preceding the release of growth hormone (Cooper *et al.,* 1972); however, the prostaglandins do not appear to stimulate LH (Zor *et al.,* 1970) and prolactin (MacLeod and Lehmeyer, 1970) release; (3) theophylline potentiates the action of prostaglandin in stimulating growth hormone release (Schofield, 1970); (4) the increased release of TSH induced by prostaglandin is blocked by the prostaglandin synthetase inhibitor, indomethacin (Dupont and Chavancy, 1972); and (5) the high [K^+]-induced release of LH (Amoss *et al.,* 1971) and TSH and ACTH (Vale *et al.,* 1971) is blocked by the prostaglandin antagonist, 7-oxa-13-prostynoic acid, as is the release of TSH induced by synthetic TRH (Vale *et al.,* 1971). The release of TSH induced by TRH is also blocked by the prostaglandin synthetase inhibitor, indomethacin, while the release induced by monobutyryl cyclic AMP is not (Dupont and Chavancy, 1972, 1973).

While these findings are fragmentary, they are consistent with an action of the prostaglandins in stimulating adenyl cyclase activity. The reported interrelation between the prostaglandins and Ca^{2+} is also consistent with this action. The effect of the prostaglandins in stimulating hormone release is dependent on a tightly bound Ca^{2+} compartment, since very rigorous Ca^{2+}-free conditions must be met before the augmented release of growth hormone, induced by prostaglandins, is inhibited (Hertelendy, 1971; Cooper *et al.,* 1972). Furthermore this rigorous Ca^{2+}-free condition does not prevent the increase in cyclic AMP content induced by the prostaglandins (Cooper *et al.,* 1972).

III. Summary

A simple model of the physicochemical events associated with the release of adenohypophysial hormones can be presented, which is consistent with much of the currently available data. The release process is initiated by an interaction between the hypothalamic releasing hormone and a specific receptor site on the plasma membrane of the adenohypophysial cell. This interaction then leads to an increase in prostaglandin synthetase activity with a subsequent increase in prostaglandin synthesis. The increase in prostaglandin then activates membrane-bound adenyl cyclase which in turn increases cyclic AMP levels within the cell. Cyclic AMP then elicits a change in the intracellular distribution of Ca^{2+} and activates a cyclic AMP-dependent protein kinase. Alternatively, the activated protein kinase alters the intracellular distribution of Ca^{2+}. This protein kinase, with Ca^{2+}, then specifically activates, by phosphorylation, a protein moiety involved in the release process. The activated protein moiety may be one element in the contractile cytoskeleton–vesicle complex, which when activated, leads to contraction with subsequent extrusion of hormone-containing granules.

ACKNOWLEDGMENTS

I acknowledge, with thanks, the valuable suggestions and criticisms of Dr. John V. Milligan during the preparation of this manuscript. The work described from the author's laboratory was supported by the Medical Research Council of Canada, Associated Medical Services Inc., the J. P. Bickell Foundation, and the CIBA Company Ltd.

REFERENCES

Amoss, M., Blackwell, R., Vale, W., Burgus, R., and Guillemin, R. (1971). Stimulation of concomitant secretion *in vitro* of LH and FSH by highly purified hypothalamic LRF: Evidence for a prostaglandin receptor for the release of LH. *Proc. Int. Union Physiol. Sci.* **9**, 17.

Borgeat, P., Chavancy, G., Dupont, A., Labrie, F., Arimura, A., and Schally, A. V. (1972). Stimulation of adenosine 3':5'-cyclic monophosphate accumulation in anterior pituitary gland *in vitro* by synthetic luteinizing hormone-releasing hormone. *Proc. Nat. Acad. Sci. U. S.* **69**, 2677–2681.

Borisy, G. G., and Taylor, E. W. (1967). The mechanism of action of colchicine. *J. Cell Biol.* **34**, 525–533.

Borle, A. B. (1971). Calcium transport in kidney cells and its regulation. *In* "Cellular Mechanisms for Calcium Transfer and Homeostasis" (G. Nichols, Jr. and R. H. Wasserman, eds.), pp. 151–174. Academic Press, New York.

Bowers, C. Y. (1971). Studies on the role of cyclic AMP in the release of anterior pituitary hormones. *Ann. N. Y. Acad. Sci.* **185**, 263–290.

Burgus, R., and Guillemin, R. (1970). Hypothalamic releasing factors. *Annu. Rev. Biochem.* **39**, 499–526.

Cehovic, G. (1969). Rôle de l'adénosine 3'-5'-monophosphate-cyclique dans la libération de TSH hypophysaire. *C. R. Acad. Sci.* **268**, 2929–2931.

Cehovic, G., Lewis, U. J., and Vander Laan, W. P. (1970). Etude de l'action de l'acide adénosine-3'-5'-monophosphate-cyclique sur la libération de l'hormone de croissance et de la prolactine *in vitro*. *C. R. Acad. Sci.* **270**, 3119–3122.

Cooper, R. H., McPherson, M., and Schofield, J. G. (1972). The effect of prostaglandins on ox pituitary content of adenosine 3':5'-cyclic monophosphate and the release of growth hormone. *Biochem. J.* **127**, 143–154.

Douglas, W. W. (1968). Stimulus-secretion coupling: The concept and clues from chromaffin and other cells. *Brit. J. Pharmacol. Chemother.* **34**, 451–474.

Douglas, W. W., Kanno, T., and Sampson, S. R. (1967). Influence of the ionic environment on the membrane potential of adrenal chromaffin cells and on the depolarizing effect of acetylcholine. *J. Physiol. (London)* **191**, 107–121.

Dupont, A., and Chavancy, G. (1972). Prostaglandins and cyclic AMP as mediators of thyrotropin-releasing hormone action. *Proc. Can. Fed. Biol. Soc.* **15**, 721 (abstr.).

Dupont, A., and Chavancy, G. (1973). Prostaglandins and cyclic AMP as mediators of thyrotropin-releasing hormone action. *Proc. Int. Congr. Endocrinol., 4th, 1972* Int. Congr. Ser. No. 256, p. 84, Abstr. 212.

Ewart, R. B. L., and Taylor, K. W. (1971). The regulation of growth hormone secretion from the isolated rat anterior pituitary *in vitro*. *Biochem. J.* **124**, 815–826.

Fleischer, N., Donald, R. A., and Butcher, R. W. (1969). Involvement of adenosine 3',5'-monophosphate in release of ACTH. *Amer. J. Physiol.* **217**, 1287–1291.

Gauthier, M., Labrie, F., Pelletier, G., and Lemay, A. (1972). Role of microtubules in basal and stimulated release of growth hormone (GH) and prolactin (LTH) in rat adenohypophysis *in vitro*. *Proc. Can. Fed. Biol. Soc.* **15**, 715 (abstr.).

Grant, G., Vale, W., and Guillemin, R. (1972). Interaction of thyrotropin releasing factor with membrane receptors of pituitary cells. *Biochem. Biophys. Res. Commun.* **46**, 28–34.

Harris, G. W. (1955). "Neural Control of the Pituitary Gland." Arnold, London.

Hertelendy, F. (1971). Studies on growth hormone secretion. II. Stimulation by prostaglandins *in vitro*. *Acta Endocrinol. (Copenhagen)* **68**, 355–362.

Hertelendy, F., Todd, H., Peake, G. T., Machlin, L. J., Johnston, G., and Pounds, G. (1971a). Studies on growth hormone secretion. I. Effects of dibutyryl cyclic AMP, theophylline, epinephrine, ammonium ion and hypothalamic extracts on the release of growth hormone from rat anterior pituitaries *in vitro*. *Endocrinology* **89**, 1256–1262.

Hertelendy, F., Peake, G., and Todd, H. (1971b). Studies on growth hormone secretion. III. Inhibition of prostaglandin, theophylline and cyclic AMP stimulated growth hormone release by valinomycin *in vitro*. *Biochem. Biophys. Res. Commun.* **44**, 253–260.

Hinman, J. W. (1972). Prostaglandins. *Annu. Rev. Biochem.* **41**, 161–178.

Horton, E. W. (1972). "Prostaglandins." Springer-Verlag, Berlin and New York.

Jutisz, M., and de la Llosa, M. P. (1968). Recherches sur le contrôle de la sécrétion de l'hormone folliculo-stimulante hypophysaire. *Bull. Soc. Chim. Biol.* **50**, 2521–2532.

Jutisz, M., and de la Llosa, M. P. (1969). L'adénosine-3',5'-monophosphate cyclique, un intermédiaire probable de l'action de l'hormone hypothalamique FRF. *C. R. Acad. Sci.* **268**, 1636–1639.

Jutisz, M., and de la Llosa, M. P. (1970). Requirement of Ca^{++} and Mg^{++} ions for the *in vitro* release of follicle-stimulating hormone from rat pituitary glands and in its subsequent biosynthesis. *Endocrinology* **86**, 761–768.

Kraicer, J., and Milligan, J. V. (1971). Effect of colchicine on *in vitro* ACTH release induced by high K^+ and by hypothalamus-stalk-median eminence extract. *Endocrinology* **89**, 408–412.

Kraicer, J., and Milligan, J. V. (1973). Effects of various secretagogues upon ^{42}K and ^{22}Na uptake during *in vitro* hormone release from the rat adenohypophysis. *J. Physiol (London)* **232**, 221–237.

Kraicer, J., Milligan, J. V., Gosbee, J. L., Conrad, R. G., and Branson, C. M. (1969a). Potassium, corticosterone, and adrenocorticotropic hormone release *in vitro*. *Science* **164**, 426.

Kraicer, J., Milligan, J. V., Gosbee, J. L., Conrad, R. G., and Branson, C. M. (1969b). *In vitro* release of ACTH: Effects of potassium, calcium and corticosterone. *Endocrinology* **85**, 1144–1153.

Labrie, F., Béraud, G., Gauthier, M., and Lemay, A. (1971a). Actinomycin-insensitive stimulation of protein synthesis in rat anterior pituitary *in vitro* by dibutyryl adenosine 3',5'-monophosphate. *J. Biol. Chem.* **246**, 1902–1908.

Labrie, F., Lemaire, S., and Courte, C. (1971b). Adenosine 3',5'-monophosphate-dependent protein kinase from bovine anterior pituitary gland. I. Properties. *J. Biol. Chem.* **246**, 7293–7302.

Labrie, F., Lemaire, S., Poirier, G., Pelletier, G., and Boucher, R. (1971c). Adenohypophyseal secretory granules. I. Their phosphorylation and association with protein kinase. *J. Biol. Chem.* **246**, 7311–7317.

Labrie, F., Barden, N., Poirier, G., and De Lean, A. (1972). Binding of thyrotrophin-releasing hormone to plasma membranes of bovine anterior pituitary gland. *Proc. Nat. Acad. Sci. U. S.* **69**, 283–287.

Lacy, P. E., Howell, S. L., Young, D. A., and Fink, C. J. (1968). New hypothesis of insulin secretion. *Nature (London)* **219**, 1177–1179.

Lemaire, S., Pelletier, G., and Labrie, F. (1971). Adenosine 3',5'-monophosphate-dependent protein kinase from bovine anterior pituitary gland. II. Subcellular distribution. *J. Biol. Chem.* **246**, 7303–7310.

Lemay, A., and Labrie, F. (1972). Calcium-dependent stimulation of prolactin release in rat anterior pituitary *in vitro* by N^6-monobutyryl adenosine 3',5'-monophosphate. *FEBS Lett.* **20**, 7–10.

Ling, G., and Gerard, R. W. (1950). External potassium and the membrane potential of single muscle fibres. *Nature (London)* **165**, 113–114.

McCann, S. M. (1971). Mechanism of action of hypothalamic-hypophyseal stimulating and inhibiting hormones. *In* "Frontiers in Neuroendocrinology" (L. Martini and W. F. Ganong, eds.), pp. 209–235. Oxford Univ. Press, London and New York.

MacLeod, R. M., and Fontham, E. H. (1970). Influence of ionic environment on the *in vitro* synthesis and release of pituitary hormones. *Endocrinology* **86**, 863–869.

MacLeod, R. M., and Lehmeyer, J. E. (1970). Release of pituitary growth hormone by prostaglandins and dibutyryl adenosine cyclic 3':5'-monophosphate in the absence of protein synthesis. *Proc. Nat. Acad. Sci. U. S.* **67**, 1172–1179.

Malawista, S. E. (1965). On the action of colchicine. *J. Exp. Med.* **122**, 361–389.

Martin, S., York, D. H., and Kraicer, J. (1973). Alterations in transmembrane potential of adenohypophysial cells in elevated potassium and calcium free media. *Endocrinology* **92**, 1084–1088.

Milligan, J. V., and Kraicer, J. (1970). Adenohypophysial transmembrane potentials: Polarity reversal by elevated external potassium ion concentration. *Science* **167**, 182–184.

Milligan, J. V., and Kraicer, J. (1971). ^{45}Ca uptake during the *in vitro* release of hormones from the rat adenohypophysis. *Endocrinology* **89**, 766–773.

Milligan, J. V., Kraicer, J., Fawcett, C. P., and Illner, P. (1972). Purified growth hormone releasing factor increases ^{45}Ca uptake into pituitary cells. *Can. J. Physiol. Pharmacol.* **50**, 613–617.

Oesterling, T. O., Morozowich, W., and Roseman, T. J. (1972). Prostaglandins. *J. Pharm. Sci.* **61**, 1861–1895.

Orczyk, G. P., and Behrman, H. R. (1972). Ovulation blockade by aspirin or indomethacin— *in vivo* evidence for a role of prostaglandin in gonadotrophin secretion. *Prostaglandins* **1**, 3–20.

Parsons, J. A. (1970). Effects of cations on prolactin and growth hormone secretion by rat adenohypophyses *in vitro*. *J. Physiol. (London)* **210**, 973–987.

Peake, G. T., Steiner, A. L., and Daughaday, W. H. (1972). Guanosine 3'5' cyclic monophosphate is a potent pituitary growth hormone secretagogue. *Endocrinology* **90**, 212–216.

Pelletier, G., Lemay, A., Béraud, G., and Labrie, F. (1972). Ultrastructural changes accompanying the stimulatory effect of N^6-monobutyryl adenosine 3',5'-monophosphate on the release of growth hormone (GH), prolactin (PRL) and adrenocorticotropic hormone (ACTH) in rat anterior pituitary gland *in vitro. Endocrinology* **91**, 1355–1371.

Poirier, G., Barden, N., Labrie, F., Borgeat, P., and De Lean, A. (1973). Partial purification and some properties of adenyl cyclase and receptor for TRH from anterior pituitary gland. *Proc. Int. Congr. Endocrinol., 4th, 1972* Int. Congr. Ser. No. 256, p. 85, abstr. 213.

Porter, J. C., Kamberi, I. A., and Grazia, Y. R. (1971). Pituitary blood flow and portal vessels. *In* "Frontiers in Neuroendocrinology" (L. Martini and W. F. Ganong, eds.), pp. 145–175. Oxford Univ. Press, London and New York.

Rasmussen, H. (1970). Cell communication, calcium ion, and cyclic adenosine monophosphate. *Science* **170**, 404–412.

Ratner, A. (1970). Stimulation of luteinizing hormone release *in vitro* by dibutyryl-cyclic-AMP and theophylline. *Life Sci.* **9**, 1221–1226.

Reinberg, A., and Stolkowski, J. (1953). Influence des variations de la concentration intracellulaire en potassium de l'antéhypophyse sur la sécrétion de l'hormone hypophysaire corticotrope. *C. R. Acad. Sci.* **236**, 1609–1611.

Robison, G. A., Butcher, R. W., and Sutherland, E. W. (1971). "Cyclic AMP." Academic Press, New York.

Samli, M. H., and Geschwind, I. I. (1968). Some effects of energy-transfer inhibitors and Ca^{++}-free or K^+-enhanced media on the release of luteinizing hormone (LH) from the rat pituitary gland *in vitro. Endocrinology* **82**, 225–231.

Schofield, J. G. (1967). Role of cyclic 3',5'-adenosine monophosphate in the release of growth hormone *in vitro. Nature (London)* **215**, 1382–1383.

Schofield, J. G. (1970). Prostaglandin E₁ and the release of growth hormone *in vitro. Nature (London)* **228**, 179.

Schofield, J. G. (1971). Effect of sulphydryl reagents on the release of ox growth hormone *in vitro. Biochim. Biophys. Acta* **252**, 516–525.

Schofield, J. G., and Cole, E. N. (1971). Behaviour of systems releasing growth hormone *in vitro. Mem. Soc. Endocrinol.* **19**, 185–199.

Schofield, J. G., and Stead, M. (1971). ATP, calcium uptake and growth hormone release. *FEBS Lett.* **13**, 149–151.

Steiner, A. L., Peake, G. T., Utiger, R. D., Karl, I. E., and Kipnis, D. M. (1970). Hypothalamic stimulation of growth hormone and thyrotropin release *in vitro* and pituitary 3'5'-adenosine cyclic monophosphate. *Endocrinology* **86**, 1354–1360.

Vale, W., and Guillemin, R. (1967). Potassium-induced stimulation of thyrotropin release *in vitro.* Requirement for presence of calcium and inhibition by thyroxine. *Experientia* **23**, 855–857.

Vale, W., Burgus, R., and Guillemin, R. (1967). Presence of calcium ions as a requisite for the *in vitro* stimulation of TSH-release by hypothalamic TRF. *Experientia* **23**, 853–855.

Vale, W., Rivier, C., and Guillemin, R. (1971). A "prostaglandin receptor" in the mechanisms involved in the secretion of anterior pituitary hormones. *Fed. Proc., Fed. Amer. Soc. Exp. Biol.* **30**, 363 (abstr.).

Wakabayashi, K., Kamberi, I. A., and McCann, S. M. (1969). *In vitro* response of the rat pituitary to gonadotrophin-releasing factors and to ions. *Endocrinology* **85**, 1046–1056.

Weeks, J. R. (1972). Prostaglandins. *Ann. Rev. Pharmacol.* **12**, 317–336.

Weisenberg, R. C., Borisy, G. G., and Taylor, E. W. (1968). The colchicine-binding protein of mammalian brain and its relation to microtubules. *Biochemistry* **7**, 4466–4479.

Wilber, J. F., and Seibel, M. J. (1973). TRH interactions with a specific anterior pituitary membrane receptor. *Proc. Int. Congr. Endocrinol., 4th, 1972* Int. Congr. Ser. No. 256, p. 85, abstr. 214.

Wilber, J. F., Peake, G. T., and Utiger, R. D. (1969). Thyrotropin release *in vitro:* Stimulation by cyclic 3′,5′-adenosine monophosphate. *Endocrinology* 84, 758–760.

York, D. H., Baker, F. L., and Kraicer, J. (1971). Electrical properties of cells in the adenohypophysis—An *in vivo* study. *Neuroendocrinology* 8, 10–16.

York, D. H., Baker, F. L., and Kraicer, J. (1973). Electrical changes induced in rat adenohypophysial cells, *in vivo,* with hypothalamic extract. *Neuroendocrinology* 11, 212–228.

Zimmerman, G., and Fleischer, N. (1970). Role of calcium ions in the release of ACTH from rat pituitary tissue *in vitro. Endocrinology* 87, 426–429.

Zor, U., Kaneko, T., Schneider, H. P. G., McCann, S. M., Lowe, I. P., Bloom, G., Borland, B., and Field, J. B. (1969). Stimulation of anterior pituitary adenyl cyclase activity and adenosine 3′:5′-cyclic phosphate by hypothalamic extract and prostaglandin E₁. *Proc. Nat. Acad. Sci. U. S.* 63, 918–925.

Zor, U., Kaneko, T., Schneider, H. P. G., McCann, S. M., and Field, J. B. (1970). Further studies of stimulation of anterior pituitary cyclic adenosine 3′,5′-monophosphate formation by hypothalamic extract and prostaglandins. *J. Biol. Chem.* **245**, 2883–2888.

ADDENDUM

A number of relevant papers have appeared since this review was written. The following are a sampling of more recent trends.

Eddy, L. J., Hershman, J. M., Taylor, R. E., and Barker, S. B. (1973). Binding of thyrotropin releasing hormone by thyrotropin-secreting cells. *Biochem. Biophys. Res. Commun.* **54**, 140–146.

Fleischer, N., and Wood, J. M. (1973). Ouabain stimulation of ⁴⁵Ca⁺⁺ accumulation in the rat pituitary. *Endocrinology* 92, 1555–1559.

Fleischer, N., Zimmerman, G., Schindler, W., and Hutchins, M. (1972). Stimulation of adrenocorticotropin (ACTH) and growth hormone (GH) release by ouabain: Relationship to calcium. *Endocrinology* 91, 1436–1441.

Grant, G., Vale, W., and Guillemin, R. (1973). Characteristics of the pituitary binding sites of thyrotropin-releasing factor. *Endocrinology* 92, 1629–1633.

Grant, G., Vale, W., and Rivier, J. (1973). Pituitary binding sites for ³H-labelled luteinizing hormone releasing factor (LRF). *Biochem. Biophys. Res. Commun.* 50, 771–778.

Ishikawa, H., and Goto, S. (1973). CRF-stimulated biosynthesis of ACTH in a cell-free system from rat anterior pituitaries. *Biochem. Biophys. Res. Commun.* 52, 884–889.

Jutisz, M., Kerdelhué, B., Bérault, A., and de la Llosa, M. P. (1972). On the mechanism of action of hypothalamic gonadotropin releasing factors. *In* "Gonadotrophins" (B. B. Saxena, C. G. Beling, and H. M. Gandy, eds.), pp. 64–71. Wiley, New York.

Labrie, F., Pelletier, G., Lemay, A., Borgeat, P., Barden, N., Dupont, A., Savary, M., Côté, J., and Boucher, R. (1973). Control of protein synthesis in anterior pituitary gland. *In* "Protein Synthesis in Reproductive Tissue" (E. Diczfalusy, ed.), pp. 301–334. Karolinska Inst., Stockholm.

Labrie, F., Gautier, M., Pelletier, G., Borgeat, P., Lemay, A., and Gouge, J.-J. (1973). Role of microtubules in basal and stimulated release of growth hormone and prolactin in rat adenohypophysis *in vitro. Endocrinology* 93, 903–914.

Spona, J. (1973). LH-RH interaction with the pituitary plasma membrane. *FEBS Lett.* **34,** 24–26.

Spona, J. (1973). LH-RH stimulated gonadotropin release mediated by two distinct pituitary receptors. *FEBS Lett.* **35,** 59–62.

Sundberg, D. K., Krulich, L., Fawcett, C. P., Illner, P., and McCann, S. M. (1973). The effect of colchicine on the release of rat anterior pituitary hormones *in vitro. Proc. Soc. Exp. Biol. Med.* **142,** 1097–1100.

Temple, R., Williams, J. A., Wilber, J. F., and Wolff, J. (1972). Colchicine and hormone secretion. *Biochem. Biophys. Res. Commun.* **46,** 1454–1461.

Wakabayashi, K., Date, Y., and Tamaoki, B.-I. (1973). On the mechanism of action of luteinizing hormone-releasing factor and prolactin inhibiting factor. *Endocrinology* **92,** 698–704.

Wilber, J. F., and Seibel, M. J. (1973). Thyrotropin-releasing hormone interactions with an anterior pituitary membrane receptor. *Endocrinology* **92,** 888–893.

Zor, U., Lamprecht, S. A., Kaneko, T., Schneider, H. P. G., McCann, S. M., Field, J. B., Tsafriri, A., and Lindner, H. R. (1972). Functional relations between cyclic AMP, prostaglandins and luteinizing hormone in rat pituitary and ovary. *Advan. Cyclic Nucleotide Res.* **1,** 503–520.

IDENTIFICATION OF ANTERIOR PITUITARY CELLS BY IMMUNOELECTRON MICROSCOPY[1]

Paul K. Nakane[2]

DEPARTMENT OF PATHOLOGY, UNIVERSITY OF COLORADO

SCHOOL OF MEDICINE, DENVER, COLORADO

I. Introduction

It has been long recognized that anterior pituitary glands of vertebrates are composed of several cell types. Each of these cell types is assigned for production of a specific hormone known to be present in the gland. These assignments were usually accomplished by observing the sequential changes in a particular type of pituitary cell following ablation of a certain peripheral endocrine gland or under certain physiological conditions when secretion of the hormone is either increased

[1] This study was supported in part by grants from the U.S.P.H.S. AI-09109 and AM-13112.

[2] A Career Development Awardee of the United States National Institutes of Health, Grant GM-46228.

45

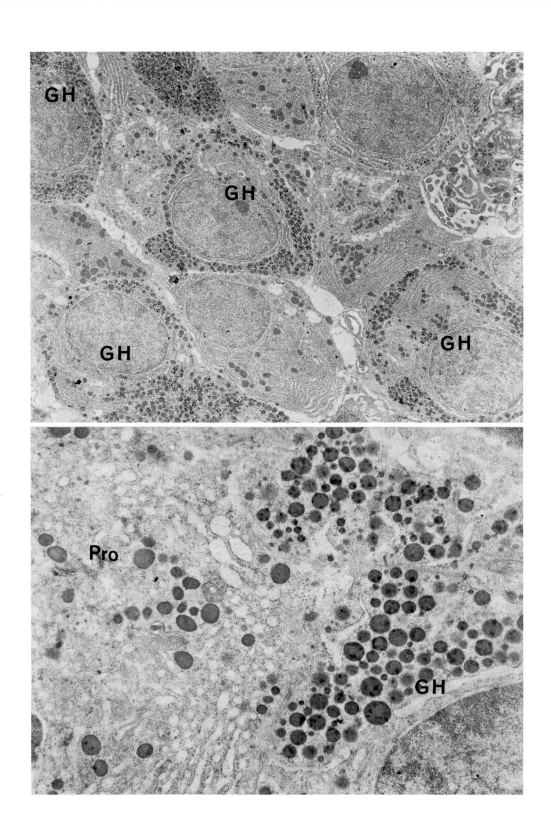

or decreased. In the course of these studies, it was necessary to assume that each of the hormones is made or stored in an identifiable cell type and that the relation between the cell type and target gland is directly mediated by the specific hormone. These assumptions appear to be reasonable and are probably correct, although experimental data to substantiate these assumptions are needed.

The recent advancements in the isolation of anterior pituitary hormones, the understanding of their molecular structure, the production of specific antibodies against these hormones, and the improvement on the technique of immunoelectron microscopy permit the direct identification and localization of the hormones within the cells of the anterior pituitary gland.

The utilization of peroxidase-labeled antibodies for the ultrastructural localization of hormones has proved so far to be the best method available (Nakane, 1968, 1970, 1971a,b, 1973). With this method, or a modification of it, several investigators have successfully identified sites of specific hormones at the light microscopic level (Baker, 1970; Baker *et al.,* 1970, 1972; Baker and Yu, 1971a,b; Phifer *et al.,* 1973) and at the ultrastructural level (Kawarai and Nakane, 1970; Mazurkiewicz and Nakane, 1972; Moriarty and Helmi, 1972).

II. Materials and Methods

A. METHODS FOR IMMUNOELECTRON MICROSCOPIC LOCALIZATION OF HORMONES

The sites of hormones were demonstrated in the anterior pituitary gland of adult albino male rats. The rats were maintained in a temperature-controlled room on Purina laboratory chow and water *ad libitum* for two weeks before use. Anterior pituitary glands of the rats were fixed for 8 hr in 2% paraformaldehyde–picric acid phosphate buffer (Stefanini *et al.,* 1967). The fixed tissues were then washed in phosphate buffered saline (PBS) overnight. Three different approaches were used to localize hormones at the ultrastructural level. Briefly:

(1) The washed tissues were frozen and sectioned at 20–30 μm in thickness. The sections were exposed to rabbit antisera against hormones, washed with PBS,

---

FIG. 1. GH cells. GH was localized directly on an ultrathin section of a pituitary gland which was embedded in methacrylate. Reaction products from 4-Cl-1 naphthol were deposited on secretion granules of GH cells.

FIG. 2. A portion of a GH cell. GH was localized directly on an ultrathin section of a pituitary gland which was fixed in paraformaldehyde–picric acid, postfixed in OsO₄, and embedded in Epon. Reaction products from 4-Cl-1 naphthol were deposited on secretion granules of GH cells, but not on the adjacent prolactin cell (Pro).

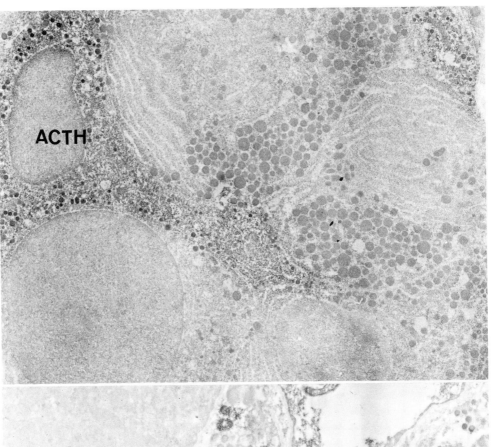

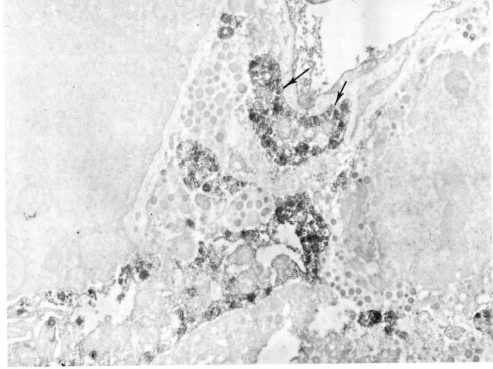

exposed to horseradish peroxidase-labeled sheep anti-rabbit immunoglobulin, and washed with PBS. The antisera-reacted tissues were postfixed in glutaraldehyde, washed with PBS, and the peroxidase was localized histochemically using the technique developed by Graham and Karnovsky (1966). The stained sections were dehydrated and embedded in Epon (Luft, 1961). Ultrathin sections of the tissues were then examined with an electron microscope without further counterstaining (Nakane, 1970; Nakane and Pierce, 1967).

(2) The washed tissues were either dehydrated and embedded in methacrylate or postfixed in OsO_4, dehydrated, and embedded in Epon (Luft, 1961). The tissues were sectioned for electron microscopy and mounted on colloidin–carbon coated grids. The embedding medium was partially removed by water saturated with xylene when the tissues were embedded in methacrylate and by 10% hydrogen peroxide when the tissues were embedded in Epon. The grids were then exposed briefly to 1% sheep serum, washed with PBS, exposed to rabbit antisera against hormone, washed with PBS, reacted with peroxidase-labeled antirabbit immunoglobulin, and washed with PBS. Finally, the peroxidase was localized histochemically using 4-Cl-1 naphthol and H_2O_2 as substrates (Nakane, 1968). The stained grids were observed with an electron microscope with or without counterstaining (Kawarai and Nakane, 1970; Nakane, 1971a).

(3) The washed tissues were dehydrated and embedded in polyethylene glycol 1000 (Sandström and Westman, 1969). The tissues were sectioned with a histological microtome and mounted on glass slides. The sections were washed with PBS, reacted with rabbit antisera against hormones, washed with PBS, and reacted with peroxidase-labeled sheep anti-rabbit immunoglobulin. The reacted sections were washed and the peroxidase was localized histochemically using diaminobenzidine and H_2O_2 as substrates. The stained sections were then dehydrated and embedded in Epon. The Epon-embedded sections were sectioned for electron microscopy. The ultrathin sections were observed with an electron microscope (Mazurkiewicz and Nakane, 1972).

The preservation of antigenicity of hormones varied from one hormone to another depending on the method used. For example, all hormones retained their antigenicity when approaches (1) or (3) were employed, but follicle stimulating hormone (FSH) lost its antigenicity when the glands were embedded in methacrylate or in Epon. Only growth hormone and prolactin retained their

---

FIG. 3. An ACTH cell (ACTH). ACTH was initially localized on 5-μm thick sections of a tissue embedded in polyethylene glycol, then re-embedded in Epon. Reaction products from diaminobenzidine were deposited in secretion granules as well as in cytoplasm.

FIG. 4. A portion of an ACTH cell reaching a wall of sinusoids. ACTH was localized on a thick section. Some secretion granules are stained more heavily at their central cores than at the periphery (arrows).

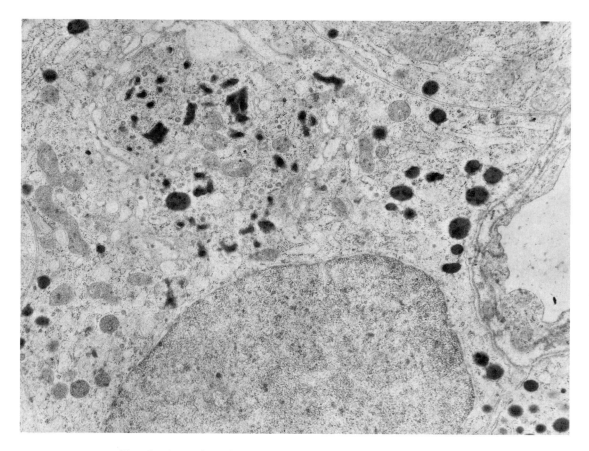

Fig. 5. A portion of a prolactin cell. Prolactin was localized directly on an ultrathin section of a pituitary gland which was fixed in paraformaldehyde–picric acid, postfixed in OsO₄, and embedded in Epon. Reaction products from 4-Cl-1 naphthol were deposited in secretion granules of prolactin cells.

antigenicity when the glands were embedded in Epon. Accordingly, a method which was proved to be useful for localization of a given hormone was used.

B. Preparation and Characterization of Rabbit Antisera against Hormones

The following rabbit antisera were used: Anti-porcine growth hormone, anti-rat growth hormone (a gift from Dr. J. Furth), anti-porcine prolactin, anti-rat prolactin (a gift from Dr. J. Furth), anti-human adrenocorticotropic hormone (ACTH) (a gift from Dr. K. Abe), anti-porcine ACTH (a gift from Dr.

P. Vague), anti-porcine ACTH, anti-17-39-ACTH (a gift from Dr. R. Phifer), anti-ovine luteinizing hormone (LH) (a gift from Dr. G. Niswander), anti-human chorionic gonadotropin (HCG) (a gift from Dr. A. Midgley), anti-ovine FSH (a gift from Dr. V. Gay), anti-human thyrotropic hormone (TSH) (a gift from Dr. W. Odell), and anti-rat TSH.

All antisera except anti-ACTH were initially reacted in agar double diffusion plates against a mixture of pituitary hormones in one well and against a relatively purified hormone in the other well. Antisera which resulted in a single continuous precipitation line were identified. From the precipitation line, the antibody was dissociated in acid solution and isolated in Sephadex G-100 columns. We were unable to obtain purified anti-LH and anti-HCG with this method because of a low dissociation efficiency. Because of this, anti-HCG which cross-reacts specifically with rat LH (a gift from Dr. A. Midgley), anti-ovine LH (a gift from Dr. G. Niswander), and anti-ovine FSH (a gift from Dr. V. Gay) were serially diluted until they were incapable of giving a positive reaction on slides. Fivefold higher concentrations of the diluted antisera were then reacted with their respective antigens in concentrations which would abolish the antibody activities. Then the same amount of antigen was used for cross-reaction (i.e., anti-FSH with LH and anti-LH with FSH). The cross-absorbed antisera were used to localize their respective hormones (Nakane, 1973).

Immunoglobulins isolated from sheep anti-rabbit gamma globulin serum by means of ammonium sulfate precipitation were usually conjugated with horseradish peroxidase. For the conjugation method in earlier experiments, difluorodinitrodiphenyl sulfone was used as the bifunctional reagent (Nakane and Pierce, 1967) and in later experiments periodate oxidized peroxidase was used (Kawaoi and Nakane, 1973). The conjugates were isolated usually in Bio-Gel P-300 or Sephadex G-200 and its nonspecific affinity to tissues was removed by adsorption with activated charcoal and acetone-washed rat liver powder.

III. Results

A. Growth Hormone Cells

Growth hormone cells were found sparsely or not at all in areas adjacent to the intermediate lobe and anterior ventral portion of the gland. In the remaining areas, the cells were dispersed evenly. The cells were usually situated along sinusoids and varied in shape from oval to pyramidal with an approximate diameter of 10–13 μm (Fig. 1). The nucleus was usually situated in the center of the cell while the growth hormone secretion granules with a diameter of about 300–350 mμ were found throughout the cytoplasm (Fig. 2). Some laminated rough endoplasmic reticulum was found frequently near the nucleus. Mitochondria in

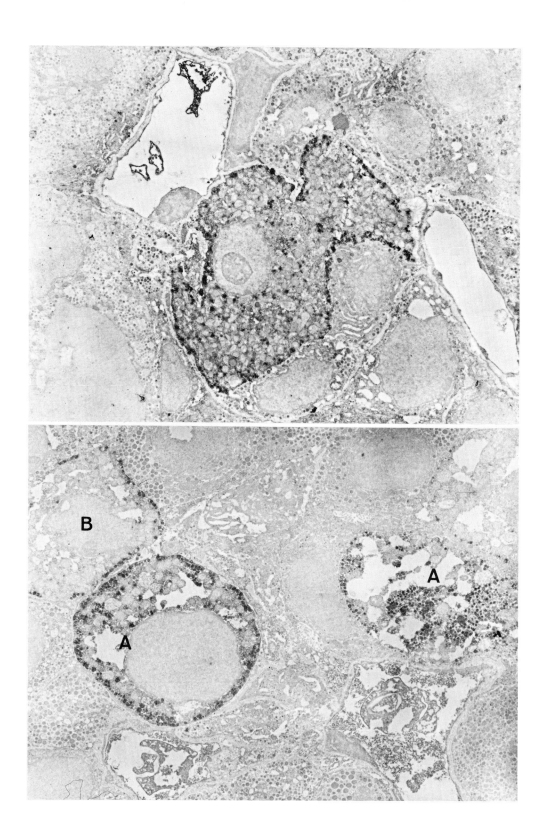

clusters were also found near the nucleus and occasionally among the secretion granules.

B. ACTH Cells

The distribution of ACTH cells was much like that of growth hormone cells. Few cells were observed in the area adjacent to the intermediate lobe and anterior ventral portion of the gland. ACTH cells appeared stellate in shape and were found frequently in juxtaposition to the growth hormone cells (Fig. 3). Usually the body of the ACTH cell was situated at the center of a cord among a cluster of growth hormone cells and extended its cytoplasm to neighboring sinusoids. ACTH-containing secretion granules were distributed mainly at the periphery of the cytoplasm. The mitochondria, endoplasmic reticulum, and Golgi bodies were clustered near the nucleus. In a portion of ACTH cells which reached the walls of sinusoids, mitochondria were usually found in the center and secretion granules with a diameter of about 200 mμ were situated near the edge of the cell (Fig. 4). Some of the secretion granules appeared solid and some had central cores.

C. Prolactin Cells

The cells containing prolactin were found sparsely in the anterior ventral portion of the gland. However, unlike growth hormone cells or ACTH cells, prolactin cells were found in the areas near the intermediate lobe. The prolactin cells had unique shapes. They were frequently cup-shaped and surrounded gonadotropic cells. The nucleus of the prolactin cell was situated centrally, *i.e.,* the base of the cup. The portion of cytoplasm which surrounded gonadotropic cells also contained hormone as well as the area near the nucleus. These cells usually had well developed rough endoplasmic reticulum and contained large secretion granules with a diameter often in excess of 800 mμ. The secretion granules varied from spherical to polymorphic (Fig. 5).

D. TSH Cells

Unlike the other types of cells, TSH cells were more frequently found in clusters in the center of the pituitary gland. The TSH cells were polygonal to stellate in shape and were usually located in the center of the cord in the manner similar to ACTH cells. The cytoplasm of TSH cells was more voluminous than

Fig. 6. TSH cell. TSH was localized on a thick section. Secretion granules containing TSH were distributed at the peripheries of the cell.

Fig. 7. FSH cells. FSH was localized on a thick section. Type A cells (A) contained secretion granules throughout their cytoplasm. Type B cells (B) contained secretion granules at the peripheries of their cytoplasm.

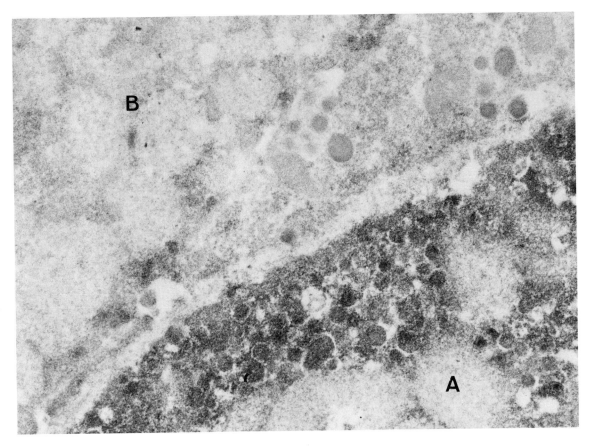

FIG. 8. FSH cells. A similar view to that of Fig. 7, but at a higher magnification.

that of ACTH cells and faced the wall of sinusoids directly. The clusters of TSH cells usually did not contain ACTH cells and vice versa. The secretion granules were more frequently situated near the plasma membrane. Large vacuoles containing little or no hormone were dispersed throughout the cytoplasm and round, oval mitochondria were found among them. Occasionally, Golgi complexes were found near the nucleus (Fig. 6).

Secretion granules were situated near the plasma membrane and were about 150–200 mμ in diameter. Some secretion granules appeared solid and some had central cores.

E. FSH CELLS

FSH cells were distributed throughout the anterior pituitary gland including the area adjacent to the intermediate lobe. When tissue sections were reacted with

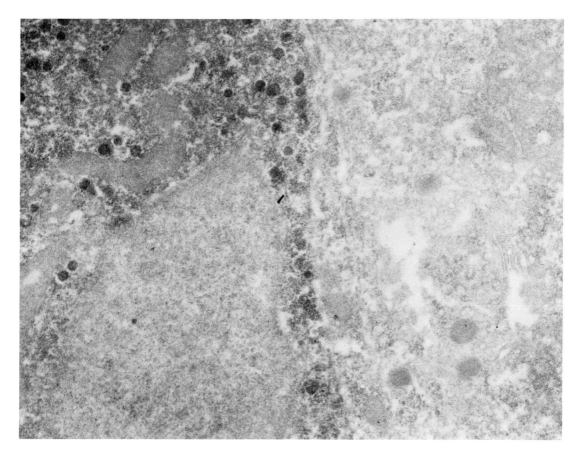

Fig. 9. A portion of an FSH cell with only one type of secretion granule.

anti-FSH, two cell types reacted with the antiserum (Figs. 7 and 8). The type A cells were oval in shape with large vacuoles dispersed throughout the cytoplasm. Hormone-containing secretion granules were found among the vacuoles. Type B cells were angulated and contained secretion granules which were situated near the plasma membrane. Dilated rough endoplasmic reticulum occupied most of the cytoplasm and mitochondria were dispersed among the endoplasmic reticulum saccules. Both cell types contained secretion granules measuring about 200–250 mμ in diameter (Fig. 9).

F. LH Cells

The distribution and cellular morphology were essentially identical to that of the type A FSH cells. LH was localized in several types of secretion granules in

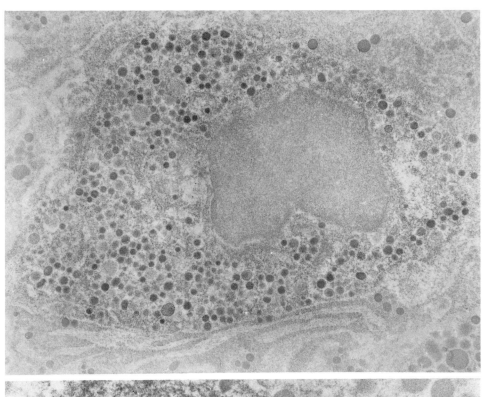

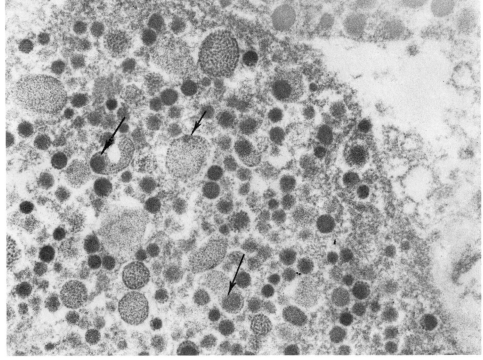

the cell (Fig. 10). LH was found in small dense secretion granules ranging from 75 to 250 mμ and in large dilated vacuoles of up to 600 mμ in diameter (Fig. 11).

IV. Discussion

Distribution and morphological characteristics of each of the cell types specialized in production of a specific hormone have been described. In general, growth hormone cells and ACTH cells were distributed similarly and were found neither in the area adjacent to the intermediate lobe nor in the most anterior and ventral portion of the gland. Prolactin, FSH, and LH cells had similar distributions; the latter two were concentrated in the sex zone. TSH cells were usually found in the middle of the gland. Within each cord, growth hormone cells and gonadotropic cells were oval, situated at the periphery, and faced sinusoids; on the other hand, ACTH cells and TSH cells were situated in the center of the cord and their cytoplasmic extension reached the wall of the sinusoids in between other cell types (Nakane, 1970).

The appearance and location of the growth hormone cells identified immunocytochemically confirm the description made for growth hormone cells by Farquhar and Rinehart (1954) and Kurosumi (1968). Since there are no other cell types which match this description, there should be little difficulty in identifying this cell type in *normal* pituitary glands independent of immunocytochemical means. Likewise, the prolactin cells may be identified without difficulty because of their unique spherical to polymorphic secretion granules as described by Smith and Farquhar (1966).

With our study, ACTH cells and TSH cells were more similar than different. Both of these cell types were situated usually in the center of the cord, were angulated, and possessed secretion granules which were smaller than in other cell types and were situated near the plasma membrane. TSH cells were usually larger than ACTH cells, and the secretion granules of TSH cells were slightly smaller than those of ACTH cells, but with a given electron micrograph, the size of the cells is not a useful criterion since it varies depending on the level of sectioning.

With conventional electron microscopy, many investigators have identified ACTH cells and differentiated them from TSH cells. The majority of investiga-

---

FIG. 10. An LH cell. LH was localized initially on 5-μm thick sections of a tissue embedded in polyethylene glycol, then re-embedded in Epon. Reaction products from diaminobenzidine were deposited in secretion granules.

FIG. 11. A portion of an LH cell. The hormone was localized as in Fig. 9. Arrows indicate the variety of secretion granules found in this type of cell.

tors identified a given population of cells in the pituitary gland as ACTH cells because these cells underwent morphological changes following alteration of physiological conditions which are known to change the ACTH content of the gland (Kurosumi and Kobayashi, 1966; Siperstein and Allison, 1965; Nakayama *et al.*, 1969). In differentiating these cells identified as ACTH-containing cells from TSH cells, the majority of investigators utilized the size, shape, and location of the cells and the size and morphological appearance of secretion granules. All investigators are in agreement that ACTH cells are stellate in shape with secretion granules situated near the plasma membrane. The size and morphological appearance of the secretion granules has been in controversy. Siperstein and her colleagues (Siperstein and Allison, 1965; Siperstein and Miller, 1970) measured the secretion granules to be approximately between 200–260 mμ and the granules are oval and solid in mature states; on the other hand, Kurosumi and Kobayashi (1966) measured the secretion granules of the ACTH cell to be between 150 and 200 mμ and most granules in the normal states are haloed granules. Since these investigators prepared the electron microscopic specimens in a slightly different manner, it has been difficult to establish uniform criteria usable in identification of ACTH cells with the electron microscope.

Our attempt to clarify this point has been met with several technical difficulties. For example, the majority of investigators utilized as an initial fixative for the gland either glutaraldehyde or osmium tetroxide, but these fixatives unfortunately inactivate the immunogenicity of both TSH and ACTH and were not usable. Not being able to use an identical technical procedure, we are unable to correlate the observations made by others to ours since the fixatives employed in our study do not result in identical morphological preservation of ACTH and TSH cells.

The localization of the two types of gonadotropic hormones have been the subject of disagreement for many years. Purves and Griesbach (1951, 1954) suggested that gonadotropic cells situated at the periphery of the gland contained FSH and those situated in the center of the gland contained LH. Conversely, Rennels (1957, 1963) concluded that the gonadotropic cells at the periphery of the gland contained LH and those at the center contained FSH, whereas Barrnett *et al.* (1956) suggested that FSH and LH were in the same cell. With the use of highly specific immunochemical techniques and resolution of electron microscopy, it is clear that FSH-containing cells and LH-containing cells are distributed in a similar manner and in many instances both of these hormones are present within a given cell. The concentration of FSH and LH varies from cell to cell. In the center of the gland, some cells contain almost exclusively either FSH or LH. However, a majority of gonadotropic cells at the periphery of the gland contained both FSH and LH (Nakane, 1973). This is in accordance with the observation that the release of FSH and LH at midcycle occurs simultaneously in human females (Midgley and Jaffe, 1969) and that FSH and LH are present in the same cell in the human pituitary gland (Leleux and Robyn, 1971; Phifer *et al.*, 1973). Isolation and characterization of a factor which is responsible for release of both LH and FSH (Schally *et al.*, 1971) also support the finding that FSH and LH may occur in the same cell.

Immunocytochemical localization of hormones in the anterior pituitary gland at the ultrastructural level presents several problems. First of all, the hormones are either protein or glycoprotein and are soluble in the usual physiological solutions and need to be insolubilized by fixatives. The rate of fixation should be fast enough so that the hormones will not diffuse from their original sites. If the diffusion should occur during the fixation, the hormone may be fixed to the secondary sites and confused with the primary sites of the hormone. At present, a method has not been devised to detect this possible false localization. Since the majority of fixatives used in electron microscopy are directed toward proteins, the loss of antigenicity of hormone is expected during the fixation. Several investigators devised a method to estimate this loss during fixation, however, the method used isolated antigens rather than antigens *in situ* (Kraehenbuhl *et al.*, 1971) and the actual effect on the antigenicity at their natural sites may not be accurately determined.

Therefore, if one is successful in localizing antigens in tissues, one must keep in mind that the positive sites represent sites of antigen but the antigen may have migrated to the secondary site and may not be present in their natural habitat. If one is unsuccessful in localizing a given antigen in expected areas, one must keep in mind that the antigen may not have been fixed and been washed away during the fixation, or the antigen may be there, but has been denatured during the fixation. Furthermore, the effect of fixative *may* vary depending on the sites of antigens. For example, an antigen in Golgi complexes may be more easily affected than those in the endoplasmic reticulum.

Although the majority of the cells in the anterior pituitary gland of the male rat may be classified into certain cell types with conventional electron microscopy, it must be remembered that not all cells have the characteristics described in this study. Some of them are in synthetic phase and may not be positively identified as to their hormone content. Whether these cell types undergo a characteristic synthetic phase which leads to the mature cell type or share a common characteristic during the synthetic phase and at the terminal stage obtain a distinct characteristic remains to be determined. How many of these characteristics may be employed for cell identification under abnormal physiological conditions remains to be seen. Until such studies are done, when one wishes to investigate a particular cell type under physiologically altered conditions, one may have to rely on immunocytochemical methods or some other method which will identify the type of hormone in the cells.

ACKNOWLEDGMENTS

It is with pleasure that I express my thanks to Drs. G. B. Pierce, Y. Kawarai, A. Kawaoi, and J. Mazurkiewicz, and to Ms. M. B. Wilson, M. V. Van Der Schouw, and L. D. King for their technical assistance during the course of this study and in preparation of the manuscript.

REFERENCES

Baker, B. L. (1970). Studies on hormone localization with emphasis on the hypophysis. *J. Histochem. Cytochem.* **18**, 1.

Baker, B. L., and Yu, Y. Y. (1971a). The thyrotropic cell of the rat hypophysis as studied with peroxidase-labeled antibody. *Amer. J. Anat.* **131**, 55.

Baker, B. L., and Yu, Y. Y. (1971b). Hypophyseal changes induced by thyroid deficiency and thyroxine administration as revealed by immunochemical staining. *Endocrinology* **89**, 996.

Baker, B. L., Pek, S., Midgley, A. R., and Gersten, B. (1970). Identification of the corticotropin cell in rat hypophyses with peroxidase-labeled antibody. *Anat. Rec.* **166**, 557.

Baker, B. L., Pierce, J. G., and Cornell, J. S. (1972). The utility of antiserums to subunits of TSH and LH for immunochemical staining of the rat hypophysis. *Amer. J. Anat.* **135**, 251.

Barrnett, R. J., Ladman, E. J., McAllaster, N. J., and Siperstein, E. R. (1956). The localization of glycoprotein hormones in the anterior pituitary gland of rats investigated by differential protein solubilities, histological stains and bio-assays. *Endocrinology* **59**, 398.

Farquhar, M. G., and Rinehart, J. F. (1954). Electron microscopic studies of the anterior pituitary gland of castrated rats. *Endocrinology* **54**, 516.

Graham, R. C., and Karnovsky, M. J. (1966). The early stages of absorption of injected horseradish peroxidase in the proximal tubules of mouse kidney: Ultrastructural cytochemistry by a new technique. *J. Histochem. Cytochem.* **14**, 291.

Kawaoi, A., and Nakane, P. K. (1973). An improved method of conjugation of peroxidase with proteins. *Fed. Proc., Fed. Amer. Soc. Exp. Biol.* **32**, 840.

Kawarai, Y., and Nakane, P. K. (1970). Localization of tissue antigens on the ultrathin sections with peroxidase-labeled antibody method. *J. Histochem. Cytochem.* **18**, 161.

Kraehenbuhl, J. P., De Grandi, P. B., and Campiche, M. A. (1971). Ultrastructural localization of intracellular antigen using enzyme-labeled antibody fragments. *J. Cell Biol.* **50**, 432.

Kurosumi, K. (1968). Functional classification of cell types of the anterior pituitary gland accomplished by electron microscopy. *Arch. Histol. Jap.* **29**, 329.

Kurosumi, K., and Kobayashi, Y. (1966). Corticotrophs in the anterior pituitary glands of normal and adrenalectomized rats as revealed by electron microscopy. *Endocrinology* **78**, 745.

Leleux, P., and Robyn, C. (1971). Immunohistochemistry of individual hypophyseal cells. *Acta Endocrinol. (Copenhagen)* **153**, 168.

Luft, J. H. (1961). Improvements in epoxy resin embedding methods. *J. Biophys. Biochem. Cytol.* **9**, 409.

Mazurkiewicz, J. E., and Nakane, P. K. (1972). Light and electron microscopic localization of antigens in tissues embedded in polyethylene glycol with a peroxidase-labeled antibody method. *J. Histochem. Cytochem.* **20**, 969.

Midgley, A. R., Jr., and Jaffe, R. B. (1969). Gonadotropins in the human male or female. *Progr. Endocrinol., Proc. Int. Congr. Endocrinol., 3rd, 1968* p. 885.

Moriarty, G. C., and Helmi, N. S. (1972). Electron microscopic study of the adrenocorticotropin-producing cell with the use of unlabeled antibody and the soluble peroxidase–antiperoxidase complex. *J. Histochem. Cytochem.* **20**, 590.

Nakane, P. K. (1968). Simultaneous localization of multiple tissue antigens using the peroxidase-labeled antibody method: A study on pituitary glands of the rat. *J. Histochem. Cytochem.* **16**, 557.

Nakane, P. K. (1970). Classifications of anterior pituitary cell types with immunoenzyme histochemistry. *J. Histochem. Cytochem.* **18**, 9.

Nakane, P. K. (1971a). Application of peroxidase-labelled antibodies to the intracellular localization of hormones. *Acta Endocrinol. (Copenhagen), Suppl.* **153**, 190.

Nakane, P. K. (1971b). Effect of thyrotropin-releasing factor on thyrotropic cells *in vitro*. *Progr. Brain Res.* **34**, 139.

Nakane, P. K. (1973). Distribution of gonadotropic cells in the anterior pituitary gland of the rat. *In* "The Regulation of Mammalian Reproduction" (S. J. Segal *et al.*, eds.), p. 79. Thomas, Springfield, Illinois.

Nakane, P. K., and Pierce, G. B., Jr. (1967). Enzyme-labeled antibodies for the light and electron microscopic localization of tissue antigens. *J. Cell Biol.* **33**, 307.

Nakayama, I., Nickerson, P. A., and Skelton, F. R. (1969). An ultrastructural study of the adrenocorticotrophic hormone-secreting cell in the rat adenohypophysis during adrenal cortical regeneration. *Lab. Invest.* **21**, 169.

Phifer, R. F., Midgley, A. R., and Spicer, S. S. (1973). Immunohistologic and histologic evidence that follicle-stimulating hormone and luteinizing hormone are present in the same cell type in the human pars distalis. *J. Clin. Endocrinol. Metab.* **36**, 125.

Purves, H. D., and Griesbach, W. E. (1951). The sites of thyrotrophin and gonadotrophin production in the rat pituitary studied by McManus-Hotchkiss staining for glycoprotein. *Endocrinology* **49**, 244.

Purves, H. D., and Griesbach, W. E. (1954). The sites of follicle stimulating and luteinizing hormone production in the rat pituitary. *Endocrinology* **55**, 785.

Rennels, E. G. (1957). Two tinctorial types of gonadotrophic cells in the rat hypophysis. *Z. Zellforsch. Mikrosk. Anat.* **45**, 464.

Rennels, E. G. (1963). Gonadotrophic cells of rat hypophysis. *In* "Cytologie de l'adénohypophse" (J. Benoit and C. Da Lage, eds.), p. 201. CNRS, Paris.

Sandström, B., and Westman, J. (1969). Non-freezing light and electron microscopic enzyme histochemistry by means of polyethylene glycol embedding. *Histochemie* **19**, 181.

Schally, A. V., Arimura, A., Kastin, A. J., Matsuo, H., Baba, Y., Redding, T. W., Nair, R. M. G., Debeljuk, L., and White, W. F. (1971). Gonadotropin-releasing hormone: One polypeptide regulates secretion of luteinizing and follicle-stimulating hormones. *Science* **173**, 1036.

Siperstein, E. R., and Allison, V. F. (1965). Fine structure of the cells responsible for the secretion of adrenocorticotrophin in the adrenalectomized rat. *Endocrinology* **76**, 70.

Siperstein, E. R., and Miller, K. J. (1970). Further cytophysiologic evidence for the cells that produce adrenocorticotrophic hormone. *Endocrinology* **86**, 451.

Smith, R. E., and Farquhar, M. G. (1966). Lysosome function in the regulation of the secretory process in cells of the anterior pituitary gland. *J. Cell Biol.* **31**, 319.

Stefanini, M., De Martino, C., and Zamboni, L. (1967). Fixation of ejaculated spermatozoa for electron microscopy. *Nature (London)* **216**, 173.

LOCALIZATION OF HORMONES IN THE PITUITARY: RECEPTOR SITES FOR HORMONES FROM HYPOPHYSIAL TARGET GLANDS AND THE BRAIN

Walter E. Stumpf, Madhabananda Sar, and Donald A. Keefer

LABORATORIES FOR REPRODUCTIVE BIOLOGY, DEPARTMENTS OF ANATOMY AND PHARMACOLOGY

UNIVERSITY OF NORTH CAROLINA, CHAPEL HILL, NORTH CAROLINA

The function of the "master gland" is dictated by chemical messengers produced elsewhere in the body, namely, "releasing" hormones from the periventricular gland (Stumpf, 1970) of the central nervous system and hormones from peripheral endocrine glands. While it seems well established that there are specialized cell types in the adenohypophysis, only recently have pituitary cells been identified as target cells for blood-borne hormones. Cells in the pituitary have for some time been identified by tinctorial, histochemical, and ultrastructural approaches with relation to the hormones produced by them. Target cells in the pituitary could only be identified after technical progress had led to the development of the dry-autoradiography technique (Stumpf and Roth, 1966). The results presently available on cellular and subcellular hormone target sites in the pituitary are restricted to the light microscopic level. The localization in the pituitary of

63

releasing or inhibiting hormones and peripheral hormones needs to be extended to the electron microscopic level. Also, the correlation between hormone target sites and hormone-producing cells needs to be studied for a better understanding of pituitary functions and hormone regulation. This can be accomplished by a combination of autoradiographic, tinctorial, and immunohistochemical approaches, including ultrastructural resolution wherever possible. This will permit us to define, for instance, on which cell type a given steroid hormone "feeds back," or, perhaps more specifically, which cells or cell types are activated genomically. Experiments under way in our laboratories promise to accomplish this goal by combining dry-autoradiographic localization of tritiated steroids with the localization of pituitary hormones using the peroxidase-labeled antibody technique.

Ultrastructural localization of steroids has not been accomplished yet. Classic fixation and embedding techniques have been demonstrated to be unsafe and unreliable since noncovalently bound small molecular weight substances can be displaced and redistributed in the tissue or even lost (Stumpf and Roth, 1966). Progress in low temperature tissue preparation (Stumpf and Roth, 1967) as well as the development of special fixation and embedding procedures (Mizuhira, 1972) offer two possible routes to pursue the goal of ultrastructural steroid localization. The present report gives an overview of the available data on hormone localization in the pituitary as related to hormones produced outside of the pituitary. The autoradiographic data available are almost exclusively from our laboratory, although a few other laboratories have made attempts to localize peripheral hypophyseotropic hormones in the pituitary. Their results, however, are conflicting and the discrepancies appear to be attributable to shortcomings in technique (Stumpf, 1969).

I. Methods

Most of the experiments were conducted with male and female Sprague-Dawley rats after removal of the endocrine organ which elaborates the hormone under investigation, so that the endogenous hormone level was low and the hormone binding sites in target cells could be demonstrated more readily. The hormones used were labeled with tritium with a high specific activity between 35 and 110 Ci/mM. Thyroid hormones were labeled with ^{125}I.

The compounds were dissolved in the following manner: steroids in 5–10% ethanol–isotonic saline; TRH in 3% ethanol–water; and thyroid hormone in 50% propylene glycol. The material was injected intravenously or subcutaneously at a dose ranging between 0.1 μg and 2.0 μg per 100 gm body weight. The animals were killed by decapitation at different time intervals after the injection ranging from 1 min to 2 hr. The pituitary was removed, placed on a tissue holder, and frozen in liquefied propane at $-180°$C. The tissue was cut at $-40°$C in the Wide Range Cryostat (Harris Manufacturing Company, North Billerica, Mass.) at 2 to

4 μm thickness. The frozen sections were freeze-dried, then dry-mounted on desiccated Kodak NTB-3 emulsion coated slides, and exposed for different lengths of time. After the exposure, the slides were photographically processed, stained with methyl green-pyronin, or Gomori's trichrome, or a modification of aldehyde fuchsin and Masson's trichrome (M. Sar and W. E. Stumpf, unpublished). The dry-autoradiography technique has been described (Stumpf, 1971a; Stumpf and Sar, 1974). Immunoautoradiograms were fixed after exposure, photographically processed, and stained according to the immunoglobulin-enzyme bridge method (Mason *et al.,* 1969; Nakane and Pierce, 1967).

II. Estrogens

[³H]estradiol localization was studied in pituitaries of rats in different hormonal states, including: intact 23- to 25-day-old, intact mature males; mature males and females after short- and long-term castration; and females, castrated during delayed implantation and lactation. Other species studied include sexually immature female tree shrews, short-term castrated lactating opossum, short-term castrated female squirrel monkey, and intact female salamanders.

In the pars distalis, in all of the rats studied, radioactivity is found to be concentrated in nuclei of a relatively large population of cells with a labeling index of over 60% (Stumpf, 1968). Among the labeled cells, a gradient of nuclear concentration can be seen with some cells being heavily labeled. It appears in the rat, that under certain physiological conditions, such as pregnancy and lactation, a topographically distinct population of cells concentrates more radioactivity compared to the other labeled cells in the pars distalis. This topographical distribution of labeled cells is depicted in Fig. 1. Morphologically densely labeled cells appear

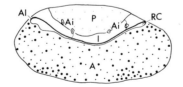

FIG. 1. Rat pituitary, coronal section, schematic, demonstrating the areas of accumulation of cells, identified as gonadotrophs, with *heavy* (large dots) nuclear concentration of radioactivity after the injection of [³H]estradiol (see text). Scattered heavily labeled cells are found in other parts of the anterior lobe (A). Many of the anterior pituitary cells show less intense moderate nuclear labeling (medium dots). A moderate uptake of radioactively labeled estrogen was observed in lining cells between anterior and intermediate lobes (AI) and in invaginated cells (Ai) found in the intermediate lobe (I) and at the border between intermediate and posterior (P) lobes. Weak nuclear labeling (small dots) has been observed in neuropituicytes, but generally not in intermediate lobe cells. Colloid in the residual cleft (RC) may concentrate varying amounts of radioactivity.

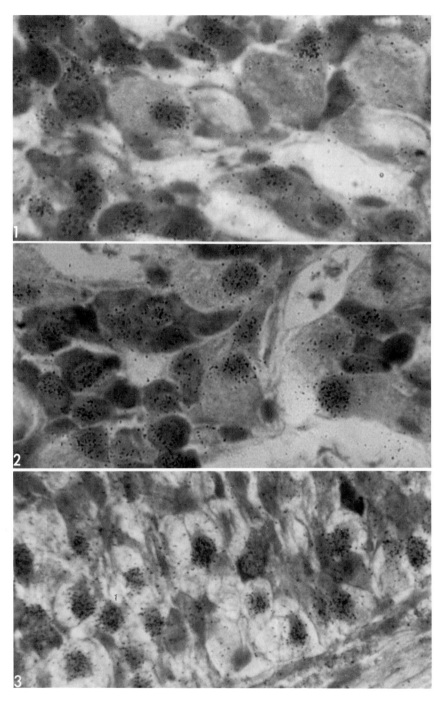

as basophils, and the distribution pattern is similar to the one described for go- nadotrophs (Purves and Griesbach, 1951). Further studies are required to pre- cisely define the quantitative differences and to identify the cell type(s) involved. While some radioactivity is contained in cytoplasm, there is never an indication of accumulation of labeled hormone at outer cell membranes or in nucleoli.

By using tinctorial identification with Gomori's trichrome or a modification of aldehyde fuchsin and Masson's trichrome (Stumpf, 1968; Sar and Stumpf, 1973a), radioactive estradiol is found in the pars distalis to be localized in different cell types, such as acidophils, basophils, and chromophobes (Plate I, Figs. 1 and 2; Plate II, Fig. 1). Among the basophils, 28-day castration cells show labeling (Plate 1, Figs. 1 and 2). It appears, that thyrotrophs also concentrate radioactivity in their nuclei. Variations of the intensity of labeling are apparent between different cells, and some of the cells of a given cell type that show nuclear concentration of radioactivity remain unlabeled. Endothelial cells do not concentrate radioactivity.

In the pars tuberalis, the small cells which stain similarly to the cells of the pars distalis show weak nuclear labeling.

In the intermediate lobe, the typical intermediate lobe cells appear essentially free of radioactivity under most of the conditions studied (Plate II, Fig. 2). Small islands or strands of cells which stain—albeit less intensively—like anterior lobe cells concentrate in their nuclei a relatively small amount of radioactivity (Plate II, Fig. 2).

In the posterior lobe, a diffuse distribution of radioactivity exists which con- trasts with the low level of radioactivity in the intermediate lobe. However, in earlier experiments no accumulation of radioactivity could be observed, while under the same conditions of exposure, distinct labeling of anterior lobe cells and other target cells existed. Failure to detect labeling in cells of the posterior pituitary may be attributable to dose, exposure time, or hormonal state of the animal. In addition to radioactively labeled "invaginated cells" (Plate II, Fig. 2) in the pos- terior lobe, in our recent studies, including animals in early pregnancy and delayed implantation, two types of cells in the posterior pituitary are seen after staining with methyl green-pyronin. Cells with large pale nuclei showed nuclear concen-

---

PLATE I.

Figs. 1 and 2. Autoradiograms of anterior pituitary of female rat (Fig. 1) and male rat (Fig. 2), castrated for 4 weeks, after injection of [³H]estradiol. Exposure time: 100 days. 4 μm, stained with aldehyde fuchsin and Masson's trichrome, ×1180. Nuclear concentration of estrogen appears in acidophils and gonadotrophs (castration cells) in female and male rat.

Fig. 3. Autoradiogram of anterior pituitary of castrated lactating opossum after injection of [³H]estradiol. Exposure time: 125 days. Stained with aldehyde fuchsin and Masson's trichrome, ×1200. Acidophils with red cytoplasm and basophils with green cytoplasm show nuclear concentration of radioactivity. Connective tissue capsule (lower right) does not show concentration of radioactivity.

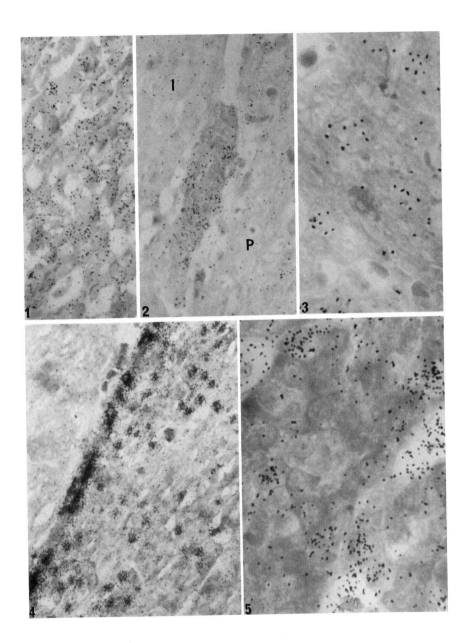

PLATE II.

Figs. 1–3. Autoradiograms of pituitary anterior lobe (Fig. 1), intermediate and posterior lobes (Fig. 2), and posterior lobe (Fig. 3) of rat with delayed implantation after the injection of [³H]estradiol. Exposure time: 222 days. 2 μm, stained with methyl green-pyronin, ×475 tration of radioactivity (Plate II, Fig. 3), whereas cells with smaller and more densely stained nuclei, as well as endothelial cells, do not show retention and concentration of radioactivity. It appears that the labeled cells in the posterior pituitary contain hormone-specific binding sites, weaker than the cells of the anterior lobe, but distinguishable from nonlabeled "nontarget" tissues. The physiological significance of this observation remains to be established.

In the posterior lobe, near the border of the intermediate lobe, so-called ectopic cells, strands, or invaginations of cells were reported for the first time to exist also in the rat (Stumpf, 1968). We have since observed such invaginations in many rat pituitaries (Plate II, Fig. 2). Although current belief is (Christ, 1966) that these "invaginated" cells originate from the intermediate lobe, their concentration of [³H]estradiol—although generally less than in labeled anterior lobe cells—and their tinctorial properties suggest that they are related to anterior lobe cells. Similarly, in the intermediate lobe such "ectopic" cells can be seen especially in the vicinity of blood vessels.

The radioactivity in the nuclei of cells in the pituitary is most likely estradiol. This can be derived from our competition experiments with chlomiphen (unpublished observation) and radioassay data from other laboratories (Eisenfeld and Axelrod, 1965; Leavitt et al., 1969). The nuclear association of estradiol, which was demonstrated for the first time in our laboratory, has been confirmed by biochemical centrifugal fractionation (Leavitt et al., 1969), including the use of pituitary cell suspensions (Leavitt et al., 1973). Earlier autoradiographic studies in other laboratories (Inman et al., 1965), including Attramadal's (1965), failed to show this subcellular distribution pattern demonstrated with the dry-mount

---

(Fig. 1), ×475 (Fig. 2), ×1180 (Fig. 3). A large population of cells of the anterior pituitary show nuclear concentration of estrogen. The labeling intensities vary among different cells. At the junction of intermediate (I) and posterior lobes (P), "an island of cells" with nuclear labeling can be seen. These cells are similar in appearance to the cells of the anterior lobe. Their nuclear concentration of radioactivity is, on the average, somewhat less than in cells of the anterior lobe, but somewhat more than in pituicytes. In the posterior lobe pituicytes with large nuclei (Figs. 2 and 3) show weak nuclear labeling, whereas other cells, including endothelial cells, remain free of labeling.

FIG. 4. Autoradiogram of pituitary of castrated lactating opossum after injection of [³H]estradiol. Exposure time: 160 days. Stained with methyl green-pyronin, ×275. Radioactivity is concentrated in nuclei of certain anterior pituitary cells (right half of picture) with varying intensities of label. Between anterior and posterior lobes (left half), a single layer of cells can be seen which shows a different distribution of radioactivity. Certain cells in this layer concentrate radioactivity in their cytoplasm, sparing the nucleus. Posterior lobe cells contain little or no radioactivity.

FIG. 5. Autoradiogram of anterior pituitary of ovariectomized squirrel monkey (age unknown) after injection of [³H]estradiol. Exposure time: 98 days. Stained with methyl green-pyronin, ×875. Radioactivity is concentrated in the nuclei of a relatively small number of anterior lobe cells. Radioactivity is also retained in blood capillaries.

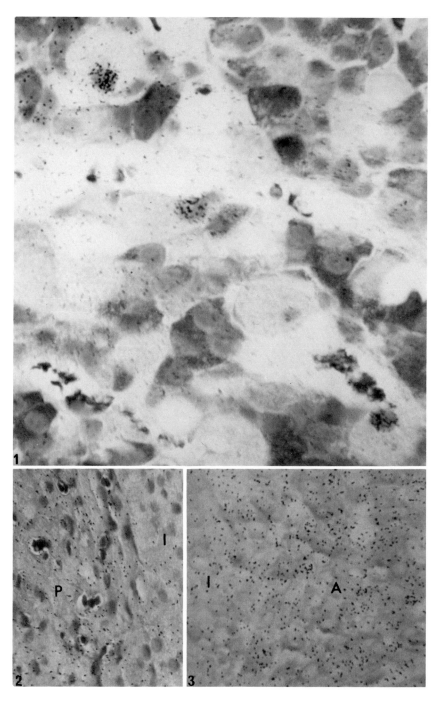

autoradiography technique. However, some of our observations have recently been confirmed by Attramadal (1970).

Vertebrate species other than the rat have been studied in our laboratory and cellular and subcellular localization of estrogen in the pituitary has been obtained in salamander, opossum, tree shrew, and squirrel monkey. The results presented here are from ongoing phylogenetic studies.

The intact, reproductively active salamander, *Triturus viridescens,* shows concentration of label in nuclei of a portion of the anterior lobe cells.

The short-term castrated, lactating opossum, *Didelphis marsupialis,* shows at least two populations of labeled cells in the anterior lobe, a large portion of basophils with a generally heavy intensity of labeling and a smaller portion of the acidophil population taking up a lesser amount of the label (Plate I, Fig. 3 and Plate II, Fig. 4).

In the opossum posterior lobe, strands of cells exhibit a weak degree of nuclear labeling, when compared to labeled anterior lobe cells. In the epithelial layer investing the posterior lobe, a heavy concentration of radioactivity is visible that is, for the most part, confined to the cytoplasm (Plate II, Fig 4). In addition, a few cells in this layer showed the typical nuclear accumulation without cytoplasmic concentration. Thus far, such pronounced cytoplasmic concentration has not been observed in other species. Dawson (1938) designated this epithelial layer as intermediate lobe on cytological grounds. However, in the intermediate lobe of the rat, such a retention pattern of labeled hormone has not been observed.

In the intact, sexually immature tree shrew, *Tupaia glis,* only a relatively small population of basophils in the anterior lobe is found to concentrate radioactivity when compared with the rat, opossum, or squirrel monkey. Since the three animals studied were not castrated, it is possible that the limited estrogen uptake is due to high endogenous estrogen levels.

The short-term castrated female squirrel monkey, *Saimiri sciureus,* shows a population of labeled cells in the anterior lobe 30 min after injection (Plate II, Fig. 5). As in the rat and other species studied, these cells exhibit varying intensities of labeling.

PLATE III.

FIGS. 1–3. Autoradiogram of anterior pituitary (A) (Figs. 1 and 3), intermediate lobe (I) (Figs. 2 and 3), and posterior lobe (P) (Fig. 2) after injection of [³H]testosterone into male castrated (Figs. 1 and 2) and intact 26-day-old female rat (Fig. 3). Exposure times: 154 days (Fig. 1), 185 days (Fig. 2), and 100 days (Fig. 3). 4 μm, stained with aldehyde fuchsin and Masson's trichrome (Fig. 1), Gomori's trichrome (Fig. 2), and methyl green-pyronin (Fig. 3), ×1280 (Fig. 1), ×480 (Figs. 2 and 3). Androgen is selectively concentrated in certain castration cells (Fig. 1), while other cells of the anterior pituitary and cells of the intermediate and posterior lobes do not concentrate radioactivity. In the female anterior pituitary, selective concentration of androgen similar to the intact immature male can be seen (Fig. 3).

In recent ontogenetic autoradiographic studies in the 2-day-old rat, estrogen target cells with the typical nuclear concentration of radioactivity have been observed in our laboratory (Sheridan *et al.,* 1973). In the neonate the number of labeled cells is low, being less than 10% of pars distalis cells, which is in contrast to 3–4 weeks old immature intact and mature castrated rats with an estrogen labeling index between 48 and 86% (Stumpf, 1968). Judged by the marginal position of the estrogen-concentrating cells in the neonate, it is likely that these are basophils only.

III. Androgens

In intact immature male and female and castrated adult male rats, a distinct cellular and subcellular distribution of radioactivity is found in the pars distalis of the pituitary (Plate III, Figs. 1 and 3). Nuclear concentration of androgen is demonstrated in a small number of cells which are identified as gonadotrophs using modified aldehyde fuchsin–Masson's trichrome, or Gomori's trichrome, as well as by long-term castration for the development of castration cells (Plate III, Fig. 1).

In the intermediate and posterior lobes, no cellular and subcellular accumulation of radioactivity exists (Plate II, Figs. 2 and 3).

The anti-androgen cyproterone acetate has been shown to reduce the nuclear uptake of radioactivity after [³H]testosterone injection (unpublished observation), and the radioactive material retained and concentrated in the pars distalis has been identified as dihydrotestosterone (Stern and Eisenfeld, 1971). The histological demonstration of the hormone in gonadotrophs is the first direct evidence for androgen target sites in the pituitary. Biochemists have failed to demonstrate specific binding of androgen in anterior pituitary cells (Leavitt *et al.,* 1969; Korach and Muldoon, 1973).

IV. Progestins

Currently, no report exists in the literature about localization of progesterone or its metabolites in the anterior pituitary, although physiological evidence suggests existence of specific progestin receptors at the level of the pituitary.

In our autoradiographic studies with [³H]progesterone, distribution of radioactivity is observed in nuclei and cytoplasm of cells of the pars distalis. In some of the cells a nuclear concentration prevails. These cells have not yet been identified further.

V. Glucocorticoids

After the administration of cortisol or corticosterone, a high level of radio-activity is observed throughout the pars distalis, with indications of nuclear concentration in certain cells. This is demonstrated for cortisol (Stumpf, 1971b) and corticosterone (unpublished). The autoradiographic studies on glucocorticoid localization are still under way.

VI. Thyroid Hormones

With the dry-mount autoradiographic technique, for the first time a distinct cellular and subcellular distribution of radioactivity is observed in the pars distalis of the pituitary, after the injection of $[^{125}I]$triiodothyronine, using 26-day-old male thyroidectomized rats (Stumpf and Sar, 1973a).

In the anterior and posterior lobes the general level of radioactivity is high with comparatively little in the intermediate lobe (Plate IV, Figs. 1 and 2).

In the anterior lobe, a large proportion of the cells shows nuclear and cytoplasmic accumulation of radioactivity that is apparent, although the radioactivity content in blood vessels is high. A nuclear labeling prevails in different tinctorial cell types, including acidophils, basophils, and chromophobes (Plate IV, Fig. 1). In the posterior lobe radioactivity appears to be concentrated in clusters, especially near the border between the intermediate and posterior lobes (Plate IV, Fig. 2). These areas of concentration appear outside of blood vessels without apparent association with certain cellular or subcellular elements at the light microscopic level. While some of the pituicytes contain radioactivity, it requires further studies to define more clearly the cellular and subcellular distribution. The difficulties are in part attributable to the limited resolution provided by ^{125}I.

Earlier autoradiographic studies of Jensen and Clark (1951) demonstrated in the rabbit a selective accumulation of radioactivity in the infundibulum and other parts of the neurohypophysis 4 hr after application of $[^{131}I]$thyroxine. Ford and Gross (1958) reported that triiodothyronine is most actively concentrated in the anterior lobe in the guinea pig, but less so in the rabbit, while with thyroxine no differences in accumulation between the anterior and posterior lobes could be seen in the rat. Recently, Bar Sella et al. (1972) found in the rat after injection of $[^{125}I]$triiodothyronine, concentrations of radioactivity in the region of transition between the intermediate and neural lobes and also over some areas of the neural lobe. Only a few silver grains were seen over the cytoplasm of pituicytes. Schadlow et al. (1972) provided biochemical evidence of the existence of a specific anterior

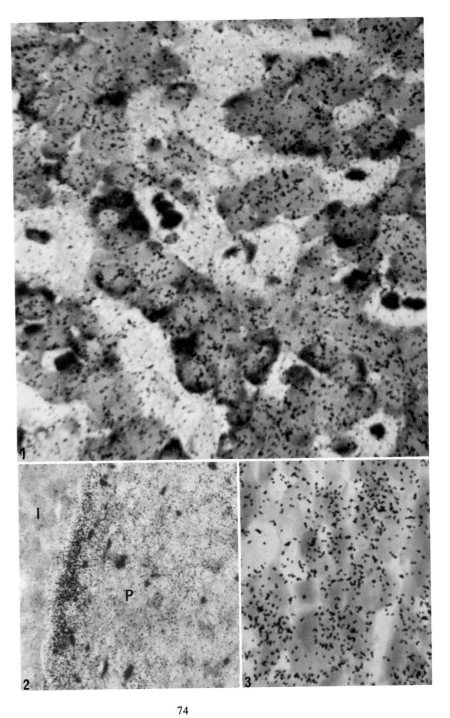

pituitary binding protein for triiodothyronine, being 9.8 times stronger than for thyroxine.

VII. TRH

After intravenous injection of TRH [L-proline-2,3-^{3}H], the rats were killed between 1 and 60 min. At the time intervals studied, radioactivity is found concentrated in nuclei as well as cytoplasm of acidophils, basophils, and chromophobes (Stumpf and Sar, 1973b). In certain cells a nuclear concentration prevails (Plate IV, Fig. 3). In a few cells radioactivity appears to accumulate at the nuclear envelope. No such concentration of radioactivity is ever seen at the outer cell membranes. In the posterior lobe of the pituitary—in contrast to the intermediate lobe—the general level of radioactivity is high, without distinct cellular accumulation.

VIII. Discussion

The autoradiographic localization of hypophyseotropic neurogenic and peripheral hormones in the pituitary has provided strong evidence for a direct action of these hormones on the pituitary. Furthermore, it appears, that each hormone binds and addresses a certain population of cells but not others, while cells of the same tinctorial group concentrate different hormones. The extent of binding, however, varies greatly between the different hormones studied. Also a given hormone, which appears to bind selectively to certain tinctorial or morphological cell type(s), may be concentrated at various intensities depending on the hormonal state of the animal. For instance, in the case of [^{3}H]estradiol, 60 to 80% of the anterior pituitary cells concentrate estrogen (Stumpf, 1968), but under certain physiological conditions, a few percent of the labeled cells concentrate several times the

---

PLATE IV.

Figs. 1 and 2. Autoradiograms of anterior pituitary (Fig. 1) and intermediate and posterior lobes (Fig. 2) 15 min after injection of [^{125}I]triiodothyronine into thyroidectomized male rats. Exposure time: 58 days. 4 μm, stained with aldehyde fuchsin and Masson's trichrome, ×1000 (Fig. 1), ×430 (Fig. 2). In Fig. 1, radioactivity appears to be concentrated in nuclei of cells with pink cytoplasm (acidophils and thyrotrophs), but also in cells with green cytoplasm (gonadotrophs). No cellular concentration of radioactivity exists in the intermediate lobe (I). In the posterior lobe (P) radioactivity is accumulated in areas at the border to the intermediate lobe, but also in the center.

Fig. 3. Anterior pituitary, 1 hr after intravenous injection of ^{3}H-TRH. Exposure time: 64 days. 4 μm, stained with methyl green-pyronin, ×1000. Nuclear concentration of radioactivity appears in certain cells.

amount of radioactivity when compared to the other cells (Fig. 1). These unpublished observations necessitate future quantitative evaluation of pituitary autoradiograms and the determination of the functional–morphological significance of these cells. Similarly, the relatively weak labeling of posterior pituitary cells with estrogen observed after long exposure times or administration of high doses of [³H]estradiol cannot be explained at this time. It is noteworthy that in the case of estrogen the labeling index is high, encompassing cells of different tinctorial types. This is at least so in the rat. The presently available data on other vertebrate species is still scanty and does not permit us to arrive at conclusions. The labeling index for estrogen in the pars distalis seems to be subject to change according to certain hormonal states. However, between castrated males and females no sex differences could be recorded (Stumpf, 1968; Korach and Muldoon, 1973).

Androgen uptake is limited to a more selective population of anterior pituitary cells of about 10 to 15% (Sar and Stumpf, 1973a), in contrast to the extensive binding of estradiol. These cells have been identified as gonadotrophs (Sar and Stumpf, 1972, 1973b) or castration cells (Sar and Stumpf, 1973a). Among the tinctorially or morphologically identified cells, basophils or castration cells, respectively, not all were labeled. It remains to be determined whether these represent different populations of FSH or LH secreting cells.

Progestogen, in contrast to estrogen and androgen, exhibits a lesser degree of nuclear concentration. While nuclear uptake of radioactivity can be demonstrated, there is also a relatively strong cytoplasmic distribution. This may be attributable to more extensive and rapid metabolism as has been shown in the uterus (Armstrong and King, 1971).

In the case of corticosterone, a nuclear hormone binding also seems to exist. More studies are required to further define the extent of localization.

As in the case of estrogen, androgen, progestin, and glucocorticoid, evidence has been provided for the first time from our laboratory about the cellular and subcellular distribution of thyroid hormones and TRH (Stumpf and Sar, 1973a,b) in the pituitary. The autoradiographic observations suggest a direct action on the pituitary of thyroxine and triiodothyronine as well as TRH. These hormones, or metabolites of them, concentrate in specific cells of the pars distalis. The nuclear localization is more pronounced when radioactively labeled triiodothyronine is used, compared to thyroxine. All of the autoradiographic studies with thyroid function-regulating hormones show both a cytoplasmic and nuclear labeling in cells of the anterior pituitary. It is remarkable that TRH is not localized in thyrotrophs exclusively and that there is no detectable accumulation of radioactivity at outer cell membranes.

All of the hormones studied in our laboratory exhibit varying degrees of selective cellular and subcellular localization in the pars distalis of the pituitary. This suggests a direct action at the pituitary level. Analogous to steroid effects on RNA synthesis in peripheral target tissues (Hamilton, 1971; Mueller et al., 1972), the prevailing nuclear concentration may indicate a nuclear receptor interaction on transcription. This seems to be the case not only for hormones from peripheral endocrine glands, but also for the TRH. Although TRH is believed by many to act through a membrane bound release mechanism (Burgus and Guillemin, 1970), the autoradiographic data suggests a different mode of action while, however, not excluding the former possibility. The autoradiographic data suggest further that TRH acts not on one cell type only, but seems to affect different cell types. For instance, the localization in acidophils agrees with its reported effect on prolactin secretion (Bowers *et al.,* 1971; Tashjian *et al.,* 1971).

It is noteworthy from the autoradiographic results that different hormones concentrate in the same tinctorially or morphologically defined cell type. For instance, radioactivity is found to concentrate in nuclei of acidophils after the injection of [^{3}H]estradiol, ^{3}H-TRH, [^{125}I]thyroxine, or [^{125}I]triiodothyronine. The autoradiographic observations are in agreement with experimental data from the literature according to which all of these hormones are capable of stimulating prolactin secretion. Desclin (1956) reported estrogen induced degranulation of acidophils in pituitary grafts. The addition of estradiol (Nicoll and Meites, 1962), thyroxine, or triiodothyronine (Nicoll and Meites, 1963) to the medium of organ culture of rat adenohypophysis has been shown to increase prolactin secretion. Cohere *et al.* (1964) also observed that both thyroxine and estrogen stimulated the development of the endoplasmic reticulum of prolactin cells of explants of rat anterior pituitary in organ culture. Intrapituitary implants of estrogen resulted in an increase of pituitary prolactin content in the rat (Ramirez and McCann, 1964).

From the localization data and the analogy with peripheral hormone target tissues, the mechanism of "feedback," thus, appears to be basically a stimulatory action (Stumpf, 1968; Stumpf and Sar, 1973c) which may lead to the production of some substance which subsequently may result in a secondary negative or positive response. Physiological evidence has been provided for a positive feedback (Hohlweg, 1934). Extranuclear effects, e.g., on the translational level, are also possible in view of the cytoplasmic radioactivity as suggested in the case of TRH and T3. Cytoplasmic radioactivity associated with subcellular anatomical components, however, cannot be resolved at the light microscopic level. Future ultrastructural localization studies can be expected to provide more conclusive information. Hormone receptor localization at the ultrastructural level, that is, immobilization and retention *in situ* of small molecular weight compounds may be accomplished in two ways: with modified classic fixation and embedding techniques (Mizuhira, 1972), or with low temperature tissue preparation as utilized in our studies and further developed (Christensen, 1969) for the ultrastructural level. At this time it is difficult to predict which way will turn out to be more successful. Both, the wet fixation-embedding and the dry low temperature approach should be pursued.

Immunoautoradiography is used here as a combination of techniques for the simultaneous localization of immunoreactive hormones by immunohistochemistry, for the identification of hormone-producing cells, and of radioactively labeled hor-

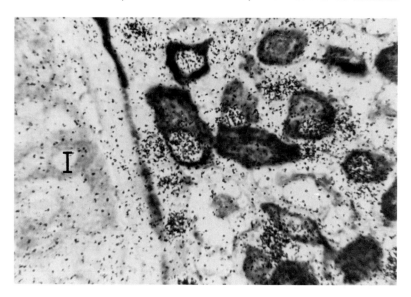

FIG. 2. Immunoautoradiogram of rat pituitary, showing simultaneous localization of peroxidase-labeled anti-hCG and [³H]estradiol. Estrogen target cells are characterized by nuclear accumulation of radioactivity. Gonadotrophs show heavy cytoplasmic immunostaining. Note that most, but not all, of the immunohistochemically defined gonadotrophs are estrogen target cells. In agreement with results obtained by tinctorial techniques, estrogen target cells also include cells other than gonadotrophs. Cells of the intermediate lobe (I) are unlabeled.

The autoradiogram was prepared by the dry-mount procedure described in Section I, 1 hr after injection of [³H]estradiol-17β into an ovariectomized rat. After post-exposure fixation and photographic processing, the autoradiogram was stained according to the immuno-globulin-enzyme bridge technique and counterstained with methyl green. Exposure time 10 months 4 μm-thick section, ×980. (Keefer, D. A., W. E. Stumpf, P. Petrusz, and M. Sar, unpublished.)

mones by autoradiography, for the identification of hormone target cells. This has been accomplished in our laboratory (Fig. 2) and promises to provide answers to many questions related to the cytology and function of the pituitary and other endocrine tissues.

IX. Conclusions

The autoradiographic results suggest:

(1) Estrogen and androgen concentrate almost exclusively in nuclei, while progestin, glucocorticoid, triiodothyronine, TRH, or their metabolites concentrate to varying degrees in nuclei and cytoplasm in certain cells of the pituitary.

(2) The pituitary is a composite hormone target tissue for central neurogenic as well as peripheral hypophyseotropic hormones.

(3) The nuclear concentration of the hormones studied corresponds to a nuclear, probably genomic, effect.

(4) Since cells of the same tinctorial cell type exhibited retention and concentration of different hormones, an individual cell may be chemically addressed not only by one hormone, but by several hormones, and, conversely, one hormone may address several cell types. Different binding affinities may exist for different hormones within a given cell, and the binding affinity for a given hormone may be subject to change depending on preceding hormonal influences. From a number of other experimental data, it seems that the responses of a given cell (type) to hormone stimulation are genetically determined, so that different hormones localized in the same tinctorial or morphological cell type may elicit uniform reactions with regard to the secretion product. For instance, all estradiol, triiodothyronine, and TRH concentrate in acidophils and all stimulate prolactin secretion. On the other hand, the possibility of hormone-induced transformation of a given cell into another cell type cannot be excluded.

(5) A difference in the intensity of estrogen labeling exists between different cell types and within a given tinctorial cell type, as is apparent from studies of animals in different hormonal states, including pregnancy and lactation. The differences in the intensity of nuclear concentration may represent different nuclear binding affinities, relating to different thresholds for (feedback) response. This gradient of binding affinities may be basic to different effects of the same hormone dependent on dose. In addition to the dose factor, hormone pretreatment may influence hormone binding and subsequent hormone related effects.

(6) No correlation seems to exist between pyroninophilia of anterior pituitary cells and the affinity to concentrate hormones.

(7) A topographical arrangement of certain cell types exists within the pars distalis of the rat.

(8) Estrogen target cells exist already in the 2-day-old neonatal rat, comprising less than 10% of the pars distalis cell population.

We propose to extend the use of the term "pituicytes" to encompass *all* of the parenchymal cells of the pituitary, distinguishing *adenopituicytes* in the adenohypophysis and *neuropituicytes* in the neurohypophysis.

ACKNOWLEDGMENTS

The studies were supported by PHS Grants No. NS09914, HD05700 and HD03007, AEC Grant AT(40-1)-4057, and a grant from the Rockefeller Foundation to the Laboratories for Reproductive Biology, University of North Carolina, Chapel Hill, North Carolina.

REFERENCES

Armstrong, D. T., and King, E. R. (1971). Uterine progesterone metabolism and progestional response: Effects of estrogens and prolactin. *Endocrinology* **89**, 191–197.

Attramadal, A. (1965). Distribution and site of action of oestradiol in the brain and pituitary gland of rat following intramuscular administration. *Proc. Int. Congr. Endocrinol., 2nd, 1964* Int. Congr. Ser. No. 83, pp. 612–616.

Attramadal, A. (1970). Cellular localization of ^{3}H-oestradiol in the hypophysis. *Z. Zellforsch. Mikrosk. Anat.* **104**, 597–614.

Bar-Sella, P., Stein, O., and Gross, J. (1972). Radioautographic localization of ^{125}I-triiodothyronine in the posterior pituitary of the rat. *Endocrinology* **91**, 302–309.

Bowers, C. Y., Friesen, H. G., Hwang, P., Guyda, H. J., and Folkers, K. (1971). Prolactin and thyrotropin release in man by synthetic pyro-glutamyl-histidyl-prolinamide. *Biochem. Biophys. Res. Commun.* **45**, 1033–1041.

Burgus, R., and Guillemin, R. (1970). Hypothalamic releasing factors. *Annu. Rev. Biochem.* **39**, 499–526.

Christ, J. F. (1966). Nerve supply, blood supply and cytology of the neurohypophysis. *In* "The Pituitary Gland" (G. W. Harris and B. T. Donovan, eds.), Vol. 3, pp. 62–130. Univ. of California Press, Berkeley.

Christensen, K. A. (1969). A way to prepare frozen thin sections of fresh tissue for electron microscopy. *In* "Autoradiography of Diffusible Substances" (L. J. Roth and W. E. Stumpf, eds.), pp. 349–362. Academic Press, New York.

Cohere, G., Bousquet, J., and Meunier, J. M. (1964). Ultrastructure d'explants de glande pituitaire de ratte adulte cultives sur milieux artificials: Action de action de variations hormonales. *C. R. Soc. Biol.* **158**, 1056–1058.

Dawson, A. B. (1938). The epithelial components of the pituitary gland of the opossum. *Anat. Rec.* **72**, 181–193.

Desclin, L. (1956). Hypothalamus et liberation d'hormone lutéotrophique. Expériences de greffe hypophysaire chez le rat hypophysectomisé. Action lutéotrophique de l'ocytocine. *Ann. Endocrinol.* **19**, 1225.

Eisenfeld, A. J., and Axelrod, J. (1965). Selectivity of estrogen distribution in tissues. *J. Pharmacol. Exp. Ther.* **150**, 469–475.

Ford, D. H., and Gross, J. (1958). The localization of I^{131}-labeled triiodothyronine and thyroxine in the pituitary and brain of the male guinea pig. *Endocrinology* **63**, 549–560.

Hamilton, T. H. (1971). Steroid hormones, ribonucleic acid synthesis and transport, and the regulation of cytoplasmic translation. *Biochem. Soc. Symp.* **32**, 49–84.

Hohlweg, W. (1934). Veränderungen des Hypophysenvorderlappens und des Ovariums nach Behandlungen mit grossen Dosen von Follikelhormon. *Klin. Wochenschr.* **13**, 92–95.

Inman, D. R., Banfield, R. E. W., and King, R. J. (1965). Autoradiographic localization of oestrogen in rat tissues. *J. Endocrinol.* **32**, 17–22.

Jensen, J. M., and Clark, D. E. (1951). Localization of radioactive *l*-thyroxine in the neurohypophysis. *J. Lab. Clin. Med.* **38**, 663–670.

Korach, K. S., and Muldoon, T. G. (1973). Comparison of specific 17β-estradiol-receptor interactions in the anterior pituitary of male and female rats. *Endocrinology* **92**, 322–326.

Leavitt, G., Kimmel, G. L., and Friend, J. P. (1973). Steroid hormone uptake by anterior pituitary cell suspensions. *Endocrinology* **92**, 94–103.

Leavitt, W. W., Friend, J. P., and Robinson, J. A. (1969). Estradiol: Specific binding by pituitary nuclear fraction *in vitro*. *Science* **165**, 496–498.

Mason, T. E., Phifer, R. F., Spicer, S. S., Swallow, R. A., and Dreskin, R. B. (1969). An immunoglobulin-enzyme bridge method for localizing tissue antigens. *J. Histochem. Cytochem.* **17**, 563–569.

Mizuhira, V. (1972). Chemical fixation of the labeled radioactive isotopes (RI) in the tissues to the electron microscopic autoradiography (EM-Aut). *Proc. Int. Congr. Histochem. Cytochem., 4th, 1972,* Kyoto, Japan, pp. 35–36.

Mizuhira, V., Uchida, K., Futaesaku, Y., and Okazaki, K. (1971). The biosynthesis of testosterone in the testicular interstitial cells of rat. *Electron Microsc., Proc. Int. Congr., 7th, 1970* Vol. III, p. 521.

Mueller, G. C., Vonderhaar, B., Kim, U. H., and Mahieu, M. L. (1972). Estrogen action: An inroad to cell biology. *Recent Progr. Horm. Res.* **28**, 1–49.

Nakane, P. K., and Pierce, G. B. (1967). Enzyme-labeled antibody for the light and electron microscopic localization of tissue antigens. *J. Cell Biol.* **33**, 307–318.

Nicoll, C. S., and Meites, J. (1962). Estrogen stimulation of prolactin production by adeno-hypophyseal explants *in vitro. Endocrinology* **70**, 272–277.

Nicoll, C. S., and Meites, J. (1963). Prolactin secretion *in vitro:* Effects of thyroid hormones and insulin. *Endocrinology* **72**, 544–551.

Purves, H. D., and Griesbach, W. E. (1951). The site of thyrotrophin and gonadotrophin production in the rat pituitary studied by McManns-Hotchkiss staining for glycoprotein. *Endocrinology* **49**, 244–264.

Ramirez, V. D., and McCann, S. M. (1964). Induction of prolactin secretion by implants of estrogen into the hypothalamo-hypophysial region of female rats. *Endocrinology* **75**, 206–214.

Sar, M., and Stumpf, W. E. (1972). Cellular localization of androgen in the brain and pituitary after the injection of tritiated testosterone. *Experientia* **28**, 1364–1366.

Sar, M., and Stumpf, W. E. (1973a). Pituitary gonadotrophs: Nuclear concentration of radioactivity after injection of [H^3] testosterone. *Science* **179**, 389–391.

Sar, M., and Stumpf, W. E. (1973b). Cellular and subcellular localization of radioactivity in the rat pituitary after injection of 1,2,^{3}H-testosterone using dry-autoradiography. *Endocrinology* **92**, 631–635.

Schadlow, A. B., Surks, M. I., Schwartz, H. L., and Oppenheimer, J. H. (1972). Specific triiodothyronine binding sites in the anterior pituitary of the rat. *Science* **176**, 1252–1254.

Sheridan, P. J., Sar, M., and Stumpf, W. E. (1973). Cellular and subcellular localization of ^{3}H-estradiol or its metabolites in the pituitary of the neonatal female rat. *Experientia* **29**, 1418–1419.

Stern, J. M., and Eisenfeld, A. J. (1971). Distribution and metabolism of ^{3}H-testosterone in castrated male rats: Effects of cyproterone, progesterone and unlabeled testosterone. *Endocrinology* **88**, 1117–1125.

Stumpf, W. E. (1968). Cellular and subcellular ^{3}H-estradiol localization in the pituitary by autoradiography. *Z. Zellforsch. Mikrosk. Anat.* **92**, 23–33.

Stumpf, W. E. (1969). Too much noise in the autoradiogram? *Science* **163**, 958–959.

Stumpf, W. E. (1970). Estrogen-neurons and estrogen-neuron systems in the periventricular brain. *Amer. J. Anat.* **129**, 207–218.

Stumpf, W. E. (1971a). Autoradiographic techniques for the localization of the hormones and drugs at the cellular and subcellular level. *Acta Endocrinol. (Copenhagen), Suppl.* **153**, 205–222.

Stumpf, W. E. (1971b). Autoradiographic techniques and the localization of estrogen, androgen and glucocorticoid in pituitary and brain. *Amer. Zool.* **11**, 725–739.

Stumpf, W. E., and Roth, L. J. (1966). High resolution autoradiography with dry-mounted, freeze-dried, frozen sections. Comparative study of six methods using two diffusible compounds, ^{3}H-estradiol and ^{3}H-mesobilirubinogen. *J. Histochem. Cytochem.* **14**, 274–287.

Stumpf, W. E., and Roth, L. J. (1967). Freeze-drying of small tissue samples and thin frozen sections below −60°C: A simple method of cyrosorption pumping. *J. Histochem. Cytochem.* **15**, 243–251.

Stumpf, W. E., and Sar, M. (1973b). ^{3}H-TRH and ^{3}H-proline radioactivity localization in pituitary and hypothalamus. *Fed. Proc., Fed. Amer. Soc. Exp. Biol.* (abstr.) **32**, 211.

Stumpf, W. E., and Sar, M. (1973a). Concentration of radioactivity in anterior pituitary cells, the posterior pituitary, ependymal cells of the median eminence and basal tuberal region after injection of ^{125}I-labeled thyroxine and triiodothyronine. *Endocrinology* 92 (*Suppl.*), A273.

Stumpf, W. E., and Sar, M. (1973c). Hormonal inputs to releasing factor cells, feedback sites. *Progr. Brain Res.* 39, 53–71.

Stumpf, W. E., and Sar, M. (1974). Autoradiographic techniques for localizing steroid hormones. *In* "Hormones and Cyclic Nucleotides. Methods in Enzymology" (B. W. O'Malley and T. G. Hardman, eds.), Vol. 20.

Tashjian, A. H., Barowsky, N. J., and Jensen, D. K. (1971). Thyrotropin releasing hormone; direct evidence for stimulation of prolactin production by pituitary cells in cultures. *Biochem. Biophys. Res. Commun.* 43, 516–523.

STRUCTURE AND FUNCTION OF THE ANTERIOR PITUITARY AND DISPERSED PITUITARY CELLS. *IN VITRO* STUDIES[1]

Marilyn G. Farquhar, Ehud H. Skutelsky, and Colin R. Hopkins

SECTION OF CELL BIOLOGY, YALE UNIVERSITY SCHOOL OF MEDICINE

NEW HAVEN, CONNECTICUT

AND

THE ROCKEFELLER UNIVERSITY

NEW YORK, NEW YORK

[1] The original research reported in this paper was supported by grants AM 15399 and AM 17780 from the National Institutes of Health, and by Rockefeller Foundation grant RF-70095 for studies in reproductive biology at The Rockefeller University.

I. Introduction

The anterior lobe of the pituitary has proved to be a less accessible tissue for the cell biologist to work with than many other glands, especially exocrine glands, such as the liver, pancreas, or parotid. In these organs there is a large amount of tissue available for analysis, a single predominating secretory cell type, and an output of secretory products which can be measured in grams/day. Therefore, cell fractionation procedures and the use of *in vitro* systems have been particularly fruitful approaches for studying the cellular mechanisms by which these tissues produce their secretory products, and as a result, we have acquired a large volume of information on the nature of their secretory processes. The situation is quite different in the case of the anterior pituitary, however, where the amount of tissue available is small by comparison, (i.e., 20–100 times smaller), and the number of secretory cell types is at least five or six (plus various vascular and connective tissue components); hence, cell fractionation, with the notable exception of the isolation of secretory granules (see below), has not been a fruitful approach for its study. To further complicate the situation the output of secretory products from the anterior pituitary is measured in microgram quantities and, until relatively recently (with the introduction of radioimmunoassay and techniques for assay by gel electrophoresis), the only means of quantitating levels of hormone output was by bioassay, with all its inherent variables.

The net result is that we do not know as much about how anterior pituitary cells function as we do about the other glandular cells mentioned. What we do know has been learned largely from application of morphological or localization techniques, i.e., light microscopy, electron microscopy, autoradiography, histochemistry, enzyme cytochemistry, and immunocytochemistry—which can be carried out on intact tissues and which, with the exception of autoradiography, give essentially static information.

In this article we will summarize what has been established to date on the structure and function of pituitary cells, focusing on our own work. Since secretion and crinophagy in pituitary cells were the subject of a recent extensive review (Farquhar, 1971), we will emphasize in this chapter more recent studies using *in vitro* systems (small tissue blocks and cell suspensions obtained by dissociation of pituitary tissue), much of which has not been published before.

II. General Background

A. CELL TYPES OF THE RAT ANTERIOR PITUITARY

The anterior lobe is known to produce six hormones in the rat. These hormones together with their abbreviations, synonyms, nature, and naming of the

TABLE I

HORMONES AND CELL TYPES OF THE RAT PITUITARY GLAND

| Cell types | Staining reactions | | | | Secretory granules (nm) | Hormones produced | | |
Name	Orange G	PAS	Erythrosin	Aldehyde-fuchsin		Name	Abbrev.	Nature
Somatotroph	+	−	−	−	350–400	Growth hormone (somatotropin)	STH	Protein
Mammotroph	+	−	+	−	600–900	Prolactin (mammotropin, lactogenic hormone)	PR (MTH LTH)	Protein
Gonadotroph	−	+	−	−	200–250	Follicle stimulating and luteinizing hormones	FSH LH	Glycoprotein Glycoprotein
Thyrotroph	−	++	−	+	140–200	Thyrotropin	TSH	Glycoprotein
Corticotroph	−	++	+	−	200–250	Adrenocorticotropin	ACTH	Polypeptide
Follicular	−	+	−	−	None present	None	—	—

cell types are given in Table I. Much of the effort in past work on the cytology of the pituitary has centered around attempts to establish the function and therefore the identity of the cell types present. In these studies advantage has been taken of the negative feedback control imposed by target endocrine organs, using target organ removal or target organ hormone to stimulate or suppress, respectively, the production of a particular trophic hormone.

Available evidence indicates that a separate cell type exists for the production of each of these hormones with the possible exception of the gonadotropins (see below); hence, the cells are named according to their secretory function. By light microscopy, the cells have been traditionally distinguished by differences in the stainability of the secretory granules. The staining properties which have been useful for distinguishing cell types in the rat pituitary are summarized in Table I and reviewed by Herlant (1964) and Purves (1961, 1966). With the electron microscope the single most helpful criterion for distinguishing the different types of cells is still secretory granule size. Initial identifications, except for the corticotroph (Siperstein and Allison, 1965), were made as a result of electron microscopic and experimental studies carried out in the mid 1950's (Farquhar and Rinehart, 1954a,b; Hedinger and Farquhar, 1957). They have subsequently been substantiated in morphological studies by others (see Kurosumi, 1968; Pooley, 1971; Costoff, 1973). Direct confirmation of these identifications has come from immunocytochemistry (Nakane, 1970, also chapter by Nakane, this volume; Moriarity, 1973) and especially as a result of the isolation and characterization of the secretory granules by McShan and his co-workers (see McShan and Hartley, 1965; Costoff and McShan, 1969; also review by Costoff, 1973). As a result there is now agreement among the majority of workers concerning the granule size range and overall cell morphology characteristic of each of the secretory cell types. The maximal diameters of the secretory granules are given in Table I, and the morphology of these cells is summarized below and is illustrated in Figs. 9–15 and 18–30.

1. Somatotrophs

The growth hormone-producing cell is the most abundant cell type present in the anterior lobe and can be readily identified by its generally ovoid or polygonal shape and presence of variable numbers of rounded or ovoid secretory granules, ~350 nm diameter, which are very osmiophilic and appear homogeneously dense (Fig. 9). In young growing animals this cell type contains a well-developed Golgi apparatus, a few lysosomes, and abundant rough endoplasmic reticulum (ER)[2]

[2] The abbreviations used in this paper are BSA, bovine serum albumin; HRP, horseradish peroxidase; KRB, Krebs–Ringer bicarbonate solution containing a complete amino acid supplement, 14 mM glucose, and either 0.3 or 0.5% BSA; LTH, prolactin; LRF or LH–RH, luteinizing hormone releasing factor (hormone); ER, endoplasmic reticulum; SBTI, soybean trypsin inhibitor; STH, growth or somatotropic hormone; TRF, thyrotropic hormone releasing factor (hormone).

which consists of stacks of elongated cisternae with a content of low density. In somatotrophs of older animals the rough ER and Golgi are simpler, and many of the cells contain numerous lysosomes with heterogeneous content.

2. Mammotrophs

The mammotroph or prolactin cell has the largest secretory granules of any anterior lobe cell type, with the mature granules being 600–900 nm, ovoid or elliptical in shape, and homogeneously dense (Fig. 15). Immature forms are smaller, and more variable in size and shape owing to the fact that they are assembled by pinching off of smaller granules from the innermost of the stacked Golgi cisternae, followed by progressive aggregation of several small granules into polymorphous forms (see Fig. 1). Mammotrophs are the predominating cell type in pituitaries of lactating females, less numerous but still abundant in pituitaries from cycling females, and less frequent in pituitaries of male rats.

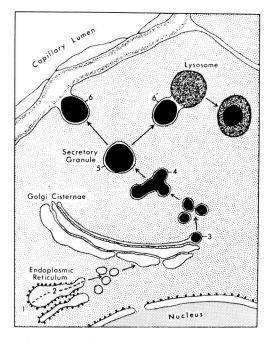

FIG. 1. Diagrammatic representation of proposed events in the secretory process of mammotrophs in the anterior pituitary gland. Prolactin is probably synthesized on ribosomes (1), segregated in and transported through the rough-surfaced ER (2), and concentrated into granules by the Golgi apparatus. Small granules arise within the inner Golgi cisterna (3) and aggregate (4), to form the mature secretory granule (5). During active secretion the latter fuse with the cell membrane (6) and are discharged into the perivascular spaces by exocytosis. When secretory activity is suppressed and the cell must dispose of excess stored hormone, some granules fuse with lysosomes (6′) and are degraded, a process referred to as "crinophagy."

3. Gonadotrophs

The gonadotroph has been known since the work of Purves and Griesbach (1951) and Halmi (1952) to be a large round cell. It contains ~200 nm secretory granules which are typically less osmiophilic than somatotroph or prolactin granules (Fig. 10). The rough ER characteristically occurs in a vesicular form of varied sizes and irregular shapes with a content of a density greater than that of the cytoplasmic matrix. After castration the ER elements become progressively distended and fuse so that individual cisternae may become huge, producing the characteristic signet-ring appearance (Farquhar and Rinehart, 1954a). In addition to the 200-nm granules, some gonadotrophs contain populations of larger (400–500 nm) granules which are notoriously difficult to fix in OsO_4, but are well preserved by aldehyde fixation. According to the immunocytochemical evidence (Mazurkiewicz and Nakane, 1972; Tougard et al., 1973; chapter by Nakane, this volume), both types of granules are secretory granules containing stored gonadotrophic hormone(s). Among the gonadotrophs, morphologically distinct subtypes have been distinguished by electron microscopy (Kurosumi, 1968; Kurosumi and Oata, 1968; Rennels et al., 1971) and by immunocytochemistry (Nakane, 1970; Moriarty, 1973; Tougard et al., 1973). Whether these are functional variants of the same cell type or different cell types, one responsible for the formation of FSH and the other for LH, remains to be established. At present, it is not clear whether both gonadotropins are produced by the same or different cell types. There is immunocytochemical evidence in the rat (Nakane, 1970) and in the human (Phifer et al., 1973) indicating the presence of LH and FSH in the same cell; however, the studies cited were carried out using antibodies raised to whole hormones which share a common polypeptide (alpha subunit). More recent studies with antibodies raised to the specific (beta) subunit (Baker et al., 1972; Tougard et al., 1973; chapter by Nakane, this volume) have not resolved this problem thus far.

4. Thyrotrophs

The thyrotroph is typically angular in shape, and contains the smallest (~140 nm) granules of any cell type (Fig. 11). Like gonadotroph granules, they have a lower density than the granules of somatotrophs and mammotrophs. It has frequently been said that thyrotrophs can be distinguished by the fact that their granules tend to line up beneath the cell membrane along the cell periphery, but, according to Nakane's (1970) immunocytochemical studies, this is not a feature specific for thyrotrophs since it is also found in corticotrophs. In the normal animal, the rough ER of thyrotrophs consists of a few simple, flattened cisternae of irregular distribution, but after thyroidectomy the ER becomes dilated and filled with a content of moderate density (Fig. 14). The situation is therefore similar to that of the gonadotroph after castration, except that in the thyroidectomy cell intracisternal granules appear within the dilated ER cavities and become more numerous with increasing time after thyroidectomy (Farquhar and Rinehart, 1954b; Kurosumi, 1968; Farquhar, 1971).

5. Corticotrophs

The identification of the cell type responsible for ACTH production has had a long and controversial history (see Costoff, 1973, for a review of this problem). However, there now appears to be agreement among interested parties, and converging evidence from electron microscopic studies on adrenalectomized rats by Siperstein and Allison (1965), Kurosumi and Kobayashi (1966), Rennels and Shiino (1968), Nakayama et al. (1969), Siperstein and Miller (1970), Pelletier and Racadot (1971), and from immunocytochemical studies by Nakane (1970), Moriarty and Halmi (1972), and Bowie et al. (1973), indicating that this cell has the following characteristics: It contains 200 nm secretory granules and it is typically angular in profile (Fig. 12). Sometimes, especially in the adrenalectomized animal, it is even spiderlike, with a number of arms radiating out from the main cell body. Because of its angular shape, the cell type with which it is most easily confused is the thyrotroph. According to most electron microscopic studies, the corticotroph can be distinguished from the thyrotroph primarily by the larger size of its granules. Other features of this cell in the normal gland are a relatively simple rough ER consisting of sparse single cisternae, abundant free ribosomes, and a small Golgi apparatus. A frequent close association with the prolactin cell was noted by Nakane (1970).

What lingering discussion or disagreement on the identification of the corticotroph that remains appears to be related to the fact that the granule content is easily extracted during fixation with OsO_4 (see Fig. 13, inset a). A remaining unresolved puzzle is why the granules stain with PAS (Bowie et al., 1974) and phosphotungstic acid (Pelletier, 1971), presumptive stains for glycoprotein (Rambourg, 1967), when the hormone (ACTH) itself contains no carbohydrate. A similar situation exists in the case of the MSH-containing granules of the intermediate lobe of the rat pituitary. In this regard it is of interest that Yalow and Berson (1971) have recently obtained some evidence suggesting that several forms of ACTH may be secreted by the human pituitary, with one form consisting of ACTH covalently linked to another peptide of unknown composition. Therefore the possibility exists that ACTH is produced by the cell and packaged into granules along with a "carrier" protein comparable to the neurophysins of the posterior lobe (Sachs, 1971).

6. Follicular Cells

These are nongranulated, presumably nonsecretory elements which were identified in early studies of the anterior pituitary by electron microscopy (Farquhar, 1957). They derive their name from the fact that they form the lining of tiny

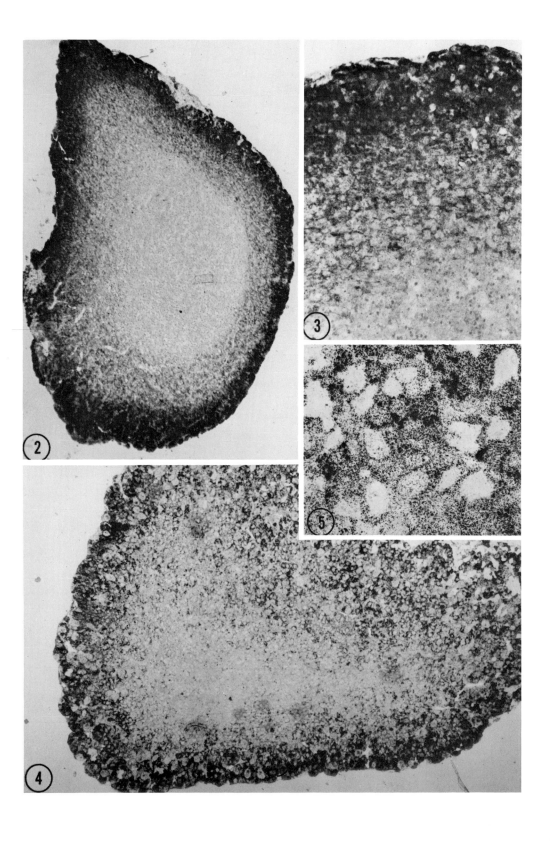

ductules or follicles. The cells possess microvilli and occasional cilia which project into the follicular lumen and long slender cytoplasmic arms that extend out between adjacent parenchymal cells (Fig. 27). Junctional complexes join one cell to another near their luminal surface, but according to the findings of Vila-Porcile (1972), the junctions do not completely seal off the intercellular spaces since horseradish peroxidase, administered i.v., gains access to the follicular lumen. The follicular cells lack secretory granules and are provided with variable but sometimes impressive numbers of polysomes and beta particles of glycogen (Fig. 30). They frequently contain lipid droplets and have modest amounts of other cell organelles (Golgi apparatus, rough ER, mitochondria, lysosomes).

The surface membrane of follicular cells differs from the secretory elements in possessing a more conspicuous glycoprotein cell coat. This cell coat material has been demonstrated by Vila-Porcile (1972), using Rambourg's (1967) PTA method, to be particularly rich on the cell surfaces facing the follicular lumina. Our own independent observations using the colloidal iron (Gasic et al., 1968) and cationized ferritin (Danon et al., 1971) methods for detection of sialic acid groups, confirm the observations of Vila-Porcile that the cell coat is thicker on the luminal than on the remainder of the follicular cell surface. However, even on the nonluminal cell surface the coat of follicular cells is considerably richer than that of secretory elements (Fig. 22).

Follicular cells have been described in the pituitaries of a number of species [discussed elsewhere (Farquhar, 1971)]. In several other species stellate cells with similar morphological features but lacking follicular lumina have been described which may represent a related or analogous cell type. Dingemans and Feltkamp (1972) and Vila-Porcile (1972) have recently published detailed fine

FIGS. 2 and 3. Radioautogram of a hemipituitary incubated for 4 hr in a mixture of tritiated amino acids (100 μCi/ml). A gradient of decreasing grain concentration is evident from the periphery to the center, indicating incomplete penetration of amino acids into these large pieces. Three zones can be distinguished: a peripheral zone (3.5–7 mm) that appears black due to a heavy concentration of exposed grains, a median zone (7–14 mm) with a lower grain concentration which appears gray, and an inner light zone with virtually no grains. Figure 3 is a higher magnification of the same specimen showing the three diffusion zones. Specimen fixed in formol–calcium and embedded in paraffin. Section (6 μm) coated with L-4 emulsion, exposed 6 weeks, and stained with hematoxylin and eosin. Fig. 2, ×14; Fig. 3, ×57.

FIGS. 4 and 5. Another radioautogram showing a smaller piece of pituitary tissue (one tenth) incubated 2 hr in [³H]leucine (200 μCi/ml). As in the case of the hemipituitary shown above, there is a gradient of decreasing grain concentration from periphery to center. The center contains virtually no grains. Figure 5 is an enlargement of the black peripheral zone to illustrate individual silver grains. Most of the cells contain heavy concentrations of grains; however, some have few or no grains and stand out as white patches against the dark background provided by the labeled cells. Specimen fixed in dilute Karnovsky's fixative (formaldehyde–glutaraldehyde) and embedded in Epon. Section (1 μm) coated with L-4 emulsion, exposed 2 weeks, and stained with toluidine blue. Fig. 4, ×142; Fig. 5, ×570.

1. Tissue chopped into small blocks

2. Incubation in 1 mg/ml Trypsin (15 min)

3. Incubation in 5 μg/ml DNase (5 min)

4. Incubation in 2 mg/ml Soya bean trypsin inhibitor (SBTI) (5 min)

5. Rinse in Ca^{2+} - free KRB

6. Incubation in 8 μg/ml neuraminidase + 1 mM EDTA in Ca^{2+} - free KRB (15 min)

7. 2 Rinses in Ca^{2+} - free KRB

8. Dispersion by pipetting in Ca^{2+} - free KRB. Cell suspension diluted with 0.2 mM KRB (+ Ca^{2+})

FIG. 6. Pituitary dissociation procedure. All incubations are carried out at 37°C.

structural studies of follicular cells in the mouse and rat, respectively. The latter author has included a thorough review in tabular form of the existing literature on these cells.

Regarding the functions of follicular cells, it was originally proposed (Farquhar, 1957), based on the observation that they underwent hypertrophy after adrenalectomy, that they might produce ACTH. It is now clear that such is not the case and that corticotrophs and follicular cells are distinct entities. Vila-Porcile and Dingemans and Feltkamp have further shown that morphological changes occur in follicular cells after a variety of experimental manipulations. Vila-Porcile discusses several possible implications of these findings, and Dingemans and Feltkamp suggest that their response is related to the incorporation and digestion of waste material. Quite recently several different investigators, have independently provided evidence for a phagocytic function of follicular cells which is described and discussed in a later section (see below).

7. Chromophobes

A word should be said about chromophobes. This term was originally introduced for cells which do not stain, i.e., lack stainable secretory granules. With routine histological methods they were said to constitute up to 65% of the cells of the anterior lobe (for a review, see Herlant, 1964; Halmi, 1973). Although some of these elements were recognized to be degranulated secretory cells, it was also believed that there was a large number of undifferentiated or "stem" cells in the pituitary gland which possessed the ability to shift their secretory activities as necessary. It is now clear that, except for follicular cells, the majority of such cells are simply secretory elements with so few granules that their stainability was below the level of sensitivity of the techniques used. With more sensitive methods, such as the PAS and aldehyde-fuchsin techniques (Purves and Griesbach, 1951; Halmi, 1952), and especially with electron microscopy, most cells can be

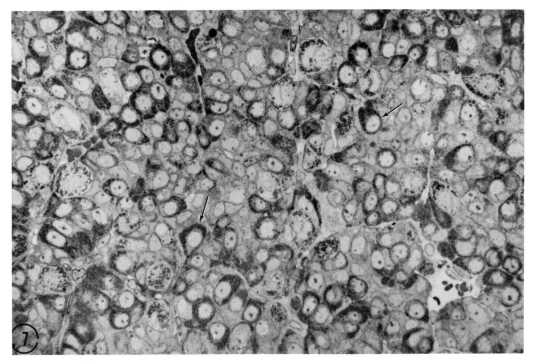

FIG. 7. Light micrograph showing a section taken from a small block of intact, undissociated anterior pituitary tissue. The cells vary in size and shape and show complicated contours. Many are angular in shape with the nucleus located at one pole of the cell (arrows). Epon section (0.5 μm) of tissue fixed in dilute Karnovsky's fixative, postfixed in 1% OsO_4 in cacodylate buffer, and stained with azure II–methylene blue. ×600.

classified into one of the cell types mentioned (secretory or follicular), and at the present time it seems unlikely that any significant number of undifferentiated "stem" cells exists in the pituitary. This conclusion is strengthened by the fact that in those situations in which increased mitotic activity has been induced experimentally in pituitary cells [e.g., by thyroidectomy (Dingemans, 1969; Stratmann et al., 1972)], mitosis has been seen to be occurring primarily within differentiated cells containing identifiable secretory granules.

B. THE SECRETORY PROCESS IN ANTERIOR PITUITARY CELLS

What is known about the secretory process in anterior pituitary cells? We have already indicated that the information we have is spotty, based primarily on static or kinetic techniques, and represents in part extrapolation from work on other systems in which more complete information is available. Since a complete description of what is known about secretion in pituitary cells together with a

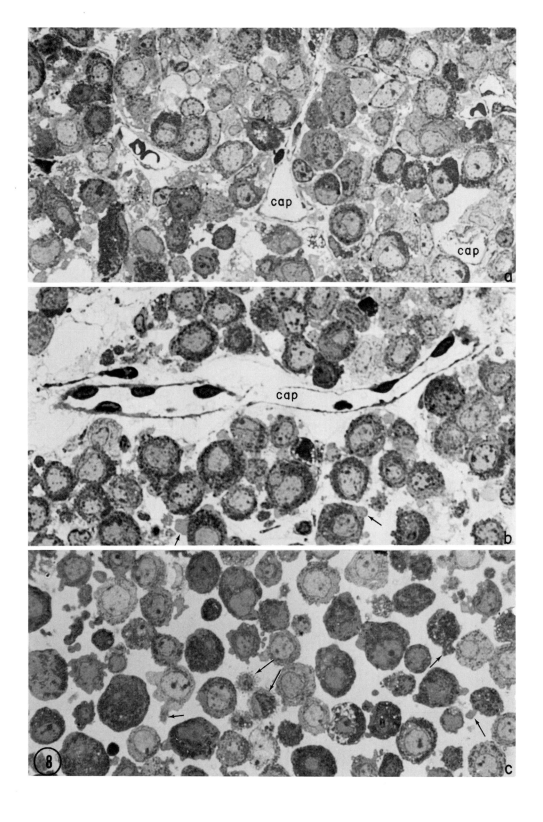

summary of what is known about this process in other systems was the subject of a recent review (Farquhar, 1971), we will simply summarize it here.

1. Mammotrophs and Somatotrophs

The most thorough studies and most extensive information are available for these two protein hormone-producing cells. In the case of the mammotroph or prolactin cell the proposed scheme is given in the diagram of Fig. 1 and is described in the legend to that figure. The later events depicted in prolactin secretion (concentration of the secretory product in Golgi cisternae, aggregation of small granules to larger, mature forms, plus exocytosis and crinophagy) are well established since they can be detected by electron microscopy (Smith and Farquhar, 1966; Farquhar, 1969, 1971). Information about the early steps in prolactin secretion (i.e., synthesis on attached polyribosomes, transfer to the cisternal space of the rough ER, and transport to the Golgi region via small vesicles) is more fragmentary, and the scheme depicted represents in part extrapolation from that worked out in the exocrine pancreas where thorough cell fractionation, autoradiographic, and morphological data are available (Palade, 1966; Jamieson, 1972). However, the morphological findings on the pituitary together with whatever data are available from cell fractionation (Meldolesi et al., 1972) and autoradiography (Tixier-Vidal and Picart, 1967) support the scheme proposed.

There are two main variations in mammotroph secretion from the scheme worked out in the exocrine pancreas: First, concentration of the secretory product takes place in the stacked Golgi cisternae (instead of within specialized condensing vacuoles). Second, an intracellular mechanism exists for the disposal of undischarged secretory granules via lysosomes. In contrast to the situation in the pancreas, a discharge option exists: When cells are stimulated to discharge their secretory product (e.g., in the lactating animal with suckling young), the granules move toward and fuse with the cell membrane and are discharged in the usual manner by exocytosis. However, when secretion is suppressed (by removal of

FIG. 8. Light micrographs of Epon sections showing progressive loosening of tissue organization during sequential steps in the dissociation procedure. (a) Central area of a small tissue block fixed after 15-min incubation in trypsin (1 mg/ml). Some removal of intercellular material and rounding-up of cells is evident mainly in the vicinity of capillaries (cap). (b) Similar small tissue block after 15-min incubation in trypsin (1 mg/ml) followed by 15-min incubation in neuraminidase (8 μg/ml) and EDTA (1 mM). The intercellular spaces are considerably enlarged, and almost all of the cells are rounded-up. Some peripheral cell blebbing can be seen (arrows). (c) Cell suspension produced by gently pipetting the tissue blocks shown in (b). Immediately following dissociation, blebbed cells are common (short arrows). However, on being returned to medium containing Ca^{2+}, blebs are much less frequent. Most of the cells present in the field are secretory elements that contain variable numbers of secretion granules, and that (except for the blebs) have relatively smooth contours. Near the center of the field (long arrows), there is a cluster of small cells with the elaborate ruffled cell margins typical of dissociated nonsecretory elements, i.e., follicular cells and endothelial cells. Preparation for microscopy as for Fig. 7. (a), $\times$808; (b) and (c), $\times$903.

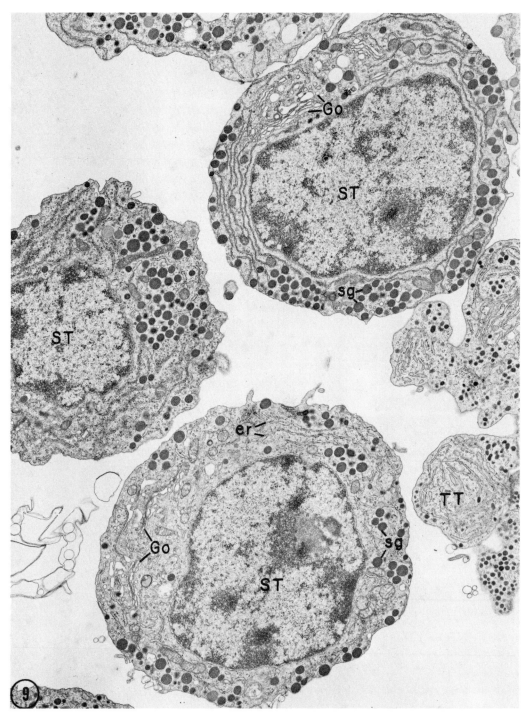

suckling young), the granules move toward and fuse with lysosomes, a process for which we have adopted the name "crinophagy," suggested by de Duve (1969).

In the case of the somatotroph, the steps in the secretory process appear to be very similar to those in the mammotroph. At least the morphological information suggests they are very similar as do the available autoradiographic (Racodot *et al.,* 1965; Hopkins and Farquhar, 1973) and cell fractionation (Howell and Ewart, 1973; Howell and Whitfield, 1973) data.

2. *Thyrotrophs, Gonadotrophs, and Corticotrophs*

Much less is known about the glycoprotein hormone-producing cells and the corticotrophs. From morphological studies it appears that the later steps in secretion: (a) packaging of hormones into granules in the stacked Golgi cisternae, (b) discharge by exocytosis, and (c) existence of crinophagy, are similar in all pituitary secretory cell types (Farquhar, 1971).

As far as the earlier steps are concerned little is known since there have been relatively few cell fractionation or autoradiographic studies on these cell types.

The glycoprotein hormones are known to be composed of two subunits (alpha and beta) (Pierce, 1971), but nothing is known about how or where the two subunits are synthesized and assembled. Regarding the intracellular sites where glycosylation takes place, work on other systems, especially the thyroid gland (Whur *et al.,* 1969; Haddad *et al.,* 1971) indicates that terminal hexose residues (galactose, fucose) are usually added in the Golgi apparatus, whereas "core" sugars (mannose) are added in the rough ER, shortly after synthesis of the polypeptide chains. Biochemical data published recently by Todd and Samli (1973) on hexose incorporation into total pituitary proteins are in keeping with this theme. Recent autoradiographic studies by Pelletier (1974) and Pelletier and Puviani (1973) suggest that fucose incorporation normally takes place in the Golgi apparatus of both gonadotrophs and thyrotrophs, but occurs in the ER (as well as in the Golgi) in gonadotrophs from castrates and in thyrotrophs from thyroidectomized animals.

3. *Is Exocytosis the Only Mechanism for Discharge of Pituitary Hormones?*

A further word should be said about exocytosis since one sometimes hears it stated or implied that exocytosis may not be the *only* or even the main mech-

---

FIG. 9. Electron micrograph of three somatotrophs or growth hormone-producing cells (ST) from a cell suspension fixed immediately after dissociation. The cells are rounded-up with the cytoplasm more or less evenly distributed around the nucleus and the Golgi apparatus (Go) located to one side. The cells can be readily identified on the basis of their well-preserved fine structural organization—especially their 350–400 nm spherical secretion granules (sg) and flattened, elongated rough ER cisternae (er). A portion of the cytoplasm of a thyrotroph (TT) containing smaller (~100–150 nm) secretion granules is also present to the right. Specimen fixed in dilute Karnovsky fixative, postfixed in OsO₄, and stained in block with uranyl acetate. Section doubly stained with uranyl acetate and lead. ×10,000.

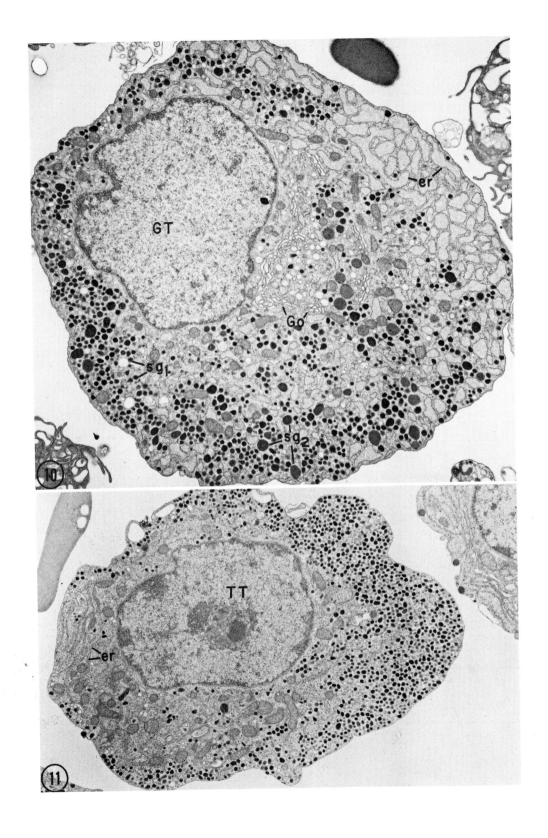

anism by which hormones are discharged from anterior pituitary cells. This kind of discussion has centered in particular on the gonadotroph where images of exocytosis were not found (Kurosumi, 1968) even under conditions in which hormone discharge was stimulated by injection of hypothalamic extracts (Pelletier *et al.,* 1971) or LH–RH (Rennels *et al.,* 1971). Subsequently, however, exocytosis was observed by Shiino *et al.* (1972a) in LH gonadotrophs after administration of purified luteinizing hormone releasing factor (LRF or LH–RF) under conditions in which LH stores were reduced (i.e., androgen-sterilized, persistent-estrous rats). Since we were curious about this problem, we recently incubated pieces of pituitary tissue from 10-day castrates *in vitro* in the presence of LRF (1.5 ng/ml for 1–30 min) and examined them by electron microscopy. We found images of exocytosis, but they were restricted to only a few of the gonadotrophs present in which such images were usually multiple. These findings indicate that discharge occurs by exocytosis, but only a few of the gonadotrophs present were apparently responsive to LRF under these conditions.

It is important to stress that in all cases of protein-secreting tissues in which the situation has been carefully probed under appropriate experimental conditions by combined morphological and biochemical analysis [e.g., exocrine pancreas (Palade, 1966); parotid (Amsterdam *et al.,* 1969); adrenal medulla (Douglas, 1966; Schneider *et al.,* 1967; Smith, 1972), and neurohypophysis (Douglas *et al.,* 1971; Sachs, 1971)], exocytosis is the *only* release mechanism so far demonstrated. *No* alternatives have been found although several (i.e., direct release from the rough ER and discharge into the cytoplasmic matrix) have been proposed.

In the case of anterior pituitary cells there is good morphological evidence for exocytosis in the case of the somatotroph (Farquhar, 1961a,b; De Virgilis *et al.,* 1968; Couch *et al.,* 1969; Coates *et al.,* 1970; Pelletier *et al.,* 1972), the mammotroph (Pasteels, 1963; Smith and Farquhar, 1966, 1970; Pelletier *et al.,* 1971, 1972; Shiino *et al.,* 1972b), the thyrotroph (Farquhar, 1969, 1971; Pelletier *et al.,* 1971; Shiino *et al.,* 1973; Moguilevsky *et al.,* 1973), the corticotroph (Rennels and Shiino, 1968; Pelletier and Racadot, 1971), and, as mentioned above, even in the case of the gonadotroph (Farquhar, 1971; Shiino *et al.,* 1972a).

FIGS. 10 and 11. A gonadotroph (GT) and a thyrotroph (TT) from a cell suspension fixed immediately after dissociation. Both cells appear intact and well-preserved. The cytoplasm of the gonadotroph is asymmetrically arranged around the nucleus with the Golgi apparatus located on the side containing the more abundant amount of cytoplasm. This cell contains two types of secretion granules: one type (sg_1), the more numerous, is relatively small ($\sim$200 nm) in size, and the other (sg_2), of which there are fewer, is much larger in diameter ($\sim$400–500 nm). As seen here, the rough ER (er) of gonadotroph is typically dilated and has a content of moderate density. The overall size of the thyrotroph is much smaller, the rough ER is more flattened, and the secretion granules are of a single type and smaller ($\sim$100–150 nm) than those of the gonadotroph. Note that after aldehyde fixation the shape of both types of gonadotroph granules and the thyrotroph granules is variable but slightly less regular than the regularly spherical somatotroph granules. $\times$9120.

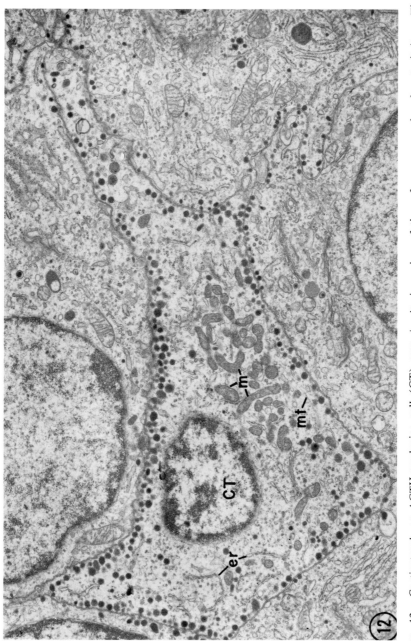

FIG. 12. Corticotroph or ACTH-producing cell (CT) seen *in situ* in a piece of tissue from a normal male rat pituitary. The cell is stellate in shape and contains ovoid ~200-nm secretory granules, the majority of which are lined up along the cell membrane. A cluster of mitochondria (m) and a few microtubules (mt) are present below the nucleus. The rough ER (er) consists of a few single, flattened cisternae. Specimen preparation as for Fig. 9. ×9500.

4. Are Microtubules and/or Microfilaments Involved in Exocytosis?

It has been proposed that in the anterior pituitary as well as in other secretory cells (see Pelletier and Bornstein, 1972; Labrie et al., 1973; and discussion by Kraicer, this volume), a microtubule–microfilament system may be involved in exocytosis. From the structural standpoint, the frequency of microtubules or microfilaments in pituitary cells is not remarkable, and no special association of such structures with secretory granules at the periphery of the cell has been noted. The hypothesis that these elements are involved in secretion is based on indirect evidence—namely, on the fact that drugs, such as colchicine and vincristine (which bind to microtubule protein) and cytochalasin B (which is supposed to affect microfilaments), inhibit secretion in various systems. This hypothesis is open to question for the following reasons:

(a) Cytochalasin B is known to inhibit glucose uptake and thereby to interfere with cell energetics (Zigmond and Hirsch, 1972; Mizel and Wilson, 1972a).

(b) Colchicine and vincristine depress ATP production in some (Jamieson, 1972) but not all (Le Marchand et al., 1973) cells. Hence inhibition of secretion might be explained in some cases by a reduction in intracellular ATP concentration.

(c) Colchicine also binds to cell membranes (Stadler and Franke, 1974), and colchicine together with several of its analogs (colcemid, lumicolchicine) drastically inhibit nucleoside uptake in certain cells (Mizel and Wilson, 1972b).

Therefore it remains to be decided by future work whether or not the drugs mentioned above inhibit secretion by interference with microtubules and microfilaments or by direct interaction with the cell membrane.

C. CONTROL OF PITUITARY SECRETION

It is not our intention to treat this topic in detail since it is discussed in this volume in the chapters by Kraicer, Hymer, and Stumpf et al., and is dealt with in other reviews (e.g., Guillemin and Burgess, 1972; Schally et al., 1973). However, no description of pituitary secretion would be complete without at least mentioning what is known about the mechanisms of control of various events in the secretory process. Briefly, there are three known control mechanisms: (1) hypothalamic releasing and inhibiting hormones, (2) hormones of target glands, and (3) in the case of excess intracellular hormonal storage, lysosomal disposal or "crinophagy."

1. Releasing and Inhibiting Hormones

As discussed by Kraicer (this volume), there is considerable evidence indicating a dual action of hypothalamic hormones on both synthesis and release of pituitary hormones. The effect on release is evident within minutes and can be detected

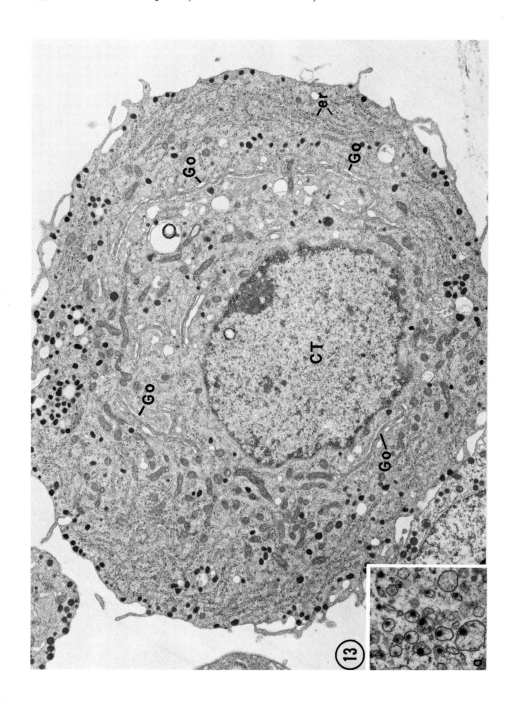

morphologically by increased frequency of granules undergoing discharge by exocytosis (Couch *et al.,* 1969; Shiino *et al.,* 1972a, 1973; Moguilevsky *et al.,* 1973). The effect on synthesis is evident only after a few hours, and according to the available biochemical data arises as a result of stimulation of protein synthesis, presumably at the translational (ribosomal) level. Both effects are presumed to be mediated by cyclic AMP and to be initiated by binding of the hypothalamic hormones to specific receptor sites on the cell membranes of pituitary cells. There are some cell fractionation data (Labrie *et al.,* 1972; Grant *et al.,* 1972; Wilber and Seibel, 1973) indicating that binding of hypothalamic hormones to plasma membrane is greater than that to other pituitary cell membranes (ER, Golgi). However, it should be mentioned that these data are not conclusive since the fractions on which the binding procedures were carried out were not analyzed in detail (for purity, yield, etc.), and since no effort was made to study equivalent sides of the membrane in all fractions. Moreover, the autoradiographic results of Stumpf *et al.* (this volume) indicate that one hypothalamic hormone (TRF) localizes to both nuclear and cytoplasmic compartments of pituitary cells, without indication of specific binding to cell membranes.

2. *Target Endocrine Hormones*

The fact that circulating levels of target hormones (steroids, thyroxine) affect the secretion of pituitary hormones is a basic tenet in endocrinology. The precise mechanisms by which this control operates is only now beginning to unfold. That the target cell hormones affect pituitary secretion indirectly through their action on the hypothalamus is well established, but now coming to light is evidence that target endocrines exert a *direct* action on pituitary cells. In the case of steroid hormones, there is now both autoradiographic (chapter by Stumpf *et al.,* this volume) and biochemical evidence (Leavitt *et al.,* 1973) for binding of steroid hormones, to pituitary cells—more precisely to nuclear components of pituitary cells. Thus the mechanism of their action on the cells of the pituitary (as well as those of the hypothalamus) is probably similar to their mechanism of action on

---

FIG. 13. Corticotroph (CT) in a cell suspension prepared from pituitaries of adrenalectomized (14 days) rats given hydrocortisone 12 hr prior to sacrifice, in order to promote accumulation of secretory granules (see Siperstein and Allison, 1965). This isolated corticotroph appears rounded-up. It is enlarged over its normal counterpart, mitochondria are more abundant, and the rough ER (er) and Golgi apparatus (Go) are more extensive. Stacks of Golgi cisternae surround the nucleus. The granules are irregular in shape and measure 200–250 nm. Specimen fixed in dilute Karnovsky's fixative as for Fig. 9. The inset on the lower left depicts a portion of a corticotroph fixed in OsO_4 alone to show the frequent haloed appearance of corticotroph granules after OsO_4 fixation. ×7500. Inset, ×20,000.

peripheral target tissues, i.e., on RNA synthesis at the transcriptional (nuclear) level (Hamilton, 1971; Mueller *et al.,* 1972; O'Malley and Means, 1974).

3. *Crinophagy*

In 1966, it was demonstrated that in pituitary mammotrophs, lysosomes play a role in regulating the secretory process by disposing of undischarged hormone (Smith and Farquhar, 1966). When hormonal secretion was suppressed, excess secretion granules were taken up and disposed of by lysosomes, a process known as crinophagy (see Fig. 1). Subsequently additional studies were undertaken to determine whether or not lysosomes had a similar function in other cell types (Farquhar, 1969, 1971). The approach used was first to stimulate hormone secretion in a given cell type (e.g., by removal of its target gland), and then to shut off secretion experimentally (by administration of the target gland hormone). It was found that in all cell types investigated—somatotrophs, gonadotrophs, and thyrotrophs—secretion granules were taken up and disposed of in lysosomes. Hence the concept of crinophagy can be assumed to be a general phenomenon among pituitary cells and to represent a mechanism by which lysosomes play a role in the regulation of the secretory process. This control mechanism, in contrast to the others described above, is a secondary level control mechanism which operates primarily when there is overproduction of secretory products. However, it is assumed that secretory granules are more or less continually funneled into the lysosomal system according to fluctuations in secretory activity. Pituitary lysosomes are known to contain proteases and peptidases capable of successively degrading pituitary hormones down to the level of inactive oligopeptides and dipeptides (McDonald *et al.,* 1971). Since the latter can diffuse through the lysosomal membrane into the cell sap, they will presumably be further degraded into their constituent amino acids which are thereby restored to the metabolic pool for reutilization within the pituitary cell (see Farquhar, 1971; McDonald *et al.,* 1971).

Crinophagy also occurs in certain other endocrine cells such as those of the adrenal medulla and the pancreatic islets. It is not known how the process is triggered in pituitary or other cells; this remains an interesting problem for the future.

FIG. 14.   Thyroidectomy cell (TT) from a cell suspension prepared from pituitaries of male rats, thyroidectomized 10 days prior to sacrifice. The isolated cells are rounded-up but otherwise show the same features as thyroidectomy cells *in situ:* Cell size is greatly increased over that of normal thyrotroph (see Fig. 11) due primarily to a striking dilation of the rough ER (er) which is engorged with a material of moderate density. The Golgi apparatus (Go) is enlarged, but relatively few secretory granules are seen. ×24,000.

FIG. 15.   Group of cells from a cell suspension prepared from normal female rat pituitaries. A mammotroph (MT) with its large (600–900 nm) secretory granules of variable shapes predominates the field. A somatotroph (ST) and thyrotroph (TT) are also present. The cytoplasm of the somatotroph contains several pale vacuoles. ×6000. Specimen preparation as for Fig. 9.

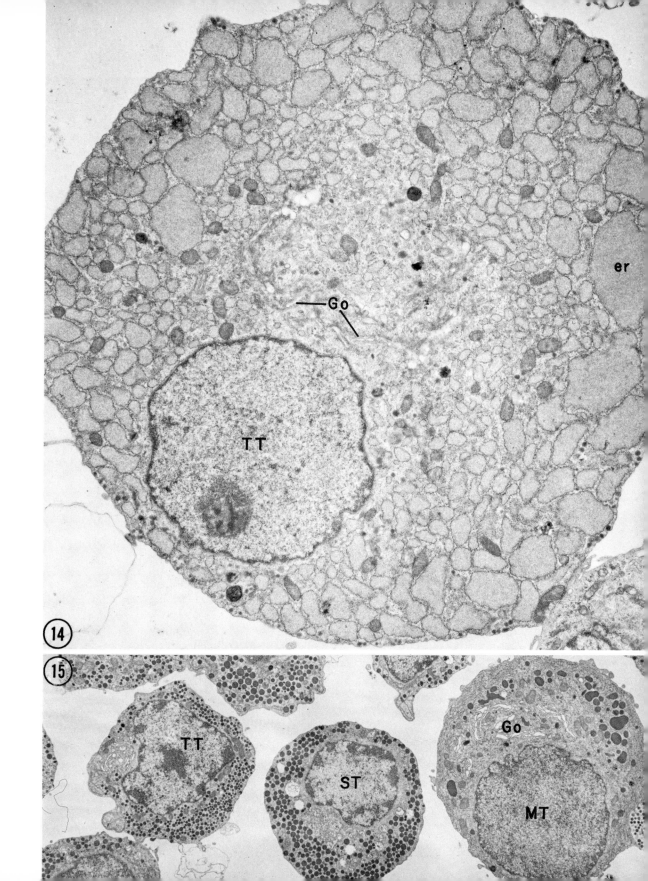

III. Evaluation of *in Vitro* Systems

Up until 2 to 3 years ago all of our studies were carried out *in vivo* on intact animals. Recently, we became interested in developing an *in vitro* system suitable for combined, morphological and biochemical studies of pituitary secretion. Although such systems are commonly used by endocrinologists, there has been relatively little information available on their fine structure, with the exception of several recent papers on incubated rat hemipituitaries (Pelletier, 1973; Pelletier *et al.,* 1972), some work on cultured whole mouse pituitaries (Yamashita, 1972a,b), and a study of slices prepared from ewe pituitaries (Tixier-Vidal *et al.,* 1971).

A. PITUITARY SLICES

Initially we tried slices of pituitary tissue (0.5–1 μm thick) comparable to those used for studies of secretion in other tissues [e.g., liver or exocrine pancreas (Jamieson and Palade, 1967a,b)] with little success. Tissue survival was very poor. After 1 hr, considerable autolysis and cell degeneration took place, and dead or dying cells were frequent, especially at the periphery of the slices near the cut surfaces.

B. HEMIPITUITARIES AND QUARTER PITUITARIES

We then attempted to use larger pieces of tissue such as hemipituitaries or quarter pituitaries, the system commonly used by endocrinologists for physiological and biochemical studies of pituitary functions. In accordance with the experience of others (e.g., Samli *et al.,* 1971), we found that when such tissue was analyzed it behaves in a satisfactory manner from a biochemical standpoint: It incorporates amino acids into protein linearly for periods up to 4 hr, the level of amino acid incorporation is high (based on local experience with other tissues), and the system can be satisfactorily pulse-labeled. When the tissue was analyzed by electron microscopy to determine the effect of incubation on cellular morphology, the preparations appeared to be in good condition with cellular fine structure reasonably well preserved up to 1 hr of incubation; however, with longer incubation times there was progressive deterioration in the condition of the tissue, in that the center of the pieces showed increasing (with time and distance from the cell surface) nuclear pycnosis, autophagy, and cell death. Autoradiographic analysis of such pieces after incubation with [^{3}H]leucine showed that there was a gradient in the number of grains present decreasing from the periphery to the center (Figs. 2 and 3). These observations indicated that penetration of even small molecular weight metabolites (i.e., amino acids) and tracers into the center of the tissue was poor, and therefore the large pieces commonly used for physiological and bio-chemical studies on pituitary function do not represent a functionally homogeneous system as commonly assumed. Even with smaller pieces (tenths), a gradient of grain distribution (decreasing from periphery to center) was seen (Figs. 4 and 5). When the size of the tissue pieces was reduced further, we again ran into the problem encountered with slices, i.e., peripheral autolysis, which was pronounced in all except the tips or end pieces of the gland that have the advantage of having only 1 cut surface. Hence, in developing a suitable *in vitro* system large pieces cannot be used because diffusion becomes limiting, and small pieces are not useful either because of the rapid autolysis of cells which occurs along cut surfaces. The autolysis seen may be due to the fact that high levels of proteases (aminopepti-dases) are known to be present in pituitary cells (McDonald *et al.,* 1971); when proteases are released from damaged (cut) cells, they may cause the death of other adjacent cells, with the phenomenon spreading from periphery to center. This assumes that the cells are more sensitive to aminopeptidases than to other proteases (e.g., trypsin, see below).

At about this time, when we were searching for alternatives to pieces and slices, an article appeared by Portanova *et al.* (1970) reporting that anterior pituitary tissue could be successfully dissociated into isolated, viable cells by using trypsin. Hence we turned our attention to attempting to develop and utilize such a system.

C. Dissociation of Pituitary Tissue: Development of a New Dissociation Procedure

1. *Trials*

Initial attempts to dissociate cells according to the method of Portanova *et al.* were discouraging since there was low cell yield and high cell fragility, and con-siderable residual undissociated tissue was left at the end of the incubation. More-over, many of the freed cells, especially the thyroidectomy cells we were working with at the time, were seen (by phase microscopy) to undergo blebbing and eventually lysis. We then undertook to develop a method for producing a prepara-tion of single dispersed pituitary cells in which morphological and functional in-tegrity were optimally preserved. In the beginning a number of different dissocia-tion procedures were tried using methods which have been applied to other tissues (involving collagenase, hyaluronidase, trypsin, EDTA, calcium-free media, or various combinations of these). In addition, we also carried out trials using neuraminidase (sialidase). Although to our knowledge this enzyme has not pre-viously been used to dissociate cells, the fact that cell surface glycoproteins are known to be involved in cell adhesion (Ambrose, 1966; Curtis, 1967) suggested that this enzyme (by its action on surface sialic acid groups) might be an agent useful for cell dissociation.[3] Various difficulties (lysis, blebbing, incomplete dissociation, reaggregation) were encountered using existing methods for tissue dissociation. Much more promising results were obtained with neuraminidase, especially when incubation with this enzyme was preceded by a brief period of incubation with trypsin.

2. *New Dissociation Procedure*

After many trials with many combinations of agents, we developed a procedure which involves incubating small tissue pieces (40–50/gland) first in 0.1 mg/ml trypsin for 15 min, and then in 8 μg/ml neuraminidase (*Clostridium perfringens*) and 1 mM EDTA for 15 min. This procedure has been given in detail elsewhere (Hopkins and Farquhar, 1973), and its principal steps are summarized in Fig. 6. All incubation media contain 0.3% BSA, 14 mM glucose, and a complete amino acid supplement in a Krebs–Ringer bicarbonate solution (KRB). After incubation the pieces can be completely dispersed by mild pipetting. The fact that the pieces hold together during the incubation and washing procedure greatly facilitates their handling. The progressive loosening of the normal tissue structure which occurs during successive steps in the dissociation procedure, as well as the resultant cell suspension is shown in Figs. 7 and 8. The procedure takes only 45 min, yields >90% single cells, and produces 1.5–2.0 $\times$ 10^6 cells/pituitary (55% yield on a DNA basis). Ninety-five percent of the cells are viable (based on trypan blue exclusion).

IV. Studies on Dispersed Pituitary Cells

A. MORPHOLOGICAL STUDIES

Electron microscopy demonstrated that the fine structure of cells obtained from normal rat pituitaries by the above dissociation procedure was well preserved. All secretory cell types—somatotrophs (Fig. 9), mammotrophs (Fig. 15), gonadotrophs (Fig. 10), thyrotrophs (Fig. 11), and corticotrophs (Fig. 12)—can be identified. In addition, several nonsecretory cell types—endothelial and follicu-

---

[3] The potential importance of such substances in adhesion of pituitary cells was also suggested by the fact that both cell junctions and the connective tissue framework are modest in anterior pituitary as compared to many other organs. As far as junctions are concerned, the total area of cell surfaces devoted to them is limited in anterior pituitary cells compared to the situation encountered in the liver, exocrine pancreas, thyroid, intestine, and other lumen-lining epithelia. Only shallow gap junctions and occasional desmosomes connect secretory elements to one another (or secretory to nonsecretory elements). The only other junctions encountered are those present between nonsecretory elements (follicular cells and endothelial cells).

lar cells (Fig. 8c)—are present. We have also carried out the procedure on pituitaries from castrates, thyroidectomized (Fig. 14), adrenalectomized (Fig. 13), and estrogen-treated rats with equally good cell yields and with equally good results in terms of preservation of fine structural organization. It is clear that the rounded-up, single cells have certain advantages over their counterparts in intact pituitary tissue (Fig. 7); Their main advantage is that they represent a functionally homogeneous cell population with equal access to metabolites, tracers, O_2, etc. In addition they have certain secondary advantages in that they present a favorable geometry with "retracted" pseudopods, a high frequency of cell profiles including all cell compartments of interest for the secretory process which facilitates autoradiography.

B. Synthesis, Transport, and Release of Secretory Products

The ability of the cells to synthesize, transport, and release secretory products was analyzed and compared with that of cells in small tissue blocks and, in some cases, with hemipituitaries. Both recently dissociated cells and cells cultured overnight were used.

1. Incorporation of Amino Acids into Total Protein

As shown in Fig. 16, dispersed cells—both recently dissociated and cultured 15 hr—incorporated linearly amino acids into total protein over a period of 4 hr. The rate of incorporation is 90% greater than that of tissue blocks and 140% greater than that of hemipituitaries. Since our autoradiographic findings indicate that amino acids do not penetrate into the center of tissue blocks and especially, into hemipituitaries, the higher rate of [³H]leucine incorporation into cell suspensions is presumed to be due largely or exclusively to the fact that they have more uniform access to the tracer. A similar relationship between the size of the tissue fragment and amino acid incorporation was noted previously by Samli et al. (1971).

2. Incorporation of [³H]Leucine into STH

The ability of the cell suspensions to incorporate amino acids into a specific secretory product was assessed by determining [³H]leucine incorporation into growth hormone (STH) by cells dissociated from pituitaries of young male rats. STH was chosen for these studies because (a) it has been established that in the young, growing male rat, STH accounts for up to 60% of the total newly synthesized protein released in vitro (Labrie et al., 1971), and (b) the amount of STH released can be readily evaluated by taking advantage of available methods for isolation of STH by polyacrylamide gel electrophoresis (Jones et al., 1965; MacLeod and Lehmeyer, 1970).

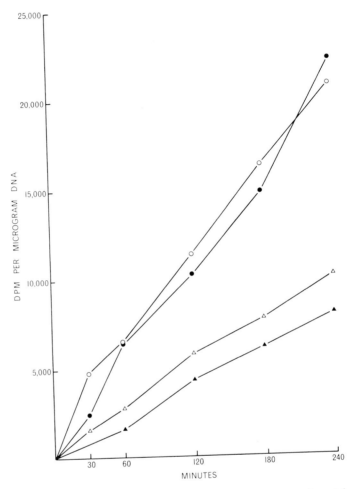

FIG. 16.   Time course of incorporation of [³H]leucine into total protein in acutely dissociated cells ○—○, cells dissociated and cultured 15 hr ●—●, small tissue blocks △—△, and hemipituitaries ▲—▲. Hemipituitaries, small tissue blocks, or 2-ml aliquots of cell suspension were placed in 25-ml Erlenmeyer flasks and incubated from 0 to 240 min in 5 ml KRB/BSA containing 10 μCi/ml [³H]leucine (52 Ci/mmole).

The results shown in Fig. 17, show that recently dissociated cells, cultured cells and tissue blocks all incorporate [³H]leucine into STH. In tissue blocks, about 25% of the total label is incorporated into STH and about 15% into prolactin. However, recently dissociated cells incorporate protein into several peaks other than those represented by these two hormones and the proportion of total label incorporated into STH is only 12–13%. In dissociated cells cultured overnight, these additional peaks of radioactivity are greatly reduced and the amounts of

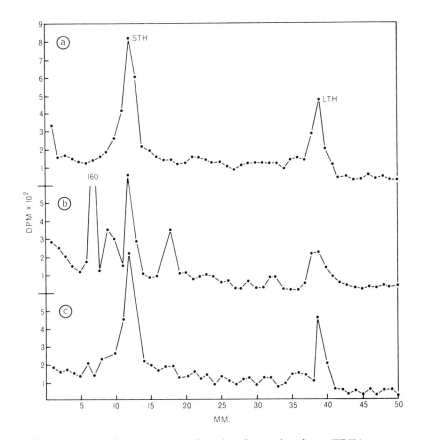

FIG. 17. Polyacrylamide gel electrophoresis of proteins from EDTA extracts of (a) tissue blocks, (b) recently dissociated cells, and (c) cells cultured 15 hr. The specimens were incubated for 4 hr in media containing 10 μCi/ml [³H]leucine then extracted in 1 mM EDTA. Free [³H]leucine was removed from the extracts by gel filtration (G-25 Sephadex). Electrophoresis was carried out on 10% slab gels at pH 9.5. Duplicate aliquots of the load samples contained ∼50,000 dpm. Growth hormone (STH) and prolactin (LTH) were identified using NIAMDD rat hormone standards. Top of gel to the left. Starting gel 1 cm long. 0 mm on axis indicates beginning of separation gel.

total label incorporated into STH and prolactin are similar to that found in tissue blocks. We have not studied further the nature of the additional labeled components elaborated in acutely dissociated cells. In view of their transient nature it seems likely that they may be related to short-term repair processes.

3. *Autoradiography*

We used electron microscope autoradiography to assess the cells' ability to transport and concentrate secretory products and to determine the intracellular route taken by newly synthesized protein. For these experiments we pulse-labeled

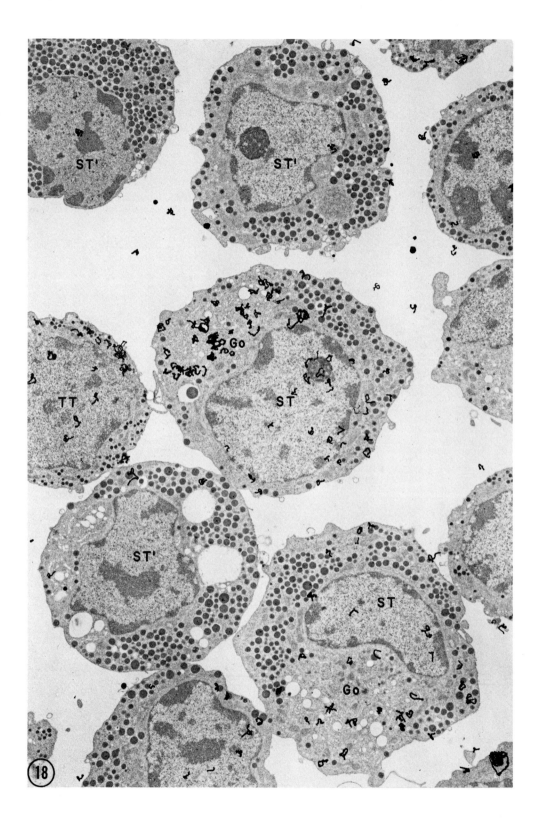

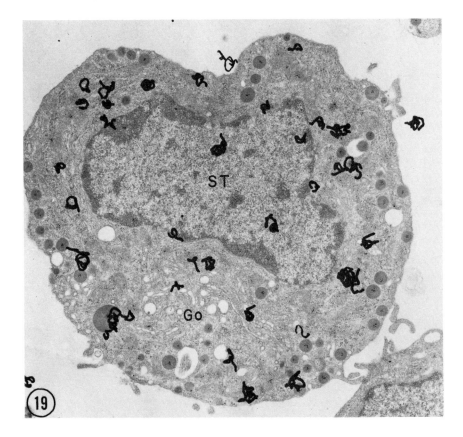

Fig. 19. Radioautogram of a somatotroph (ST), cultured for 15 hr after dissociation and given a 5-min leucine pulse, followed by a 1-min chase incubation. The radioautographic grains are rather evenly distributed over nucleus and cytoplasm with the majority being associated with the rough ER. Go = Golgi apparatus. ×13,000.

cultured cells, since the incorporation data (Fig. 17) showed that the amount of leucine incorporated into STH is twice that found in recently dissociated cells. It is of interest that only about half of the somatotrophs are heavily labeled; that is, only about half are actively engaged in incorporating amino acids into secretory (or other) proteins (Fig. 18). A similar heterogeneity of amino acid uptake was

Fig. 18. Electron microscopic radioautogram showing a group of somatotrophs cultured for 15 hr and given a 5-min pulse of [³H]leucine, followed by a 15-min chase incubation. Two of the somatotrophs (ST) show a heavy concentration of grains, primarily over the Golgi (Go) region, whereas three others (ST′) show very few grains. Except for one cell on the lower left which has large vacuoles, the unlabeled cells do not show any obvious differences from labeled cells. ×5940.

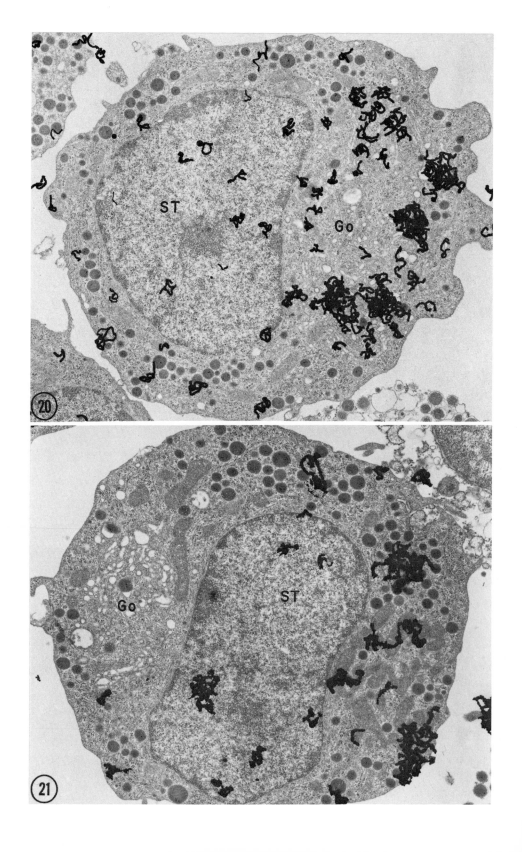

found among somatotrophs in small tissue blocks. It can therefore be concluded that somatotrophs of young male rats normally vary in their ability to incorporate labeled precursors, and under our conditions only about half the cells are turned on in this respect. This is in striking contrast to the situation in the exocrine pancreas where uptake of radioactive tracer is uniformly high in virtually all cells (Amsterdam and Jamieson, 1972).

At the end of the 5-min pulse, most of the autoradiographic grains are diffusely distributed over the nucleus and cytoplasm of the somatotroph (Fig. 19). At 15–30 min postpulse, there is a striking concentration of grains over the Golgi apparatus (Fig. 20) whereas at 60 min postpulse the proportion of grains over the Golgi is reduced, and increased numbers are related to secretory granules. By 120 min postpulse, the Golgi is virtually free of label and the majority of grains are related to secretory granules (Fig. 21).

The findings indicate that the cells are capable of transporting newly synthesized, labeled secretory protein from the rough ER to the Golgi apparatus where it is packaged into granules which are stored in the cytoplasm.

4. *Ability to Release STH*

The data shown in Tables II and III show that recently dissociated cells do not respond to secretagogues (high K^+ and dibutyryl cyclic AMP). However, cells cultured overnight do respond by releasing significantly increased amounts of STH at a level comparable to those reported previously (Schofield, 1967; MacLeod and Lehmeyer, 1970) for intact pituitary tissue. Vale *et al.* (1972) and Tixier-Vidal *et al.* (1973) also found that a period of recovery (3 and 5 days, respectively) was necessary to obtain a normal secretory response after dissociation by another procedure. Since trypsin and neuraminidase are known to affect a number of important reactions at the cell surface (Marcus and Schwartz, 1968; Currie *et al.*, 1968; Burger and Goldberg, 1967; Burger, 1969), including hormone binding to specific receptor sites in other tissues (cf. Cuatrecasas and Illiano, 1971; Rosenthal and Fain, 1971), the most reasonable interpretation of these findings is that the enzymes used in the dissociation procedure partially degrade the cell surface components and that a period of recovery is necessary for their regeneration or repair. The additional proteins synthesized by acutely dissociated cells and isolated by gel electrophoresis may be related to these repair processes.

Figs. 20 and 21. Radioautograms of two somatotrophs (ST) prepared and pulsed as in Fig. 19, and fixed after a 15-min (Fig. 20) or 120-min (Fig. 21) chase incubation. Note that after a 15-min chase most of the radioautographic grains are concentrated over the Golgi region (Go), whereas after a 120-min chase most are concentrated over the secretion granules, and the Golgi region (Go) is free of grains. At both chase intervals, and indeed at all intervals studied, there is considerable nuclear label which consistently accounts for 15–20% of the total grains. Fig. 20, $\times 11,400$; Fig. 21, $\times 13,300$.

TABLE II

EFFECT OF HIGH K+ AND DIBUTYRYL cAMP ON RELEASE OF
[³H]STH FROM ACUTELY DISSOCIATED CELLS[a]

	Cell content[b]	Medium content[b]	% Total [³H]STH released
KRB control	142,547 ± 7,127	20,363 ± 509	12.5
57 mM K+	145,378 ± 10,176	21,723 ± 1,303	13
5.0 mM DbcAMP	142,213 ± 15,643	27,088 ± 1,083	16

[a] Cell suspensions were incubated 6 hr with 10 μCi/ml [³H]leucine, and STH was isolated using gel electrophoresis as previously described (Hopkins and Farquhar, 1973). Data derived from 5 experiments. Values are mean ± standard error.

[b] dpm ³H-labeled STH/10⁶ cells.

C. SURFACE PROPERTIES OF DISPERSED CELLS

In view of the findings discussed in the preceding section, especially those indicating that recently dissociated cells (in contrast to cultured cells) do not respond to secretagogues, we carried out cytochemical tests to determine whether or not differences in surface properties would be detected between the two cell populations. Since the neuraminidase and trypsin used in the dissociation procedure are expected to remove surface sialyl residues and glycopeptides, we applied the colloidal iron method of Gasic et al. (1968) and the cationized ferritin procedure of Danon et al. (1971) to investigate the distribution and frequency of acidic groups, primarily sialyl groups, on the cell surface. Staining was demonstrated on

TABLE III

EFFECT OF HIGH K+ AND DIBUTYRYL cAMP ON RELEASE OF
[³H]STH FROM CELLS CULTURED 15 HR[a]

	Cell content[b]	Medium content[b]	% Total [³H]STH released
KRB control	148,970 ± 7,448	17,871 ± 893	11
KRB at 5°C	186,933 ± 17,863	7,788 ± 389	4
57 mM K+	138,520 ± 16,622	33,245 ± 4,986	24
0.1 mM DbcAMP	164,520 ± 18,040	20,500 ± 1,845	11
0.5 mM DbcAMP	140,383 ± 7,019	39,595 ± 2,373	22
5.0 mM DbcAMP	93,951 ± 4,697	101,780 ± 13,702	52

[a] Cell suspension preincubated for 15 hr in KRB, and for the last 10 hr in the presence of 100 μCi/ml [³H]leucine. The cells were then washed, incubated for 3 hr as shown above, and the STH isolated by gel electrophoresis as previously described (Hopkins and Farquhar, 1973). Data derived from 3 experiments. Values are means ± standard error.

[b] dpm ³H-labeled STH/10⁶ cells.

all pituitary cells with both techniques. There were some differences in staining among the various cell types, but there were no appreciable differences between recently dissociated and cultured cells of the same type. The degree of staining (indicating the number of sialic groups present) was distinctly greater on follicular than on secretory cells (Fig. 22). Among the latter elements, however, there were minor differences (e.g., with gonadotrophs staining more intensely than somatotrophs). No attempt was made to quantitate these differences. Similar studies were conducted to evaluate the binding of concanavalin A (Con A) (a plant lectin which binds to α-D-mannose or α-D-glucose residues) to the surfaces of pituitary cells using the ferritin-conjugated Con A procedure of Nicolson and Singer (1971). The results were similar to those obtained with methods for detection of sialic acid, in that no appreciable differences were detected in the level of labeling of recently dissociated versus cultured cells. The results are significant in that they indicate that sialyl and other (i.e., mannosyl or glucosyl) carbohydrate residues were not completely removed by the enzyme treatment during the dissociation procedure. The failure to detect any differences in surface groups between recently dissociated and cultured cells could be real, or could simply mean that the effect of the dissociation agents on the frequency and distribution of such groups is below the level detectable by the qualitative methods used.

D. PEROXIDASE UPTAKE BY DISSOCIATED SOMATOTROPHS

It is evident that at the time of granule discharge by exocytosis there is considerable relocation of the membranes surrounding discharged granules to the cell surface. Since the granule membranes were originally derived from the innermost Golgi cisternae this amounts to a relocation of considerable Golgi membrane to the cell membrane. In other systems (Amsterdam et al., 1969) there is evidence that during the period of recovery following granule discharge by exocytosis, membrane is removed from the cell surface to compensate for that added, thus keeping the overall cell size reasonably constant. It is not known, however, what kind of membrane is removed or what is its fate. Various hypotheses have been proposed: (1) recovery of the same membrane and recirculation back to the Golgi (Palade, 1959), (2) dismantling to macromolecular components which are subsequently reassembled (Fawcett, 1962), or (3) complete degradation (Amsterdam et al., 1971). In the case of neurosecretory neurons (Nagasawa et al., 1971) and neuromuscular junctions (Heuser and Reese, 1973; Ceccarelli et al., 1973), evidence has been obtained that following stimulation of the cells and discharge (by exocytosis) of neurosecretory granules and synaptic vesicles, respectively, areas of the cell membrane are recovered in the form of small vesicles.

We have used horseradish peroxidase as a tracer to investigate circulation of membranes during secretion in dissociated pituitary cells and to define possible

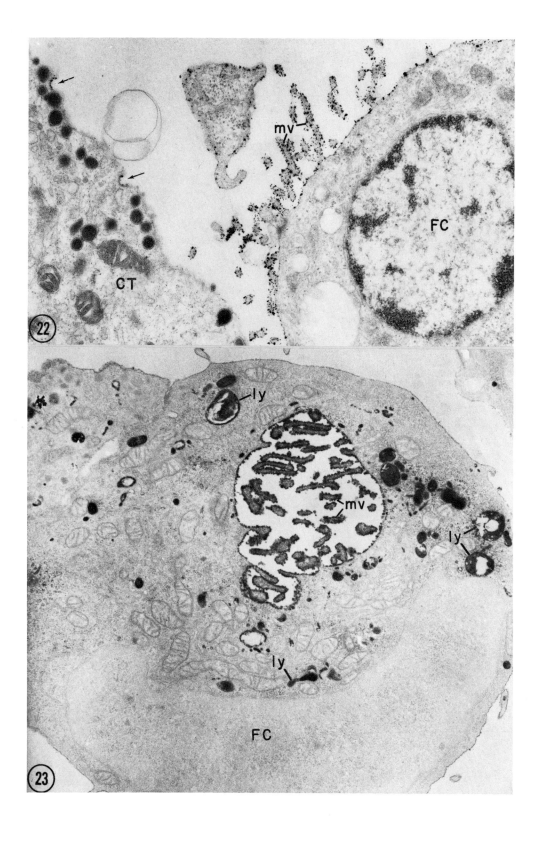

sites of relocation of recirculated membrane. Toward this end, dissociated pituitary cells cultured overnight were incubated for 1–5 hr in the presence of horseradish peroxidase (HRP) (1 mg/ml, chromatographically pure, obtained from Worthington Biochemical Corp.) and 5 mM dibutyryl cyclic AMP to stimulate exocytosis (Pelletier *et al.,* 1972). Some specimens were also incubated with HRP and cyclic AMP for 2 hr followed by incubation for 1 hr with cyclic AMP alone. It was anticipated that HRP would label the cell membranes and thus give an indication as to possible intracellular sites where such membranes might be relocated during the recovery period following granule discharge. In controls incubated without cyclic AMP, there was limited uptake of HRP in somatotrophs although uptake was pronounced in follicular cells (Fig. 23). The HRP present in somatotrophs was found primarily in small vesicles and lysosomes, with the number of labeled lysosomes increasing with time. In specimens incubated in the presence of HRP and cyclic AMP, much more HRP was found within the cells (Figs. 24–26). Most was located within lysosomes or related tubular structures or vesicles found near the cell membrane or in the Golgi region. However, and most interestingly, traces of HRP were also evident in the stacked Golgi cisternae of some cells (Figs. 25–26). It was not found in all the cisternae but was restricted to the innermost cisternae along the concave surface of the stack.[4] It is significant that, although condensing granules are not evident in these preparations incubated with cyclic AMP (Hopkins and Farquhar, 1973), this is the same cisterna from which secretory products normally emerge in these cells (see Farquhar, 1971).

[4] Very similar observations have been recently reported by Pelletier (1973) in intact pituitary fragments incubated *in vitro* in the presence of HRP and cyclic AMP.

FIG. 22. Portions of two cells from a cell suspension cultured overnight and stained by the colloidal iron technique (for demonstration of sialic acid groups). The follicular cell (FC) on the right shows numerous iron particles distributed along its cell membrane, indicating a heavy concentration of sialic acid groups on the surface of these cells. The particles are especially concentrated along the membranes of the microvilli (mv). The amount of stain seen along the membrane of the adjoining secretory cell, believed to be a corticotroph (CT), is much more limited. It is of interest that, in the latter, the heaviest deposits are seen in two pits (arrows) which occur along the cell surface. These could represent vesicles in the process of either pinching off or fusing with the cell membrane. Specimen fixed 30 min in Karnovsky's fixative, stained 30 min by Gasic's (1968) procedure, and postfixed in OsO$_4$. Specimen lightly stained in lead alone. ×19,000.

FIG. 23. Follicular cells (FC) from a cell suspension incubated 3 hr with horseradish peroxidase (HRP). The amount of tracer taken up is much greater than that taken up by secretory elements in the same cell suspension. Externally the HRP is concentrated on the microvilli (mv) which are located along the former luminal surface of the cell. Internally the tracer is concentrated in lysosomes (ly) of variable sizes and shapes. Specimen fixed 30 min in Karnovsky's fixative, incubated 30 min in the diaminobenzidine (DAB) medium of Graham of Karnovsky, and postfixed in OsO$_4$. Section lightly stained with lead alone. ×13,300.

It can be concluded that these findings provide evidence in support of the suggestion of Palade (1959) that membrane relocated to the cell surface during exocytosis is recaptured intact and recirculated back to the Golgi. The findings indicate further that the recirculation is restricted to *specific* Golgi cisternae—i.e., the cisternae along the concave Golgi surface—the same cisternae from whence the pieces of membrane originally came.

E. COMMENTS ON PITUITARY DISSOCIATION METHODS

A number of other investigators have recognized the desirability of having available a suspension of pituitary cells for studies on hormone secretion and have attempted to dissociate pituitary tissue and to obtain such preparations (Ishikawa, 1969; Portanova *et al.,* 1970; Bala *et al.,* 1971; Vale *et al.,* 1972; Kudo *et al.,* 1972; Hymer *et al.,* 1973; Leavitt *et al.,* 1973; Tixier-Vidal *et al.,* 1973). These studies have been discussed by us elsewhere (Hopkins and Farquhar, 1973) and are reviewed by Hymer in this volume, hence they are not detailed here. As far as our own results are concerned, it is difficult at this time to compare them directly with those of others, since the data are incomplete and since none of the previous authors have provided a thorough structural and functional analysis of the resultant cell suspension. It is clear in our system that the function and morphological integrity of the cells is preserved and that the dispersed system has a number of advantages over intact tissue for the studies of pituitary secretion and its control. With dispersed cell suspensions it is possible to: (1) overcome individual variations between glands, since the cell population from a number of glands is randomized, (2) eliminate the diffusion problems encountered in even small tissue fragments, (3) provide rounded-up single cells with a simplified topography (which facilitates certain procedures such as autoradiographic counting), and (4) provide a suitable starting material for the subfractionation of the heterogeneous

---

FIGS. 24–26. Somatotrophs (ST) from a cell suspension incubated for 3 hr in horse-radish peroxidase (HRP) (1 mg/ml) plus dibutyryl cyclic AMP (5 mM). In Fig. 24, the tracer is seen lining the cell membrane and concentrated in lysosomes (ly) and numerous vesicular (v) and tubular (t) structures, which are undoubtedly also lysosomal in nature. In addition, smaller concentrations are seen within stacked Golgi cisternae (Go) in two places (arrows). Figures 25 and 26 are enlargements of stacks of Golgi cisternae containing HRP. In Fig. 25, which is an enlargement of the part of Fig. 24 outlined, seven successive Golgi cisternae (1–7) can be counted. HRP is restricted to the penultimate cisterna (2) along the concave Golgi surface. It is also seen within three rounded bodies, believed to be lysosomes (ly), present nearby. In Fig. 26, HRP is restricted to the innermost of the stacked Golgi cisternae and to two curved tubular structures which occur in close proximity. Another fenestrated Golgi cisterna (fc) is seen in full face to the right. A large accumulation of fibrillar material (fi) surrounded by rough ER (er) is present on the left of the field shown in Fig. 24. Such accumulations are common in cell suspensions incubated with cyclic AMP. Specimen preparation as for Fig. 23. Fig. 24, ×18,000; Figs. 25 and 26, ×40,000.

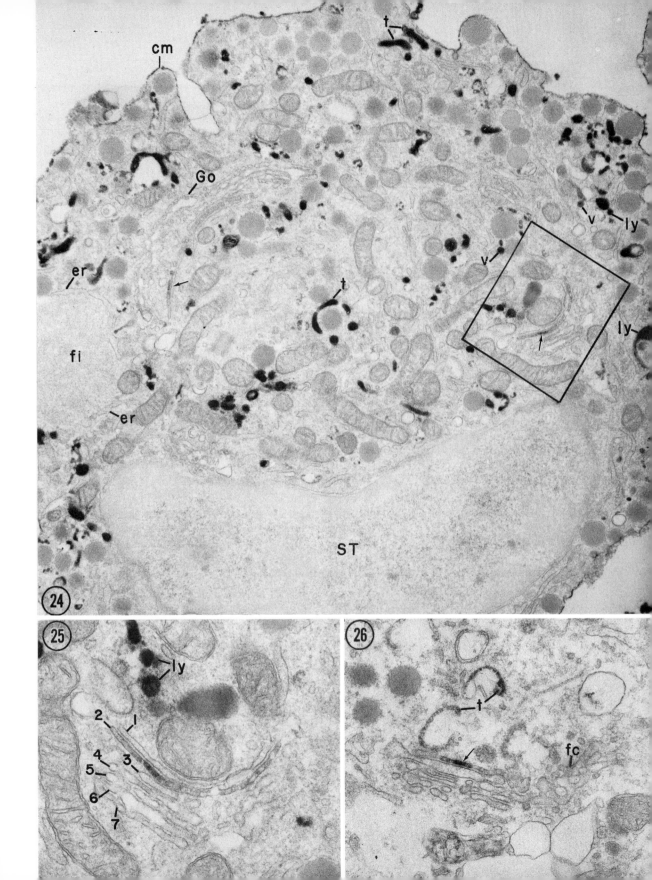

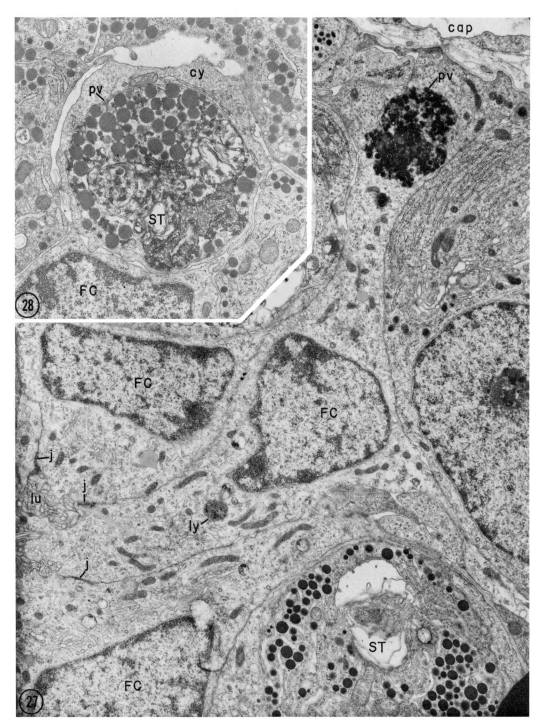

pituitary cell population into preparations consisting of a single cell type. Some examples of applications of the first three points listed above have been outlined in this chapter. At the moment, efforts in our and a number of other laboratories are directed toward trying to separate pituitary cells according to specific cell types utilizing dispersed cells as a starting material. In this respect, Hymer and his co-workers (1972) have already achieved some enrichment of cell types using a combination of gravity and equilibrium sedimentation.

V. *In Vitro* Studies on Follicular Cells: Their Role in Phagocytosis of Cells and Cell Debris

It was previously reported (Farquhar, 1971) that in pituitary tissue incubated *in vitro,* follicular cells function in the phagocytosis of cells and cell debris. This phenomenon is particularly evident in blocks of pituitary tissue. As already mentioned, since metabolites do not fully penetrate hemipituitaries or large tissue blocks, many secretory cells, especially in the center of such pieces, undergo cell death. Apparently the follicular cells incorporate and dispose of such cells. Reconstructing the situation from electron micrographs, it appears that what happens is that the long arms of the follicular cell, which normally extend between and surround the secretory elements, simply encircle the damaged or dead cell, the arm collapses on it and subsequently internalizes it into the follicular cell cytoplasm in a large phagocytic vacuole (Figs. 27–28). The process resembles that whereby an octopus pulls in prey by way of its tentacles. When this process is studied in electron micrographs of tissue blocks, the relationship between the phagocytized object and the thin layer of surrounding follicular cell cytoplasm is not always clear (Fig. 28). If, however, the blocks are dissociated after an *in vitro* incubation of 2 hr, the follicular cells can be recovered (by appropriate manipulations) in groups still attached by their junctional complexes. In such preparations virtually every cell has phagocytic vacuoles with recognizable cell

FIGS. 27 and 28. Figure 27 shows a cluster of follicular cells from a small block of pituitary tissue incubated 2 hr *in vitro*. The follicular lumen (lu) is seen to the left and a capillary (cap) above. Junctional complexes (j) are present along the lateral cell membranes near their luminal surfaces between adjoining follicular cells (FC). The follicular cell which dominates the field is stellate in shape and is seen to have one foot on the follicular lumen below and another on the capillary above. A lysosome (ly) is seen between the follicular lumen and the cell's nucleus. The capillary pole of this cell contains a large phagocytic vacuole (pv) with recognizable secretory granules derived from an ingested secretory cell. Figure 28 is a similar field showing another phagocytic vacuole (pv) containing a partially digested somatotroph (ST) within a follicular cell. Only a thin rim of follicular cell cytoplasm (cy) surrounds the vacuole. Preparation as for Fig. 9. Fig. 27, ×10,200; Fig. 28, ×18,700.

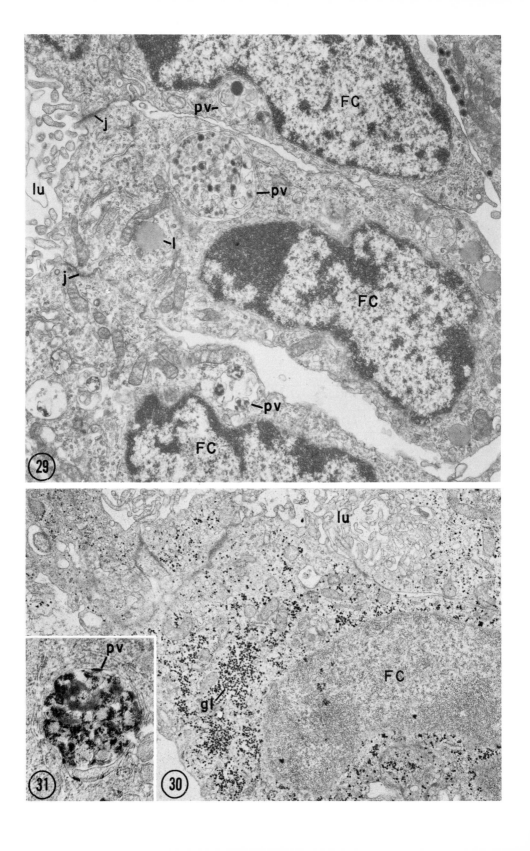

debris (clumped secretory granules) in varying stages of digestion (Figs. 29–31). Acid phosphatase tests carried out on such specimens (Fig. 31) indicate that the phagocytic vacuoles contain lysosomal enzymes capable of digesting the phagocytosed material.

One can ask, why is it that the follicular cells themselves do not succumb to the conditions prevailing in the centers of tissue blocks incubated *in vitro?* The answer probably lies in their large stores of glycogen (Fig. 30) which make available an energy source through anaerobic glycolysis. Thus these cells, like phagocytes in general, would not be adversely affected by low oxygen tension as long as their energy requirement could be satisfied anerobically at the expense of their glycogen reserves or of glucose present in the medium.

One can ask further, does this phagocytic function of follicular cells take place *in vivo?* The answer is in all probability, yes, since one can occasionally find lysosomes containing tissue debris in follicular cells in tissue taken from intact animals. Moreover, Yamashita (1969) reported (but did not illustrate the process) that follicular cells phagocytize necrotic cells in the differentiating postnatal rat pituitary in which some cell involution is evident. Dingemans and Feltkamp (1972) found evidence of phagocytosis by follicular cells or "nongranulated cells" in mouse pituitaries after a variety of experimental procedures (radiothyroidectomy, castration, adrenalectomy, and transplantation of the pituitary to the kidney capsule). Beyond this, Vila-Porcile *et al.* (1973) have obtained evidence that under certain experimental conditions (i.e., the postlactating rat) lysosomal residues are released into intercellular spaces from prolactin cells. We have seen similar images (i.e., of extracellular discharge of lysosomal residues) in thyroidectomized animals and have further obtained evidence that such residues may be picked up and disposed of by follicular cells. Thus the follicular cells apparently constitute a scavanger system for disposing of cells and cell debris. An indication of the

---

FIG. 29. Cluster of three follicular cells (FC) from a piece of pituitary tissue which had been incubated 2 hr *in vitro* and then dissociated with trypsin alone. The cells remain attached to one another by cell junctions (j) located near their luminal surfaces (lu). Phagocytic vacuoles (pv) are seen in all three cells. The phagosome in the center cell contains partially digested but still recognizable secretory granules. l = lipid droplets. Specimen preparation as for Fig. 9. ×15,200.

FIG. 30. Portions of three follicular cells (FC) from a specimen prepared in such a way as to demonstrate glycogen. The former lumen (lu) is present above. It can be seen that all three cells contain abundant beta particles of glycogen (gl). Specimen fixed in OsO_4 in phosphate buffer. Specimen stained with lead alone. ×14,250.

FIG. 31. Phagocytic vacuole (pv), similar to that in Fig. 29, from a preparation incubated for acid phosphatase. Dense reaction product for this lysosomal enzyme is seen within the phagocytic vacuole along with partially digested, rounded secretion granules. Specimen fixed in glutaraldehyde, incubated for 90 min in a modified (Barka-Anderson) acid phosphatase medium at pH 5.0, postfixed in OsO_4, and stained in block with uranyl acetate. Specimen lightly stained with lead alone. ×19,000.

rapidity of the process is given by the fact that the material ingested *in vitro* is incorporated and partially digested within 2–3 hr. As already pointed out by Vila-Porcile *et al.* (1973), the phagocytosis of dead cells by follicular cells is reminiscent of the role played by Sertoli cells of the testis in phagocytosis and digestion of forming sperm after suppression of spermatogenesis [e.g., following irradiation (Hugon and Borgers, 1966)].

Finally, one can ask if phagocytosis is the only or even the main function of the follicular cells? This question cannot be answered at the present time. One is reminded again of the analogy with the Sertoli cell which functions as a "nurse" cell during spermiogenesis and as a phagocyte to dispose of developing sperm on suppression of spermiogenesis. The ubiquity of the follicular cells and their intimate relationship to the parenchymal elements suggest that these unusual cells could have additional functions that remain to be discovered.

VI. Summary

After reviewing available information on the nature of the various cell types in the anterior pituitary gland and what is known about their secretory processes, new data is presented on the fine structure and function of cells dissociated from rat anterior pituitaries.

Available information indicates there are five secretory cell types—somatotrophs, mammotrophs or prolactin cells, gonadotrophs, thyrotrophs, and corticotrophs—plus several nonsecretory cell types such as follicular cells and vascular and connective tissue elements (endothelial and adventitial cells, respectively). There is now general agreement on the fine structural characteristics of all of these cells types. The main problem (pertaining to the identification of cell types) that needs to be resolved is whether or not the two gonadotropins are produced by a single cell type (as suggested by the available immunocytochemical data) or by separate cell types.

As far as knowledge of secretory processes in the various cell types is concerned, there are many gaps in our information owing primarily to the fact that the pituitary, with its large number of cell types, is not as suitable a tissue as some others on which to carry out cell fractionation studies. The information available suggests that the main events of secretion in pituitary cells parallel the theme worked out in other, better-studied protein-secreting tissues. There are, however, some variations on this theme (e.g., the existence of crinophagy or a mechanism for disposal of excess secretion granules via lysosomes). It is stressed that in the pituitary as in other systems, exocytosis appears to be the main mechanism for hormone discharge and the only established mechanism. Evidence for the involvement of microtubules and/or microfilaments in exocytosis is indirect and therefore inconclusive.

Results are presented on our attempts to use existing *in vitro* systems for combined morphological and biochemical studies on pituitary secretion. Hemipituitaries (as well as tenths) were unacceptable owing to poor penetration of nutrients and radioactive (amino acid) tracers into the pieces. This resulted in a gradient of increasing autolysis and decreasing grain concentration from the periphery to the center of the pieces. Slices (200–400 μm) were equally unacceptable due to autolysis of cells along the cut surfaces which spread from the periphery to the centers. Therefore, neither hemipituitaries or slices were acceptable.

Dispersed pituitary cells gave more promising results. A new procedure was developed for dissociating anterior pituitary tissue and producing a viable suspension of single cells. The procedure involves incubation of small tissue blocks in 1 mg/ml trypsin (15 min), followed by incubation in 8 μg/ml neuraminidase and 1 mM EDTA (15 min), followed by mechanical dispersion. Cell yields are ~55%, based on recovered DNA. By electron microscopy all five types of secretory cells plus endothelial and follicular cells can be identified and are morphologically well preserved up to 20 hr after dissociation.

The ability of the dissociated cells to synthesize, transport, and release secretory products was assessed. Throughout this period, the cells incorporate linearly [³H]leucine into protein for up to 4 hr at a rate 90% greater than hemipituitaries, and they synthesize, transport intracellularly, and release the two major pituitary secretory products, growth hormone and prolactin. Immediately after dissociation the cells' ability to respond to secretogogues (high K$^+$ and dibutyryl cyclic AMP) is impaired, but after a 6–12-hr culture period, the cells apparently recover and discharge 24 and 52%, respectively, of their content of prelabeled growth hormone over a 3-hr period in response to these two secretogogues. This represents a stimulation of 109 and 470% over that released by cells incubated in control medium. The results demonstrate that function and morphological integrity are preserved in this cell system.

Cytochemical tests were carried out to determine the cell surface properties (primarily distribution of sialyl residues) of dissociated cells. No differences were detected between recently dissociated cells and cells cultured overnight. Staining (for acidic groups) was present in low concentration on the plasma membranes of all secretory cell types and in higher concentration on the plasma membranes of follicular cells especially on their luminal surfaces.

Cultured cells were also incubated in the presence of cyclic AMP (to stimulate exocytosis) and horseradish peroxidase (to label the plasma membrane) in order to determine possible sites of relocation of membrane recirculated following exocytosis. Most of the HRP was found in lysosomes; however, confirming the findings of Pelletier (1973), some was detected in the Golgi apparatus, primarily in the innermost cisterna along the concave Golgi face. The results indicate that at least some of the membrane relocated to the cell surface during exocytosis is recaptured intact and recirculated back to the Golgi apparatus—in fact back to the very same cisterna which normally provides the granule membranes.

Studies were also carried out on follicular cells *in vitro*. Follicular cells represent an unusual cell type apparently peculiar to the anterior pituitary. They possess no known secretory function and for many years, partially for lack of other known functions, were considered to represent supportive elements. Evidence is now accumulating to indicate that these cells are phagocytes. Evidence is presented which indicates that *in vitro,* these cells function in the disposal and removal of dead cells and cell debris which they phagocytize and subsequently digest within lysosomes. It remains to be determined whether or not this is the only or the main function of such cells.

ACKNOWLEDGMENTS

The authors gratefully acknowledge the skilled technical assistance in all aspects of this work provided by Mrs. Nancy Dwyer. Figures 10 and 13 were prepared by Dr. Toku Kanaseki. Figure 1 is from Smith and Farquhar (1966) and Figs. 8 to 11 and 16 to 21 are from Hopkins and Farquhar (1973) and are reproduced with the permission of the Rockefeller University Press.

REFERENCES

Ambrose, E. J. (1966). Electrophoretic behavior of cells. *Progr. Biophys. Mol. Biol.* **16**, 241–265.

Amsterdam, A., and Jamieson, J. D. (1972). Structural and functional characterization of isolated pancreatic exocrine cells. *Proc. Nat. Acad. Sci. U. S.* **69**, 3028–3032.

Amsterdam, A., Ohad, I., and Schramm, M. (1969). Dynamic changes in the ultrastructure of the acinar cell of the rat parotid gland during the secretory cycle. *J. Cell Biol.* **41**, 753–773.

Amsterdam, A., Schramm, M., Ohad, I., Salomon, Y., and Selinger, Z. (1971). Concomitant synthesis of membrane protein and exportable protein of the secretory granule in rat parotid gland. *J. Cell Biol.* **50**, 187–200.

Baker, B. L., Pierce, J. G., and Cornell, J. S. (1972). The utility of antiserums to subunits of TSH and LH for immunochemical staining of the rat hypophysis. *Amer. J. Anat.* **135**, 251–268.

Bala, R. M., Burgus, R., Ferguson, K. A., Guillemin, R., Kudo, C. F., Olivier, G. C., Rodger, N. W., and Beck, J. C. (1970). Control of growth hormone secretion. *In* "The Hypothalamus" (L. Martini, M. Motta, and F. Fraschini, eds.), pp. 401–448. Academic Press, New York.

Bowie, E. P., Williams, M. G., and Rennels, E. G. (1974). Evidence for PAS positive reaction of the corticotroph granules of the rat adenohypophysis. *Histochemistry* **38**, 281–284.

Bowie, E. P., Williams, G., Shiino, M., and Rennels, E. G. (1973). The corticotroph of the rat adenohypophysis: A comparative study. *Amer. J. Anat.* **138**, 499–519.

Burger, M. M. (1969). A difference in the architecture of the surface membrane of normal and virally transformed cells. *Proc. Nat. Acad. Sci. U. S.* **62**, 994–1001.

Burger, M. M., and Goldberg, A. R. (1967). Identification of a tumor-specific determinant on neoplastic cell surfaces. *Proc. Nat. Acad. Sci. U. S.* **57**, 359–366.

Ceccarelli, B., Hurlbut, W. P., and Mauro, A. (1973). Turnover of transmitter and synaptic vesicles at the frog neuromuscular junction. *J. Cell Biol.* **57**, 499–524.

Coates, P. W., Ashby, E. A., Krulich, L., Dhariwal, A. P. S., and McCann, S. M. (1970). Morphologic alterations in somatotrophs of the rat adenohypophysis following administration of hypothalamic extracts. *Amer. J. Anat.* **128**, 389–412.

Costoff, A. (1973). "Ultrastructure of the Rat Adenohypophysis: Correlation with Function," pp. 1–220. Academic Press, New York.

Costoff, A., and McShan, W. H. (1969). Isolation and biological properties of secretory granules from rat anterior pituitary glands. *J. Cell Biol.* **43**, 564–574.

Couch, E. F., Arimura, A., Schally, A. V., Saito, M., and Sawano, S. (1969). Electron microscope studies of somatotrophs of rat pituitary after injection of purified growth hormone releasing factor (GRF). *Endocrinology* **85**, 1084–1091.

Cuatrecasas, P., and Illiano, G. (1971). Membrane sialic acid and the mechanism of insulin action in adipose tissue cells. *J. Biol. Chem.* **246**, 4938–4946.

Currie, G. A., van Doorninck, W., and Bagshawe, K. D. (1968). Effect of neuraminidase on the immunogenicity of early mouse trophoblast. *Nature (London)* **219**, 191–192.

Curtis, A. S. G. (1967). "The Cell Surface: Its Molecular Role in Morphogenesis." Academic Press, New York.

Danon, D., Goldstein, L., Marikovsky, Y., and Skutelsky, E. (1971). Cationized ferritin used for labelling of negative charges on cell surfaces. *Electron Microsc., Proc. Int. Congr., 7th, 1970* Vol. 3, pp. 33–34.

de Duve, C. (1969). The lysosome in retrospect. *In* "Lysosomes in Biology and Pathology" (J. T. Dingle and H. B. Fell, eds.), Vol. 1, pp. 3–40. North-Holland Publ., Amsterdam.

De Virgiliis, G., Meldolesi, J., and Clementi, F. (1968). Ultrastructure of growth hormone-producing cells of rat pituitary after injection of hypothalamic extracts. *Endocrinology* **83**, 1278–1284.

Dingemans, K. P. (1969). The relation between cilia and mitoses in the mouse adenohypophysis. *J. Cell Biol.* **43**, 361–367.

Dingemans, K. P., and Feltkamp, C. A. (1972). Nongranulated cells in the mouse adenohypophysis. *Z. Zellforsch. Mikrosk. Anat.* **124**, 387–405.

Douglas, W. W. (1966). The mechanism of release of catecholamines from adrenal medulla. *Pharmacol. Rev.* **18**, 471–480.

Douglas, W. W., Nagasawa, J., and Schulz, R. (1971). Electron microscopic studies on the mechanism of secretion of posterior pituitary hormones and significance of microvesicles ('synaptic vesicles'): Evidence of secretion by exocytosis and formation of microvesicles as a by-product of this process. *Mem. Soc. Endocrinol.* **19**, 353–377.

Farquhar, M. G. (1957). 'Corticotrophs' of the rat adenohypophysis as revealed by electron microscopy. *Anat. Rec.* **127**, 291.

Farquhar, M. G. (1961a). Origin and fate of secretory granules in cells of the anterior pituitary gland. *Trans. N. Y. Acad. Sci.* [2] **23**, 346–351.

Farquhar, M. G. (1961b). Fine structure and function in capillaries of the anterior pituitary gland. *Angiology* **12**, 270–292.

Farquhar, M. G. (1969). Lysosome function in regulating secretion: Disposal of secretory granules in cells of the anterior pituitary gland. *In* "Lysosomes in Biology and Pathology" (J. T. Dingle and H. B. Fell, eds.), Vol. 2, pp. 462–482. North-Holland Publ., Amsterdam.

Farquhar, M. G. (1971). Processing of secretory products by cells of the anterior pituitary gland. *Mem. Soc. Endocrinol.* **19**, 79–122.

Farquhar, M. G., and Rinehart, J. F. (1954a). Electron microscopic studies of the anterior pituitary gland of castrate rats. *Endocrinology* **54**, 516–541.

Farquhar, M. G., and Rinehart, J. F. (1954b). Cytologic alterations in the anterior pituitary gland following thyroidectomy: An electron microscope study. *Endocrinology* **55**, 857–876.

Fawcett, D. W. (1962). Physiologically significant specializations of the cell surface. *Circulation* **26**, 1105–1132.

Gasic, G. J., Berwick, L., and Sorrentino, M. (1968). Positive and negative colloidal iron as cell surface electron stains. *Lab. Invest.* **18**, 63–71.

Grant, G., Vale, W., and Guillemin, R. (1972). Interaction of thyrotropin releasing factor with membrane receptors of pituitary cells. *Biochem. Biophys. Res. Commun.* **46**, 28–34.

Guillemin, R., and Burgus, R. (1972). The hormones of the hypothalamus. *Sci. Amer.* **227**, 24–33.

Haddad, A., Smith, M. D., Herscovics, A., Nadler, N. J., and Leblond, C.P. (1971). Radioautographic study of *in vivo* and *in vitro* incorporation of fucose-^{3}H into thyroglobulin by rat thyroid follicular cells. *J. Cell Biol.* **49**, 856–882.

Halmi, N. S. (1952). Two types of basophils in the rat pituitary: "Thyrotrophs" and "gonadotrophs" vs. beta and delta cells. *Endocrinology* **50**, 140–142.

Halmi, N. S. (1973). The hypophysis (pituitary gland). *In* "Histology" (R. O. Greep and L. Weiss, eds.), Chapter 27, pp. 891–914. McGraw-Hill, New York.

Hamilton, T. H. (1971). Steroid hormones, ribonucleic acid synthesis and transport, and the regulation of cytoplasmic translation. *Biochem. Soc. Symp.* **32**, 49–84.

Hedinger, C. E., and Farquhar, M. G. (1957). Elektronenmikroskopische Untersuchungen von Zwei Typen acidophiler Hypophysenvorderlappenzellen bei der Ratte. *Schweiz. Z. Pathol. Bakteriol.* **20**, 766–768.

Herlant, M. (1964). The cells of the adenohypophysis and their functional significance. *Int. Rev. Cytol.* **17**, 299–382.

Heuser, J. E., and Reese, T. S. (1973). Evidence for recycling of synaptic vesicle membrane during transmitter release at the frog neuromuscular junction. *J. Cell Biol.* **57**, 315–358.

Hopkins, C. R., and Farquhar, M. G. (1973). Hormone secretion by cells dissociated from rat anterior pituitaries. *J. Cell Biol.* **59**, 276–303.

Howell, S. L., and Ewart, R. B. L. (1973). Synthesis and secretion of growth hormone in the rat anterior pituitary. II. Properties of the isolated growth hormone storage granules. *J. Cell Sci.* **12**, 23–35.

Howell, S. L., and Whitfield, M. (1973). Synthesis and secretion of growth hormone in the rat anterior pituitary. I. The intracellular pathway, its time course and energy requirements. *J. Cell Sci.* **12**, 1–21.

Hugon, J., and Borgers, M. (1966). Ultrastructural and cytochemical changes in spermatogonia and Sertoli cells of whole-body irradiated mice. *Anat. Rec.* **155**, 15–32.

Hymer, W. C., Kraicer, J., Bencosme, S. A., and Haskill, J. S. (1972). Separation of somatotrophs from the rat adenohypophysis by velocity and density gradient centrifugation. *Proc. Soc. Exp. Biol. Med.* **141**, 966–973.

Hymer, W. C., Evans, W. H., Kraicer, J., Mastro, A., Davis, J., and Griswold, E. (1973). Enrichment of cell types from the rat adenohypophysis by sedimentation at unit gravity. *Endocrinology* **92**, 275–287.

Ishikawa, H. (1969). Isolation of different types of anterior pituitary cells in rats. *Endocrinol. Jap.* **16**, 517–523.

Jamieson, J. J. (1972). Transport and discharge of exportable proteins in pancreatic exocrine cells: *In vitro* studies. *Curr. Top. Membranes Transp.* **3**, 273–338.

Jamieson, J. D., and Palade, G. E. (1967a). Intracellular transport of secretory proteins in the pancreatic exocrine cell. I. Role of the peripheral elements of the Golgi complex. *J. Cell Biol.* **34**, 577–596.

Jamieson, J. D., and Palade, G. E. (1967b). Intracellular transport of secretory proteins in the pancreatic exocrine cell. II. Transport to condensing vacuoles and zymogen granules. *J. Cell Biol.* **34**, 597–615.

Jones, A. E., Fisher, J. N., Lewis, U. J., and Vanderlaan, W. P. (1965). Electrophoretic comparison of pituitary glands from male and female rats. *Endocrinology* **76**, 578–583.

Kudo, C. F., Rubenstein, D., McKenzie, J. M., and Beck, J. C. (1972). Hormonal release by dispersed pituitary cells. *Can. J. Physiol. Pharmacol.* **50**, 860–867.

Kurosumi, K. (1968). Functional classification of cell types of the anterior pituitary gland accomplished by electron microscopy. *Arch. Histol. Jap.* **29**, 329–362.

Kurosumi, K., and Kobayashi, Y. (1966). Corticotrophs in the anterior pituitary glands of normal and adrenalectomized rats as revealed by electron microscopy. *Endocrinology* **78**, 745–758.

Kurosumi, K., and Oota, Y. Electron microscopy of two types of gonadotrophs in the anterior pituitary glands of persistent estrous and diestrous rat. (1968). *Z. Zellforsch. Mikrosk. Anat.* **85**, 34–46.

Labrie, F., Béraud, G., Gauthier, M., and Lemay, A. (1971). Actinomycin-insensitive stimulation of protein synthesis in rat anterior pituitary *in vitro* by dibutyryl adenosine 3',5'-monophosphate. *J. Biol. Chem.* **246**, 1902–1908.

Labrie, F., Barden, N., Poirer, G., and De Lean, A. (1972). Binding of thyrotropin-releasing hormone to plasma membranes of bovine anterior pituitary gland. *Proc. Nat. Acad. Sci. U. S.* **69**, 283–287.

Labrie, F., Gauthier, M., Pelletier, G., Borgeat, P., Lemay, A., and Gouge, J.-J. (1973). Role of microtubules in basal and stimulated release of growth hormone and prolactin in rat adenohypophysis *in vitro*. *Endocrinology* **93**, 903–914.

Leavitt, W. W., Kimmel, G. L., and Friend, J. P. (1973). Steroid hormone uptake by anterior pituitary cell suspensions. *Endocrinology* **92**, 94–103.

Le Marchand, Y., Singh, A., Assimacopoulos-Jeannet, F., Orci, L., Rouiller, C., and Jeanrenaud, B. (1973). A role for the microtubular system in the release of very low density lipoproteins. *J. Biol. Chem.* **248**, 6862–6870.

McDonald, J. K., Callahan, P. X., Ellis, S., and Smith, R. E. (1971). Polypeptide degradation by dipeptidyl aminopeptidase I (cathepsin C) and related peptidases. *In* "Tissue Proteinases" (A. J. Barrett and J. T. Dingle, eds.), pp. 69–107. North-Holland Publ., Amsterdam.

MacLeod, R. M., and Lehmeyer, J. E. (1970). Release of pituitary growth hormone by prostaglandins and dibutyryl adenosine cyclic 3':5'-monophosphate in the absence of protein synthesis. *Proc. Nat. Acad. Sci. U. S.* **67**, 1172–1179.

McShan, W. H., and Hartley, M. W. (1965). Production, storage and release of anterior pituitary hormones. *Ergeb. Physiol., Biol. Chem. Exp. Pharmakol.* **56**, 264–296.

Marcus, P. I., and Schwartz, V. G. (1968). Monitoring molecules of the plasma membrane: Renewal of sialic acid-terminating receptors. *In* "Biological Properties of the Mammalian Surface Membrane" (L. A. Mason, ed.), p. 143. Wistar Inst. Press, Philadelphia, Pennsylvania.

Mazurkiewicz, J. E., and Nakane, P. K. (1972). Light and electron microscopic localization of antigens in tissues embedded in polyethylene glycol with a peroxidase-labeled antibody method. *J. Histochem. Cytochem.* **20**, 969–974.

Meldolesi, J., Marini, D., and Marini, M. L. D. (1972). Studies on *in vitro* synthesis and secretion of growth hormone and prolactin. I. Hormone pulse labeling with radioactive leucine. *Endocrinology* **91**, 802–808.

Mizel, S., and Wilson, L. (1972a). Inhibition of the transport of several hexoses in mammalian cells by cytochalasin B. *J. Biol. Chem.* **247**, 4102–4105.

Mizel, S. B., and Wilson, L. (1972b). Nucleoside transport in mammalian cells. Inhibition by colchicine. *Biochemistry* **11**, 2573–2578.

Moguilevsky, J. A., Cuerdo-Rocha, S., Christot, J., and Zambrano, D. (1973). The effect of thyrotrophic releasing factor on different hypothalamic areas and the anterior pituitary gland: A biochemical and ultrastructural study. *J. Endocrinol.* **56**, 99–109.

Moriarty, G. C. (1973). Adenohypophysis: Ultrastructural cytochemistry. A review. *J. Histochem. Cytochem.* **21**, 855–894.

Moriarty, G. C., and Halmi, N. S. (1972). Electron microscopic study of the adrenocorticotrophin-producing cell with the use of unlabeled antibody and the soluble peroxidase-antiperoxidase complex. *J. Histochem. Cytochem.* **20**, 590–603.

Mueller, G. C., Vonderhaar, B., Kim, U. H., and Mahieu, M. L. (1972). Estrogen action: An inroad to cell biology. *Recent Progr. Horm. Res.* **28**, 1–49.

Nagasawa, J., Douglas, W. W., and Schulz, R. A. (1971). Micropinocytic origin of coated and smooth microvesicles ("synaptic vesicles") in neurosecretory terminals of posterior pituitary glands demonstrated by incorporation of horseradish peroxidase. *Nature (London)* **232**, 341–342.

Nakane, P. K. (1970). Classifications of anterior pituitary cell types with immunoenzyme histochemistry. *J. Histochem. Cytochem.* **18**, 9–20.

Nakayama, I., Nickerson, P. A., and Skelton, F. R. (1969). An ultrastructural study of the adrenocorticotrophic hormone-secreting cell in the rat adenohypophysis during adrenal cortical regeneration. *Lab. Invest.* **21**, 169–178.

Nicolson, G. L., and Singer, S. J. (1971). Ferritin-conjugated plant agglutinins as specific saccharide stains for electron microscopy: Application to saccharides bound to cell membranes. *Proc. Nat. Acad. Sci. U. S.* **68**, 942–945.

O'Malley, B. W., and Means, A. R. (1974). Female steroid hormones and target cell nuclei. *Science* **183**, 610–620.

Palade, G. E. (1959). Functional changes in the structure of cell components. *In* "Subcellular Particles" (T. Hayashi, ed.), pp. 64–80. Ronald Press, New York.

Palade, G. E. (1966). Structure and function at the cellular level. *J. Amer. Med. Ass.* **198**, 815–825.

Pasteels, J. L. (1963). Recherches morphologiques et expérimentales sur la sécrétion de prolactine. *Arch. Biol.* **74**, 439.

Pelletier, G. (1971). Détection des glycoprotéines dans les cellules corticotropes de l'hypophyse du rat. *J. Microsc. (Paris)* **11**, 327–330.

Pelletier, G. (1973). Secretion and uptake of peroxidase by rat adenohypophyseal cells. *J. Ultrastruct. Res.* **43**, 445–459.

Pelletier, G. (1974). Autoradiographic studies of synthesis and intracellular migration of glycoproteins in the rat anterior pituitary gland. *J. Cell Biol.* **62**, 185–197.

Pelletier, G., and Bornstein, M. B. (1972). Effect of colchicine on rat anterior pituitary gland in tissue culture. *Exp. Cell Res.* **70**, 221–223.

Pelletier, G., and Puviani, R. (1973). Detection of glycoproteins and autoradiographic localization of ^{3}H-fucose in the thyroidectomy cells of rat anterior pituitary gland. *J. Cell Biol.* **56**, 600–605.

Pelletier, G., and Racadot, J. (1971). Identification des cellules hypophysaires sécrétant l'ACTH chez le rat. *Z. Zellforsch. Mikrosk. Anat.* **116**, 228–239.

Pelletier, G., Peillon, F., and Vila-Porcile, E. (1971). An ultrastructural study of sites of granule extrusion in the anterior pituitary of the rat. *Z. Zellforsch. Mikrosk. Anat.* **115,** 501–507.

Pelletier, G., Lemay, A., Béraud, G., and Labrie, F. (1972). Ultrastructural changes accompanying the stimulatory effect of N^6-monobutyryl adenosine 3′,5′-monophosphate on the release of growth hormone (GH) prolactin (PRL) and adrenocorticotropic hormone (ACTH) in rat anterior pituitary gland *in vitro. Endocrinology* **91,** 1355–1371.

Phifer, R. F., Midgley, A. R., and Spicer, S. S. (1973). Immunohistologic and histologic evidence that follicle-stimulating and luteinizing hormones are present in the same cell types in the human pars distalis. *J. Clin. Endocrinol. Metab.* **36,** 125–141.

Pierce, J. G. (1971). Eli Lilly lecture: The subunits of pituitary thyrotropin—their relationship to other glycoprotein hormones. *Endocrinology* **89,** 1331–1334.

Pooley, A. S. (1971). Ultrastructure and size of rat anterior pituitary secretory granules. *Endocrinology* **88,** 400–411.

Portanova, R., Smith, D. K., and Sayers, G. (1970). A trypsin technic for the preparation of isolated rat anterior pituitary cells. *Proc. Soc. Exp. Biol. Med.* **133,** 573–576.

Purves, H. D. (1961). Morphology of the hypophysis related to its function. *In* "Sex and Internal Secretion" (W. C. Young, ed.), 3rd ed., Vol. 1, pp. 161–238. Williams & Wilkins, Baltimore, Maryland.

Purves, H. D. (1966). Cytology of the adenohypophysis. *In* "The Pituitary Gland" (G. W. Harris and B. T. Donovan, eds.), Vol. 1, p. 147. Univ. of California Press, Berkeley.

Purves, H. D., and Griesbach, W. E. (1951). The site of thyrotrophin and gonadotrophin production in the rat pituitary studied by McManus-Hotchkiss staining for glycoprotein. *Endocrinology* **49,** 244–264.

Racadot, J., Olivier, L., Porcile, E., and Droz, B. (1965). Appareil de Golgi et origine des grains de sécrétion dans les cellules adenohypophysaires chez le rat. Etude radioautographique en microscopie électronique après injection de leucine tritiée. *C. R. Acad. Sci.* **261,** 2972–2974.

Rambourg, A. (1967). Détection des glycoprotéines en microcopie électronique: Coloration de la surface cellulaire et de l'appareil de Golgi par un mélange acide chromique-phosphotungstique. *C. R. Acad. Sci.* **265,** 1426–1428.

Rennels, E. G., and Shiino, M. (1968). Ultrastructural manifestations of pituitary release of ACTH in the rat. *Arch. Anat., Histol. Embryol.* **51,** 575–590.

Rennels, E. G., Bogdanove, E. M., Arimura, A., Saito, M., and Schally, A. V. (1971). Ultrastructural observations of rat pituitary gonadotrophs following injection of purified porcine LH-RH. *Endocrinology* **88,** 1318–1326.

Rosenthal, J. W., and Fain, J. N. (1971). Insulin-like effect of clostridial phospholipase C, neuraminidase, and other bacterial factors on brown fat cells. *J. Biol. Chem.* **246,** 5888–5895.

Sachs, H. (1971). Secretion of neurohypophysial hormones. *Mem. Soc. Endocrinol.* **19,** 965–973.

Samli, M. H., Lai, M. F., and Barnett, C. A. (1971). Protein synthesis in the rat anterior pituitary. I. Interrelationship of amino acid uptake to protein synthesis. *Endocrinology* **88,** 540–547.

Schally, A. V., Arimura, A., and Kastin, A. J. (1973). Hypothalmic regulatory hormones. *Science* **179,** 341–350.

Schneider, F. H., Smith, A. D., and Winkler, H. (1967). Secretion from the adrenal medulla: Biochemical evidence for exocytosis. *Brit. J. Pharmacol. Chemother.* **31,** 94–104.

Schofield, J. G. (1967). Role of cyclic 3′,5′-adenosine monophosphate in the release of growth hormone *in vitro. Nature (London)* **215,** 1382–1383.

Shiino, M., Arimura, A., Schally, A. V., and Rennels, E. G. (1972a). Ultrastructural observations of granule extrusion from rat anterior pituitary cells after injection of LH-releasing hormone. *Z. Zellforsch. Mikrosk. Anat.* **128,** 152–161.

Shiino, M., Williams, M. G., and Rennels, E. G. (1972b). Ultrastructural observation of pituitary release of prolactin in the rat by suckling stimulus. *Endocrinology* **90,** 176–187.

Shiino, M., Williams, M. G., and Rennels, E. G. (1973). Thyroidectomy cells and their response to thyrotrophin releasing hormone (TRH) in the rat. *Z. Zellforsch. Mikrosk. Anat.* **138,** 327–332.

Siperstein, E. R., and Allison, V. F. (1965). Fine structure of the cells responsible for secretion of adrenocorticotrophin in the adrenalectomized rat. *Endocrinology* **76,** 70–79.

Siperstein, E. R., and Miller, K. J. (1970). Further cytophysiologic evidence for the identity of the cells that produce adrenocorticotrophic hormone. *Endocrinology* **86,** 451–486.

Smith, A. D. (1972). Storage and secretion of hormones. *Sci. Basis Med.* pp. 74–102.

Smith, R. E., and Farquhar, M. G. (1966). Lysosome function in the regulation of the secretory process in cells of the anterior pituitary gland. *J. Cell Biol.* **31,** 319–347.

Smith, R. E., and Farquhar, M. G. (1970). Modulation in nucleoside diphosphatase activity of mammotrophic cells of the rat adenohypophysis during secretion. *J. Histochem. Cytochem.* **18,** 237–250.

Stadler, J., and Franke, W. W. (1974). Characterization of the colchicine binding of membrane fractions from rat and mouse liver. *J. Cell Biol.* **60,** 297–303.

Stratmann, I. E., Ezrin, C., Sellers, E. A., and Simon, G. T. (1972). The origin of thyroidectomy cells as revealed by high resolution radioautography. *Endocrinology* **90,** 728–734.

Tixier-Vidal, A., and Picart, R. (1967). Etude quantitative par radioautographie au microscope électronique de l'utilisation de la DL-leucine-³H par les cellules de l'hypophyse du Canard en culture organotypique. *J. Cell Biol.* **35,** 501–519.

Tixier-Vidal, A., Kerdelhué, B., Bérault, A., Picart, R., and Jutisz, M. (1971). Action *in vitro* du facteur hypothalamique de libération de l'hormone lutéinisante (LRF) sur l'antéhypophyse d'agnelle. II. Etude ultrastructurale des tissue incubés. *Gen. Comp. Endocrinol.* **17,** 33–59.

Tixier-Vidal, A., Kerdelhué, B., and Jutisz, M. (1973). Kinetics of release of luteinizing hormone (LH) and follicle stimulating hormone (FSH) by primary cultures of dispersed rat anterior pituitary cells. Chronic effect of synthetic LH and FSH releasing hormone. *Life Sci.* **12,** 499–509.

Todd, J. M., and Samli, M. H. (1973). The incorporation of [³H]glucosamine, [³H]fucose and [¹⁴C]mannose into protein in the rat anterior pituitary incubated *in vitro. Biochim. Biophys. Acta* **297,** 11–21.

Tougard, C., Kerdelhué, B., Tixier-Vidal, A., and Jutisz, M. (1973). Light and electron microscopic localization of binding sites of antibodies against ovine luteinizing hormone and its two subunits in rat adenohypophysis using peroxidase-labeled antibody technique. *J. Cell Biol.* **58,** 503–521.

Vale, W., Grant, G., Amoss, M., Blackwell, R., and Guillemin, R. (1972). Culture of enzymatically dispersed anterior pituitary cells: Functional validation of a method. *Endocrinology* **91,** 562–572.

Vila-Porcile, E. (1972). Le réseau des cellules folliculo-stellaires et les follicules de l'adénohypophyse du rat (Pars distalis). *Z. Zellforsch. Mikrosk. Anat.* **129,** 328–369.

Vila-Porcile, E., Olivier, L., and Racadot, O. (1973). Exocytose polarisée des corps résiduels lysosomiaux des cellules à prolactine dans l'adénohypophyse de la Ratte en post-lactation. *C. R. Acad. Sci.* **276,** 355–357.

Whur, P., Herscovics, A., and Leblond, C. P. (1969). Radioautographic visualization of the incorporation of galactose-^{3}H and mannose-^{3}H by rat thyroids *in vitro* in relation to the stages of thyroglobulin synthesis. *J. Cell Biol.* 43, 289–311.

Wilber, J. F., and Seibel, M. J. (1973). Thyrotropin-releasing hormone interactions with an anterior pituitary membrane receptor. *Endocrinology* 92, 888–893.

Yalow, R. S., and Berson, S. A. (1971). Size heterogeneity of immunoreactive human ACTH in plasma and in extracts of pituitary glands and ACTH-producing thymoma. *Biochem. Biophys. Res. Commun.* 44, 439–445.

Yamashita, K. (1969). Electron microscopic observations on the postnatal development of the anterior pituitary of the mouse. *Gunma Symp. Endocrinol.* 6, 177–196.

Yamashita, K. (1972a). Fine structure of the mouse anterior pituitary maintained in a short-term incubation system. *Z. Zellforsch. Mikrosk. Anat.* 124, 465–478.

Yamashita, K. (1972b). A simple method for the cultivation of anterior pituitary tissues: Electron microscopical evaluation of the cultured tissues. *Z. Anat. Entwickl.-Gesch.* 137, 106–113.

Zigmond, S. H., and Hirsch, J. G. (1972). Cytochalasin B: Inhibition of D-2-deoxyglucose transport into leukocytes and fibroblasts. *Science* 176, 1432–1434.

SEPARATION OF ORGANELLES AND CELLS FROM THE MAMMALIAN ADENOHYPOPHYSIS

W. C. Hymer

DEPARTMENT OF BIOLOGY, THE PENNSYLVANIA STATE UNIVERSITY

UNIVERSITY PARK, PENNSYLVANIA

I. Introduction

The extreme cellular heterogeneity of the mammalian adenohypophysis provides a major obstacle to the study of this interesting gland. Not surprisingly, cytologists have speculated about the localization of hormones within the different cell types for many years. Baker (1970) recently observed ". . . few areas of endocrine research have been as troubled by our inability to localize hormones accurately as has been the case with pituitary cytology. Consequently, much research has been misdirected and ineffectual." While few would disagree with these comments, the progress that endocrinologists have made in the clarification of this problem

137

through use of direct experimental approaches has been quite remarkable indeed (see for example, chapters by Herlant and Nakane).

The role that hypothalamic factors play in the regulation of hormone secretion from the different adenohypophysial cell types has received considerable attention in the last ten years or so. The results thus far indicate that, with the possible exception of gonadotropin releasing factor (Schally *et al.,* 1971), each factor causes hormone release only from its target cell type (McCann and Porter, 1969). How is specificity conferred? What intracellular events occur prior to hormone secretion? Attractive working models have been proposed to answer such questions, but most rely on the indirect experimental approach for their answers. Once again cellular heterogeneity becomes a major factor in interpretation of experimental results.

It is precisely at this point that "structure–function" studies assume considerable importance. It is the purpose of this chapter to review some of the information which has been gathered when this approach has been used. Results of studies dealing with isolation of subcellular organelles from the mammalian adenohypophysis will be reviewed in the first portion of this chapter. As far as is possible they will be considered in terms of (a) the isolation methods employed, (b) the morphological characteristics of the isolated organelle, and (c) their functions as they relate to the problem of control of hormone turnover within the gland. Finally, progress which has been made to date on the challenging problem of separation of the different adenohypophysial cell types will be summarized in the second portion of this chapter.

II. Isolation of Subcellular Organelles from Adenohypophysial Tissue

A. Secretory Granules

The results from the early work of Catchpole (1948), McShan and his colleagues (McShan and Meyer, 1949, 1952; McShan *et al.,* 1953), and LaBella and Brown (1958, 1959), clearly show that pituitary hormones are associated with cytoplasmic elements which can be concentrated by the relatively simple procedure of differential centrifugation. From these early studies, it is also apparent (1) that growth hormone is associated with a particle that sediments at approximately the same centrifugal force as that required to sediment mitochondria (9×10^4 g-min) while gonadotropic hormones sediment at higher forces (1.2×10^6 g-min), and (2) that gonadotropin is easily dissociable from its particle form. This latter observation accounts for the homogenization technique developed and used by McShan and his students in which rat pituitaries are homogenized "in a cold, sharp-pointed glass homogenizing tube and mashed rather than ground with a pestle so as to minimize breaking the granules and solubilizing the hormones" (Costoff and McShan, 1969). Since the granule fractions obtained by differential centrifugation are heavily contaminated with other organelles, additional methods were developed to purify them.

It was reasoned that if pure populations of hormonally active secretion granules could be obtained, an independent method of pituitary cell identification would be theoretically possible. Moreover, such preparation might provide clues as to how the hormone is packaged within its granule form. The extent to which these expectations have been realized is considered below.

1. *Methodology for Granule Isolation*

The assumption basic to all granule isolation studies is that specific hormones are packaged in granules of varying size and/or density. Examination of the methodologies used to obtain pituitary granules (Table I) reveals a general procedural sequence of (a) low speed centrifugation of homogenates prepared in isotonic sucrose to remove nuclei and unbroken cells, (b) differential centrifugation, (c) filtration of the particles through Millipore filter membranes of different pore sizes, and, finally, (d) density gradient centrifugation. The filtration step, originally introduced by Hartley *et al.* (1960), is useful for removing large contaminating particles or aggregates from the initial granule suspension. Furthermore, this step makes it possible to exclude large granules from small ones.

Those involved with isolating populations of growth hormone granules are likely to face the problem of excessive mitochondrial contamination. Various methods have been used to remove mitochondria. These involve (a) separation of the "fluffy" layer from a crude granule pellet (LaBella *et al.*, 1971), (b) removal of mitochondria by chromatography through columns of diatomaceous earth (Hymer and McShan, 1963), or (c) filtration through Nucleopore filters (Costoff and McShan, 1969). These methods, shown by both electron microscopy and assay of the granule fractions for mitochondria-specific enzymes, are highly effective in removing mitochondria.

The conditions for density gradient centrifugation vary widely (Table I). Essentially homogeneous populations of granules have been obtained on either continuous or discontinuous density gradients, although the discontinuous gradients are subject to "turnover" effects. With the exception of the study LaBella *et al.* (1971) in which Ficoll was used, the gradients to date have been prepared in sucrose. In some cases Diodrast, the diethanolamine salt of 3,5-diiodo-4-pyridone-N-acetic acid, has been used to increase the density.

2. *Hormone Contents of Isolated Granules*

A comparison of the results of the seven fractionation studies in Table I has been done on the basis of the approximate average density at which the major portion of recovered hormone is found within the density gradient. It should be emphasized that these values are estimates calculated from published line drawings of the gradients after centrifugation. They are therefore subject to error and should be considered, at best, as only approximations. It is interesting to note, however, that a definite pattern of hormone distribution within the gradient emerges. Thus, in all cases where ACTH and TSH activities have been measured, a major portion

TABLE I

METHODS USED TO ISOLATE ADENOHYPOPHYSIAL SECRETION GRANULES

Investigators	Starting material	Density gradient parameters	Approximate average density at which major portion of recovered hormone is found					
			ACTH	TSH	FSH	LH	STH	LTH
Hartley et al. (1960)	Rat, 44,000 g/min supernatant; filtration	6–45% sucrose + 17.5% diodrast Continuous 1.11–1.26 12 × 10⁶ g/min	—	—	1.21(H)[a]	1.21(H)	—	—
Perdue and McShan (1962, 1966)	Rat, 44,000 g/min supernatant; filtration	15–44% sucrose + 17.5% diodrast Discontinuous 1.14–1.26 1.32 × 10⁶ g/min	1.14(H)	1.14(H)	1.14(H)	1.14(H)	—	—
Hymer and McShan (1963)	Rat, 2750 g/min supernatant; filtration; chromatography through celite	30–80% sucrose Discontinuous 1.15–1.41 6 × 10⁶ g/min	1.17(H)	1.19(H)	1.28(H)	1.28(H)	1.28(L)[b]	1.28(L)

Reference	Source; preparation	Gradient						
Tesar et al. (1969)	Bovine, 8000 g/min pellet; filtration	50–65% sucrose Discontinuous 1.23–1.32 6.8 × 10⁶ g/min	—	—	—	—	1.30(L)	1.30(L)
	Bovine, 280,000 g/min pellet; filtration	10–40% sucrose × 17.5% diodrast Continuous 1.13–1.24 13.5 × 10⁶ g/min	—	1.20(H)	1.20(H)	1.20(H)	—	—
Costoff and McShan (1969)	Rat, 2750 g/min supernatant; filtration	6–45% sucrose + 17.5% diodrast Continuous 1.11–1.26 12 × 10⁶ g/min	1.13(H)	1.13(H)	1.17(H)	1.17(H)	1.20(L)	1.23(L)
LaBella et al. (1971)	Bovine, 140,000 g/min pellet	30–55% Ficoll Discontinuous 1.15–1.26 0.4 × 10⁶ g/min	—	—	—	—	1.18	1.25

[a] High speed centrifugation used to sediment granules (6 × 10⁶ g/min) from band in gradient into pellet form.
[b] Low speed centrifugation used to sediment granules (2 × 10⁶ g/min) from band in gradient into pellet form.

of these hormones is recovered in granules which are of relatively low density. On the other hand, prolactin is clearly associated with denser particles.

How pure are these granule preparations? In virtually every study the isolated particles have been monitored for the presence of mitochondrial enzymes (e.g., succinic dehydrogenase), lysosomal enzymes (e.g., acid phosphatase), or microsomal enzymes (e.g., glucose-6-phosphatase). In all cases, the concentration of these enzymes is either negligible or absent. Sections of granule pellets, examined

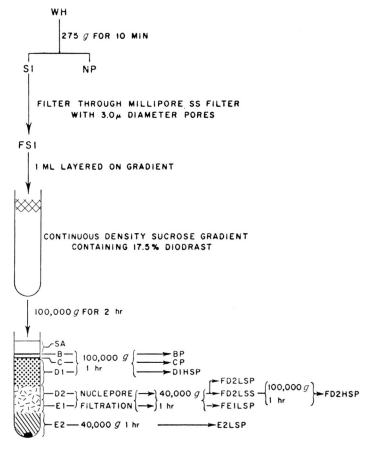

FIG. 1. Procedure for isolation of rat pituitary secretory granules. Designation of fractions: WH, whole homogenate (50 mg FT/ml); NP, nuclear pellet; S1, supernatant from nuclear pellet; FS1, filtered supernatant; SA to E2, zones obtained on the continuous density gradient; SA, soluble A zone; BP, ribosomal pellet, CP, microsomal pellet; D1HSP, D1 high speed pellet; FD2LSP, filtered D2 low speed pellet; FD2LSS, filtered D2 low speed supernatant; FD2HSP, filtered D2 high speed pellet; FE1LSP, filtered E1 low speed pellet, E2LSP, E2 low speed pellet, E2LSP, E2 low speed pellet. From Costoff and McShan (1969).

by electron microscopy, also indicate excellent purity. On the other hand, the results of the hormone assays indicate that considerable cross contamination of granules has occurred in all instances.

The thorough study by Costoff and McShan (1969) is representative of the quality of the results obtained to date. Their method for isolation of the different secretion granules (Fig. 1) is claimed to give highly repeatable patterns. Electron micrographs of thin sections through pellets of granules that were recovered from different positions in the density gradient (Fig. 2a–d) show (1) that the preparations are essentially free of other contaminating organelles and (2) that the size of the granules increase the further down into the gradient they are found. In most studies of this kind the pellets are sectioned at different levels to determine if stratification of organelles has occurred during their preparation for electron microscopy.

With the exception of two recent studies (Hodges and McShan, 1970; LaBella et al., 1971), hormone contents of the granule preparations have been estimated using conventional bioassay procedures. The reader is referred to the original papers for details of the bioassay used in any given study. It is worthwhile noting, however, that the granule pellets are usually suspended in 0.9% saline and administered to the test animal in this fashion. This means that granules per se are probably being injected (at least in the case of growth hormone-containing particles). The manner in which breakdown of granules occurs within the test animal has not been studied.

TABLE II

PERCENTAGES OF HORMONAL ACTIVITIES RECOVERED FROM GRADIENT FRACTIONS

Fractions[a]	Percentages[b]					
	TSH	ACTH	FSH	LH	STH	LTH
NP	0.0	1.0	0.0	0.5	0.0	0.0
SA	4.0	19.1	13.5	23.1	19.3	8.6
BC + CP	0.2	0.0	0.0	0.0	0.0	0.0
D1HSP	82.0[c]	53.4	3.1	1.5	0.0	0.0
FD2LSP	0.0	2.4	10.4	7.7	6.0	0.0
FD2HSP	1.5	10.8	55.7	51.4	0.0	0.0
FE1LSP	0.0	0.0	0.0	0.0	66.6	1.0
E2LSP	0.0	0.0	0.0	0.0	2.4	89.3
Total recoveries	87.7	86.7	82.7	84.2	94.3	98.9

[a] See Fig. 1 for fraction designations.

[b] Based on amounts of the FS1 fraction added to the gradient.

[c] Maximum percentage recoveries for each of the hormones are underlined. From Costoff and McShan (1969).

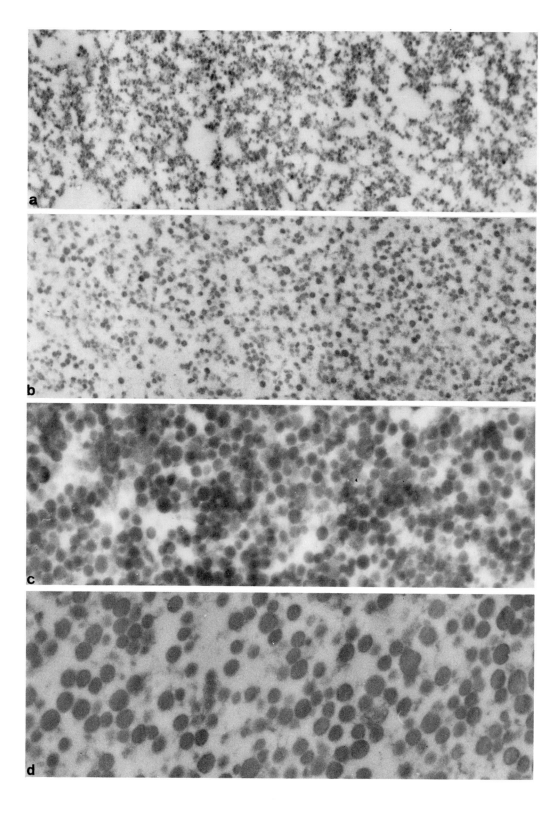

The percentages of hormonal activities recovered from the different gradient fractions in Costoff's study are given in Table II. Recoveries of hormone, based on amounts of "cytoplasm" added to the gradient, range from 83 to 99%. In all cases, a major part of each hormone is associated with some granule band. However, considerable overlap in activities of TSH and ACTH, as well as FSH and LH, indicate that granules containing these hormones are of a similar size and/or density and are not separated by the density gradient technique. The data in Table I would also appear to argue in favor of this generalization. The observation that a significant quantity of hormone is detected in the soluble (SA) fraction is of interest. As suggested by Costoff and McShan (1969), hormone in this fraction could arise from (a) granules which are broken during homogenization or (b) cisternal elements of the endoplasmic reticulum. This finding could also indicate that a small percentage of hormone exists within the adenohypophysial cells in nongranular form. Some investigators have considered the possibility that hormone which is rapidly released from the cell in response to a physiological secretagogue might come from such a soluble pool. The electron microscope radioautographic studies of Racadot et al. (1965) appear to argue against this possibility, but if the pool is small, it might not be detectable by this technique. More study is required to answer this important question.

For many years, probably the single most useful criterion for the ultrastructural identification of the different pituitary cell types has been the size of the secretory granules in their cytoplasm. These granules, which are electron opaque and limited by a single membrane, show ranges in diameters which agree closely with measurements made on the isolated granules (see Table III). These measurements, when considered with the bioassay data, appear to support the validity of using granule size as a useful criterion for cell type identification. Similar findings have been reported in other studies (Perdue and McShan, 1962; Tesar et al., 1969). Close inspection of the data in Table III, however, shows that considerable overlap in the size measurements occurs in some instances. This problem has been examined in detail in a statistical study by Pooley (1971). His measurements, made in serial sections as well as by sections negatively stained by phosphotungstic acid, indicate that up to fourfold differences in diameters of granules within individual cells can be found. Thus, concludes Pooley, the results ". . . indicate that the different cell types cannot always be identified solely on the basis of granule size, though this is probably the most useful criterion."

Electrophoresis has also been used to detect hormones. Proteins in homogenates

---

FIG. 2a–d. (a) The D1HSP secretory granule fraction with which TSH and ACTH are associated. (b) A section from the FD2HSP pellet with which the gonadotropins are associated. (c) An electron micrograph of the FE1LSP secretory granule pellet containing growth hormone activity. (d) The E2LSP fraction of the large acidophilic LTH granules. ×15,000. From Costoff and McShan (1969).

TABLE III

DIAMETERS OF SECTIONS OF GRANULES FROM RAT ANTERIOR PITUITARY GLANDS[a]

Sections of	Cell types					
	TSH	ACTH	FSH	LH	STH	LTH
Intact cells						
Mean	85 ± 9[b]	106 ± 10	133 ± 12	145 ± 13	240 ± 17	318 ± 22
Range	40–145	50–180	75–200	75–235	100–390	250–870

	Fractions from gradient[c]					
	D1HSP	D1HSP	FD2HSP	FD2HSP	FE1LSP	E2LSP
Isolated granule pellets						
Mean	89 ± 9[b]	89 ± 9	150 ± 15	150 ± 15	242 ± 16	361 ± 22
Range	35–150	35–150	75–218	75–218	180–380	220–880

[a] Diameters are given in millimicrometers.

[b] Standard error of the mean.

[c] Fractions (Fig. 1) from which maximum amounts of the hormones were recovered. From Costoff and McShan (1969).

prepared from rat, mouse, human, and cow adenohypophyses have been separated into a number of protein bands by starch and polyacrylamide gel electrophoresis (see, for example, Hodges and McShan, 1970; Jones *et al.,* 1965; Macleod *et al.,* 1969; Yamamoto *et al.,* 1970). Growth hormone and prolactin were among the first hormones to be separated by this technique. In a recent study from McShan's laboratory (Hodges and McShan, 1970), hormones associated with isolated secretory granules have also been separated by polyacrylamide gel electrophoresis. Using the procedure developed by Costoff to obtain the granules, hormones were extracted by several freeze–thaw cycles and homogenization in distilled water. After electrophoresis in 7.5% polyacrylamide gels the distribution of hormones within the gels, as determined by bioassay, agreed well with patterns previously reported (Kragt and Meites, 1966). Thus, growth hormone and prolactin are associated with the major cathodal and anodal discs, respectively. LH and TSH activities have similar mobilities and are located in a broad zone above the STH disc. FSH is detected in a narrow segment a few millimeters below STH. These results are of importance because they strongly suggest that the hormone in its stored form (i.e., the granule) has a mobility similar to that found in the crude pituitary extracts. This electrophoretic technique has also proved to be an extremely useful one to monitor hormone synthesis.

3. Biochemical Properties of Isolated Granules

a. *Enzymes.* In two separate studies (Perdue and McShan, 1962; Tesar *et al.,* 1969), fractions with granules containing high concentrations of FSH, LH, and

TSH were shown to contain alkaline protease. Since these preparations were virtually homogeneous, it is probable that this enzyme is associated with the granule element per se. Adsorption of the enzyme onto the granule surface appears less likely since this enzyme (which is assayed by hydrolysis, at pH 8.3, of amino acids from a substrate of denatured hemoglobin) was not present in the other granule fractions. The relative specific activity of this enzyme, in terms of enzyme activity/milligram protein, was 9 and 15 times higher than that found in the starting homogenate (Perdue and McShan, 1962; Tesar et al., 1969, respectively). Both investigators tentatively postulate that the association of this enzyme with the granule is somehow related to its breakdown and subsequent release of hormone into the circulation during exocytosis. However, these experiments should be extended, especially in light of the observation that nucleosides can interfere with this assay to give spuriously high values Pickup and Hope (1971).

Other enzyme activities tested in preparations of isolated secretory granules have been acid protease, acid and alkaline phosphatase, glucose-6-phosphatase, and succinic dehydrogenase. In all cases these enzymes are either absent or present in negligible quantities.

Histochemistry at the level of the electron microscope provides a sensitive tool to study the subcellular distribution of enzymes in the adenohypophysis. In the elegant study of Smith and Farquhar (1970) it was shown that in addition to Golgi and cell membrane areas, reaction product of the enzyme nucleoside diphosphatase (NDPase) is also found in the space between the dense content of the prolactin granule and its limiting membrane (Fig. 3) in mammotropic cells of the rat adenohypophysis. Interestingly, enzyme activity is greatest after estrogen treatment, moderate during lactation, and minimal in the postlactating animal, thereby paralleling rates of prolactin secretion known to occur in these different endocrine states. Smith and Farquhar conclude, "unfortunately nothing is known about the function of this enzyme activity, but its location along that face of the Golgi involved in the formation of secretory granules, together with its functional modulations during secretion, suggests that it may be associated with the concentration and packaging of secretory products."

b. *Granule Stability Studies.* The early studies of McShan and his co-workers showed that (a) gonadotropin associated with a small granule fraction is readily dissolved when this fraction is treated with water, and (b) that purified "basophilic" granules are easily disrupted by treatment with physiological saline with a resulting solution of gonadotropin (Hartley et al., 1960). In marked contrast is the observation that extraction of purified growth hormone-containing granules with distilled water leaves particles which are still recognizable as granules (Hymer and McShan, 1963). The relative stability of the STH granules is appreciated by workers in the field, and in many studies care is taken to assure that these granules are indeed broken. For example, the results from radioimmunoassay show that homogenization of rat adenohypophysial tissue in 0.01 M NaOH liberates virtually all of the growth hormone from its sedimentable form (Birge et al., 1967). We

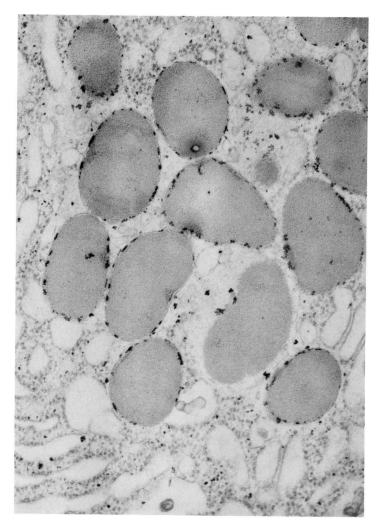

Fig. 3. Mature or fully condensed secretory granules from a prolactin cell in the pituitary of a postlactating rat. Clumps of nucleoside diphosphatase reaction product are seen outlining the granule in the space between the dense content and the limiting membrane. ×48,000. From Smith and Farquhar (1970).

recently reported (Snyder and Hymer, 1971) that lysis of GH granules with the non-ionic detergent Triton X-100 is required in order to recover complete biological activity from polyacrylamide gels. If, however, pituitaries are incubated for 1 hour in Hanks balanced salt solution depleted of Mg^{2+} and Ca^{2+} ions, STH-containing granules isolated from such glands contain the hormone in a form that does not require detergent treatment prior to electrophoresis. These results are somewhat consistent with our earlier observations (Hymer and McShan, 1963) that incubation of isolated STH granules for ½ hour in 1 mM EDTA results in considerable disruption of these particles. One tentative interpretation of these findings is that divalent cations may somehow participate in granule structure. However, more definitive studies are required to establish this point with certainty.

c. *Fractionation of Granule Proteins.* The discovery of proinsulin prompted considerable interest in the possible form(s) that adenohypophysial hormones might take within their different secretion granules. The only study which has appeared to date on this important subject is summarized in this section. LaBella and his co-workers (1971) obtained, with the combined techniques of (1) differential centrifugation, (2) removal of the mitochondrial "fluffy" layer, and (3) density gradient centrifugation with Ficoll (cf. Table I), highly purified preparations of 200–600 mμ diameter granules from bovine pituitaries. Comparative biological and radioimmunoassay (RIA) tests on these granule preparations showed that they contained only STH and prolactin (PL). Neither TSH nor LH were detectable by RIA. Interestingly, all or most of the protein in these granules could be accounted for by the two hormones. (This finding is based on immunoassayable contents of GH and PL/mg granule fraction relative to theoretical potency estimates of the pure bovine hormones.) To determine what form the hormone might take with the granule, preparations of hormonally active granules (140 mg protein) were solubilized in 0.02 M borate buffer (pH 8.0) containing 0.5% lauryl sulfate prior to application on DEAE-cellulose chromatography columns. Proteins eluted from this column were either (1) assayed for amino acid composition, (2) assayed for STH and PL activity by RIA, or (3) lyophilized and passed through a G-75 Sephadex column to obtain an estimate of the molecular weight of the protein component. An elution profile of the granule proteins reveals four major components, two of which show good correspondence to NIH hormone standards (Fig. 4). The amino acid composition of component 2 is remarkably similar to that of NIH-GH, while components 3 and 4 are more like NIH-PL (Table IV). Immunoassayable STH was associated almost exclusively with component 2, whereas PL activity predominated in components 3 and 4 (Table V). As before, hormone accounts for virtually all of the Lowry protein in these components. Finally, estimates of the molecular weight of the GH component suggest that the molecule may be packaged in a monomeric form of ~25,000 molecular weight. Prolactin, on the other hand, may be packaged in aggregate form (Table V) although the possibility that the extraction procedures produced aggregate forms could not be ruled out.

It is quite clear that this study was made possible by use of the sensitive procedure of radioimmunoassay. As this technique becomes more routine, additional studies of this type will certainly be carried out.

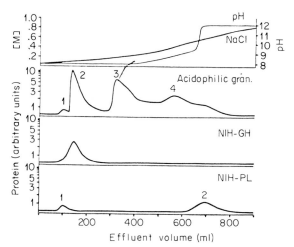

FIG. 4. Elution profile of proteins contained in solubilized acidophilic granules after chromatography on DEAE-cellulose. The elution patterns of purified GH and PL are given for comparison. From LaBella *et al.* (1971).

TABLE IV

AMINO ACID COMPOSITION OF THE PROTEIN FRACTIONS OBTAINED BY DEAE-CELLULOSE CHROMATOGRAPHY OF "ACIDOPHILIC" GRANULES[a]

Amino acid	AG₃	AG₄	NIH-PL (residues/1000 total residues)	AG₂	NIH-GH
ASP	125	123	129	92	91
THR	46	48	44	62	63
SER	80	77	81	70	78
GLU	128	129	139	146	145
PRO	68	60	59	36	36
GLY	60	64	60	61	63
ALA	54	59	54	85	80
VAL	46	47	47	35	36
CYS[b]	—	17	—	—	10
MET	25	24	25	14	13
ILEU	49	48	49	35	34
LEU	106	98	110	122	107
TYR	36	34	35	36	28
PHE	32	35	31	69	60
LYS	49	49	50	59	61
HIS	35	30	37	17	17
ARG	61	59	60	69	71

[a] The mean of duplicate amino acid analyses are shown for a typical chromatographic fractionation. Several fractions from each peak were pooled, dialyzed, and lyophilized prior to amino acid determination.

[b] Calculated as cysteine plus cysteic acid in acid hydrolysates. From LaBella *et al.* (1971).

TABLE V

DETERMINATION OF PROTEIN, GH, PL, AND MOLECULAR WEIGHTS OF FRACTIONS OBTAINED
FROM CHROMATOGRAPHY OF ACIDOPHILIC GRANULES ON DEAE-CELLULOSE

Protein component[a]	Fraction no.	Protein (μg/ml by Lowry)	GH (μg/ml by RIA)	PL (μg/ml by RIA)	Mol. wt.
AG$_1$	13	5	<0.04	1.2	25,000
AG$_2$	23	320	276	Not detected	25,000
	24	290	267	Not detected	
AG$_3$	50	476	12	616	70,000
	52	500	10	587	
AG$_4$	90	439	28	218	40,000
	95	386	17	177	70,000
	105	262	10	204	>70,000

[a] See Fig. 4. From LaBella *et al.* (1971).

B. RIBOSOMES: RNA AND PROTEIN SYNTHESIS

The techniques used to measure RNA and protein biosynthesis in mammalian tissues are now fairly standard. It is therefore not surprising to learn that some effort has been made in the last six years to measure these biosynthetic activities in the mammalian adenohypophysis. In this section the results of several studies are considered in terms of the methodology used to measure adenohypophysial RNA and protein biosynthesis. The picture that emerges thus far is one of a tissue with rigorous activity which, in some instances, may be altered as a result of endocrine imbalances induced by e.g., castration, thyroidectomy, etc.

1. *Autoradiographic Studies*

In a careful study by Städtler and his colleagues (1970), RNA synthesis has been evaluated in the different pituitary cell types after intraperitoneal administration of [³H]cytidine to 300-gm male rats. The following values, expressed as mean grain density (i.e., silver grain/μm² nuclear area) were reported: small chromophobes 4.79, large chromophobes 5.43, thyrotrophs 4.10, gonadotrophs 2.95, and somatotrophs 2.20. Eight weeks after castration, silver grain density in the small chromophobes increased by a factor of 4, while it doubled in the other cell types. In preliminary studies utilizing the approach of cell separation, we obtained approximately similar results in cells prepared from normal male rats (see Section III,B,3). Taken together, it would appear the chromophobes constitute the most active RNA-synthesizing cell compartment in the pituitary.

Similar approaches have been used to estimate relative protein synthetic activities in the different cell types. Siperstein (1963) measured *in vivo* uptake of [³H]glycine into pituitary cellular protein and reported that the large chromophobe appeared to be the most active cell type, while the somatotroph was considerably less active. Kobayashi *et al.* (1965) measured uptake of [³H]leucine into glands after incubation for 2 hr and found that chromophobes and acidophils had approximately equal activity (mean grain counts per cell). Apart from the fact that a different amino acid was used in these studies, it is conceivable that the different results might be explained by a hypothalamic inhibitory effect on protein synthesis in the acidophils. Additional quantitative studies are obviously required to settle this important question.

2. Biochemical Studies

a. *RNA Metabolism.*   We have examined the rate at which single rat adeno-hypophyses incorporate [³H]uridine into RNA *in vitro* (Hymer and Stere, 1967). After a short lag period, synthesis, which is linear in glands incubated up to 10 hr in Medium 199, occurs at the expense of radioactive precursor in the intracellular pool (Fig. 5). More than 98% of the total acid precipitable counts are in RNA. To establish the nature of the RNA species synthesized, total cellular RNA extracted 0.5 to 4 hr after incubation was separated on linear gradients of 5–20% sucrose. Radioactivity patterns were fairly typical of those reported in other mammalian systems. Thus, in the early pulse periods, predominant labeling was found in material heavier than 28 S ribosomal RNA. It is likely that a portion of these molecules are of the ribosomal precursor type. However, a majority of

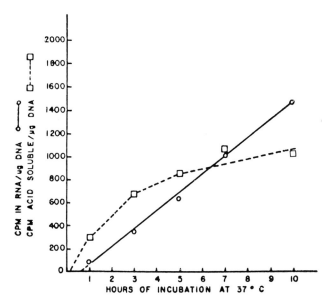

Fig. 5.   RNA synthesis in anterior pituitary glands incubated *in vitro*. Results represent the average specific activities of 2 separate experiments in which each flask contained the equivalent of a single pituitary gland (in halves). From Hymer and Stere (1967).

the counts after a 4-hr pulse were recovered in ribosomal RNA. These data imply that ribosomal RNA is being synthesized during the incubation period. This was subsequently confirmed in an experiment in which ribosomes were isolated from glands after 4 hr incubation *in vitro*. RNA liberated from the ribosomes by treatment with 0.5% sodium dodecyl sulfate was analyzed on a sucrose gradient. The profile (Fig. 6) clearly shows that newly synthesized ribosomal RNA is indeed present in cytoplasmic ribosomes.

The physicochemical characteristics of ribonucleoprotein particles isolated from the bovine anterior pituitary have been studied in detail by Adiga *et al.* (1968b). Ribosomes have an RNA:protein ratio of 0.92, while polyribosomes have a slightly higher ratio (1.15). Analytical ultracentrifugation of ribosome preparations show components with sedimentation coefficients ($s_{20,w}$) of 55, 75, 108, 137, and 160 S. Removal of Mg^{2+} from the preparation by dialysis at 4°C against a Tris-KCl buffer leads to a gradual disappearance of the heavier species and an increase in the monomer peak (75 S). When treated with EDTA, suspensions of ribosomes show two major components with sedimentation coefficients of 44 and 27 S. These data, together with the UV absorption spectra, show that pituitary ribosomes are physically similar to liver ribosomes.

b. *Protein Metabolism.* The kinetics of incorporation of labeled amino acids into the *in vitro* pituitary approximate those previously discussed for labeled

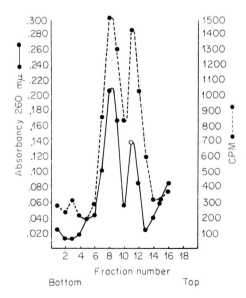

FIG. 6. Sucrose gradient sedimentation pattern of RNA extracted from a pituitary ribosome fraction isolated from glands previously incubated 4 hr in Medium 199 containing [³H]uridine. The 2 major peaks of radioactivity correspond to newly synthesized 28 S and 18 S ribosomal RNA. From Hymer and Stere (1967).

uridine (Fig. 5). Thus, uptake of labeled amino acid into pituitary cells proceeds in exponential fashion with time until a constant level is reached after 60–90 min of incubation. Moreover, incorporation of amino acid into protein, after a slight lag, proceeds linearly during 4–6 hr incubation (Grieshaber and Hymer, 1968; Samli *et al.,* 1971). Of the total labeled amino acids taken up by the tissue, a major fraction is incorporated into protein (after 3 hr it may reach 70–90%, depending on the concentration of unlabeled amino acids in the incubation media). Such an incubation system has been used by many investigators to estimate protein synthesis rates in adenohypophyses from, e.g., castrated or thyroidectomized animals or to evaluate levels of activity in tissues stimulated to release hormone.

Since, however, these protein synthetic activities may have little bearing on the actual rates of hormone production, additional protein separation procedures are obviously required. The rigorous study by Rao *et al.* (1967) provides a good case in point. Using micro-modifications of procedures described in the literature for the large-scale isolation of growth hormone and prolactin, labeled bovine growth hormone and prolactin were apparently successfully isolated from tissue slices previously incubated in radioactive amino acids. Identity of the isolated hormones was established on the basis of amino acid composition, N and C terminal analysis, chromatography on Amberlite and DEAE-cellulose columns, polyacrylamide gel electrophoresis, and immunoassays on Ouchterlony plates using purified antisera. Clearly, this approach is limited not only by quantities of tissue, but by laboratory facilities as well. It is therefore not surprising to find that the most widely used technique to measure synthesis of new hormone has been that of polyacrylamide gel electrophoresis. This technique permits separation of small quantities of newly synthesized proteins present in rat hemipituitaries, dissociated cells, or subcellular organelles. Localization of hormone in the gel is usually accomplished by bioassay or radioimmunoassay. Indeed, electrophoretic mobilities are reproducible enough so that most investigators use R_f as the sole criterion for hormone identification. Radioactivity in comparable gel discs then provide an estimate of new hormone synthesis. As emphasized by Kohler *et al.* (1971), however, different percentage gels and various buffer systems should be used to test conformation and charge of the hormone in question. Furthermore, since the amount of protein in any given disc is so small, it has not as yet been possible to demonstrate complete homogeneity of any protein hormone. MacLeod has reported that 75% of the counts in the prolactin band are precipitated by an antibody to rat prolactin (MacLeod and Lehmeyer, 1972), a finding which supports the usefulness of the acrylamide gel technique. It therefore seems that the gel electrophoresis technique, when used in conjunction with radioimmunoassay, will probably continue to be the most useful way of measuring small amounts of hormones synthesized in tissue slices, isolated cells, or cell-free systems.

3. *Cell-Free Synthesizing Systems*

The report of Adiga *et al.* (1966) appears to be the first in which a cell-free system capable of sustained protein synthesis was specifically applied to pituitary tissue. Since then only a handful of papers have appeared (indeed most by Adiga and co-workers, 1968a,b, 1971) utilizing this important technique.

As originally described by Adiga *et al.* (1966), the basic components of the system include ribosomes, pH 5 enzyme, an ATP-generating source, amino acids, and other co-factors. Ribosomes are usually prepared by (1) liberation from microsomes with 1% deoxycholate in the presence of a small amount of 140,000 g supernatant fraction (included to enhance activity) followed by (2) centrifugation through 1.0 M sucrose. The pH 5 enzyme has been prepared from the 140,000 g supernatant fraction after passage through G-25 Sephadex either by pH adjustment or ammonium sulfate fractionation. The optimum conditions for protein biosynthesis have been established by Adiga and will not be detailed here, but it is worthwhile noting that an approximately linear rate of synthesis takes place over 1.5 hr of incubation. Furthermore, using a battery of techniques, e.g., column chromatography, isoelectric precipitation, and paper and gel electrophoresis, these investigators have also been able to show in fairly convincing fashion that ACTH, growth hormone and prolactin are all synthesized in the cell-free system.

One of the more interesting applications of the method was an experiment in which pituitary polysomes of differing aggregate size were incubated in the cell-

TABLE VI

HORMONE AND PROTEIN SYNTHESIS WITH POLYSOME FRACTIONS OF VARYING SIZE

		Isotope incorporated in 1 hr with cell-free assay system (cpm)		
Tube no.[a]	Polysome aggregate no.	Growth hormone	Prolactin	Protein
9–11	55 S subunit	260	500	6,900
12–14	1	700	2,200	15,000
15–17	1–2	1,100	3,400	25,600
18–20	2	1,600	6,800	47,400
21–23	3	3,500	11,200	60,100
24–26	4	4,600	20,300	80,200
27–29	5	7,000	27,800	98,800
30–32	6–7	8,200	33,500	107,180
33–35	7–9	5,900	21,000	82,200
36–38	Polydisperse region (above 9)	4,900	14,500	35,800
38–41	Bottom of gradient	700	2,500	25,200

[a] From the sucrose gradient. From Adiga *et al.* (1968a).

free system for 1 hr prior to analysis for new hormone synthesis (Adiga *et al.,* 1968a). As seen in Table VI, radioactivity recovered in growth hormone and prolactin rose progressively with increasing size, reaching a maximum at the hexamer to heptamer level. It is of considerable interest that the predicted polysome size requirement for synthesis of proteins with molecular weights of approximately 20,000 correlated rather well with these experimental results (bovine prolactin and growth hormone have molecular weights of 21,000 and 26,000, respectively).

In future work it can be anticipated that the cell-free protein-synthesizing system will play an important role in studying ways in which certain physiologically active secretagogues, e.g., releasing factors, cyclic AMP, ions, might act on the protein synthetic machinery of the pituitary cell. Indeed, a recent report by Adiga *et al.* (1971) offers suggestive evidence that the cyclic nucleotide can directly stimulate translation by pituitary polysomes, but more definitive work is required to establish this important point. With the development of sensitive radioimmunoassay techniques and the establishment of an active cell-free protein synthesizing system by the rat adenohypophysis (Reagan *et al.,* 1972), new and exciting information can be expected in the next few years.

C. Adenohypophysial Enzyme Systems

1. *Hormone Secretion*

The *precise* sequence of intracellular events which occur in a pituitary cell that has been stimulated to release hormone is by no means defined, yet a few of the steps have been recently elucidated. Since 1970, evidence for the involvement of a pituitary adenyl cyclase and protein kinase in the release process has accumulated. A brief progress report follows.

a. *Binding a Releasing Factor to the Cell Membrane.* Commercial availability of a radiolabeled synthetic releasing factor which retains full biological activity, viz., [³H]TRF (thyrotropin releasing factor), has enabled investigators to look into ways in which this peptide may act on the pituitary thyrotroph. In one of the first published reports on this subject, Labrie and his co-workers (1972) measured binding of [³H]TRF to adenohypophysial plasma membranes isolated by a modified Neville procedure (1960). Half-maximal binding was obtained after 5-min incubation at $0°C$. The equilibrium constant for the reaction was $4 \times 10^{-7} M$; the concentration of TRF which gave half-maximal binding was 23 nM (a level which is physiologically active). It was of considerable interest that Labrie noted that addition of unlabeled TRF induced dissociation of the [³H]TRF–receptor complex with a half-life of approximately 14 min. This result suggests a high degree of specificity; a suggestion further supported by the finding that binding was unaffected by addition of MSH-releasing inhibiting hormone, lysine–vasopressin, ACTH, STH, LTH, insulin, or glucagon to the incubation medium.

Grant *et al.* (1972) also studied binding of [³H]TRF to cell membrane preparations isolated from TSH-secreting mouse pituitary tumors. In general, the characteristics of binding and dissociation were approximately similar to those observed in the study of Labrie *et al.* (1972). Binding of [³H]TRF to cultured pituitary cells obtained from thyroidectomized animals was lower than that obtained with the tumor preparations, but the number of thyrotrophs in these cultures was not assessed.

Wilber and Seibel (1973), using a slightly different approach, have incubated [³H]TRF with intact pituitary glands. After incubation, homogenization and differential centrifugation, the membrane fraction with the highest 5′-nucleotidase specific activity bound 320% more TRF/mg protein than corresponding fractions from numerous other tissues. Moreover, binding was enhanced 400% in pituitaries taken from animals that had received propylthiouracil.

Finally, the binding affinity of differing TRF analogs is apparently directly correlated with their biological activity. For example, the inactive analog pGlu-His-OMe does not compete for the [³H]TRF bound even at 200-fold concentration (Grant *et al.,* 1972).

Taken together, these results strongly indicate that there is a specific TRF-binding site on the thyrotroph cell membrane. As other labeled releasing factors become available, it will be of interest to see if receptor molecules on the plasma membranes of the other pituitary cell types can also be identified.

b. *Activation of Adenyl Cyclase.* Adenyl cyclase and adenosine 3′,5′-monophosphate (cAMP) both play an important role in the control of hormone release from the pituitary (see chapter by Kraicer). For example, it has been shown by Zor *et al.* (1970) that crude hypothalamic extracts stimulate pituitary adenyl cyclase activity and significantly elevate the intracellular concentration of cAMP. Similar kinds of effects have recently been noted when synthetic TRF or FRF–LRF are used as stimulating agents. From the preliminary observations of Poirier *et al.* (1973) and those of McKeel and Jarett (1970), it now seems fairly certain that a large proportion of the pituitary adenyl cyclase activity is associated with the plasma membrane fraction. There is as yet no direct evidence that the adenyl cyclase and receptor protein are complexed in the pituitary cell membrane as they appear to be in the adrenal cell (Pastan *et al.,* 1970).

Zor *et al.* (1970), MacLeod and Lehmeyer (1970), and DuPont and Chavancy (1973) have all observed that prostaglandins (especially PGE_1) increase adenyl cyclase activity (and [cAMP]) in the pituitary as well as cause secretion of ACTH, TSH, and STH. These kinds of observations have prompted some to suggest that physiologically induced activation of the adenyl cyclase system may be mediated by certain prostaglandins. If this turns out to be the case, prostaglandin could serve as a "connecting-link" between the membrane receptor protein and the adenyl cyclase component.

c. *Activation of cAMP-Dependent Protein Kinase.* The results from a series of papers published from Labrie's laboratory in 1971 (Labrie *et al.,* 1971a,b; Lemaire *et al.,* 1971) invite speculation that a next probable step in the secretion process involves activation of a protein kinase. This enzyme (whose assay involves incubation of enzyme with histone protein in the presence of $[\gamma\text{-}^{32}P]ATP$ and subsequent estimation of the rate of histone phosphorylation) was partially purified from beef pituitary tissue. Addition of $5 \times 10^{-7} M$ cAMP causes a fourfold increase in activity, i.e., the enzyme is cAMP dependent. Approximately half of the total enzyme activity is recovered in the 200,000 g supernatant (i.e., soluble fraction), with the remaining activity being widely distributed among the other subcellular fractions. The finding that the enzyme associated with the particle fractions exhibits relatively low cAMP dependence, coupled with the observation that over 80% of the enzyme available to exogenous substrate resides in the soluble fraction, prompted Labrie to speculate that the primary effect of cAMP might be exerted through the cytosol enzyme. Thus, according to Labrie,

> . . . an increase of the intracellular level of cyclic AMP through activation of adenylate cyclase activity, inhibition of cyclic phosphodiesterase, or exogenous treatment with cyclic AMP, would result in increased binding of the cyclic nucleotide to the receptor moiety of the inactive protein kinase complex in the cytosol, with resultant release of catalytic subunit and its subsequent binding and phosphorylation of the various substrates available in a soluble or particulate form in the cell.

What kinds of proteins might be phosphorylated in the pituitary cell that has been stimulated to secrete hormone? One answer appears to be that of the granule protein itself. Purified bovine pituitary secretory granules are phosphorylated in the presence of $[\gamma\text{-}^{32}P]ATP$ by endogenous protein kinase. The radioactive phosphate is incorporated largely as ester phosphate of serine and threonine residues. Since washing of the granule fraction in solutions of high ionic strength removes only negligible activities, Labrie feels that either structural protein in the granule itself or protein in its membrane envelope are most likely candidates for phosphorylation sites.

d. *Hormone Release.* Essentially no data of a biochemical nature are available to help explain the mechanism(s) that may be involved in the final stages of release of hormone from the cell. If a rapid and dynamic variation in the level of phosphorylation of secretion granule membranes and plasma membranes are required in the membrane fusion process (Labrie *et al.,* 1971b), the scheme presented in the above section appears to be at least partially consistent. If, however, microtubules and/or ions are directly involved in the release process (see chapter by Kraicer), more pieces of the puzzle need filling in before the complete picture emerges.

2. Hormone Degradation

Several pituitary peptidases have now been separated and characterized (McDonald *et al.*, 1966, 1968a,b). Some of these are endopeptidases, viz., acid and alkaline proteinase. Others are able to split one amino acid from the NH_2-terminus of polypeptides; these are the aminopeptidases. Since several synthetic β-naphthylamide peptide derivatives effectively serve as substrate for the aminopeptidases, these enzymes are usually termed aminoacyl arylamidases. Still another class of pituitary peptidases catalyze the hydrolysis of dipeptides from their β-naphthylamide derivatives; these are the dipeptidyl arylamidases. The preferred substrates, activators, pH optima, and subcellular distribution of each of these pituitary enzymes has been conveniently summarized by Vanha-Perttula (1969) (Table VII).

Several years ago Meyer and Clifton (1956) showed that pituitary alkaline proteinase activity was increased after long-term treatment with estrogen. More recently Vanha-Perttula (1969) found that activity of all four aminoacyl arylamidases, dipeptidyl arylamidase I and III, acid proteinase, and acid phosphatase increased significantly in pituitaries of female rats that had been injected with estrogens for 20 days. Fluctuations in activity of specific enzymes were also noted after testosterone treatment and castration.

McDonald *et al.* (1968b) have proposed an interesting model to account for the fluctuations of some of these different enzymes as related to regulation of hormone turnover. Under conditions where hormone secretion is diminished and secretory granules are taken up by lysosomes (see chapter by Farquhar), acid proteinase within the lysosome could catalyze hydrolysis of protein hormone to

TABLE VII

PROTEOLYTIC ENZYMES OF THE ADENOHYPOPHYSIS

Enzyme	Preferred substrate	Activator	pH optimum	Subcellular distribution
AAP I[a]	Leu-βNA	Co^{2+}, Mn^{2+}	7.5	Microsomes
AAP II	Arg-βNA	Co^{2+}, Mn^{2+}	7.0	Soluble
AAP III	Lys-βNA	–SH	7.5	Soluble
AAP IV	Arg-βNA	–SH	7.0	Soluble
DAP I[b]	Ser-Tyr-βNA	–SH, Cl^-	4.0	Lysosomes
DAP II	Lys-Ala-βNA		5.5	Lysosomes
DAP III	Arg-Arg-βNA	–SH	9.0	Soluble
Acid proteinase	Hemoglobin		3.8	Lysosomes
Alkaline proteinase	Hemoglobin		8.2	Soluble
Plasmin-like peptidase	CBZ-Gly_2-Arg-NA	Urokinase	8.5	?

[a] Aminoacyl arylamidase.
[b] Dipeptidyl arylamidase. From Vanha-Perttula (1969).

inactive polypeptides. Since 75 and 100% of the biological activity of growth hormone and prolactin are lost by cleavage of only seven peptide bonds in each of these hormones after treatment with this enzyme (Ellis, 1960), this hypothesis is not an unreasonable one. The resulting peptide products could then be degraded to amino acids by the lysosomal dipeptidyl arylamidase III and other peptidases of the cell sap. The resulting amino acids would then presumably be restored to a metabolic pool for reutilization.

The reader should be aware that interpretation of biochemical data in terms of physiological function is extremely difficult since the *cellular* localization of these proteolytic enzymes is at present largely unknown. However, the recent observations on the localization of dipeptidyl aminopeptidase II by electron microscope histochemistry (McDonald *et al.*, 1971) clearly indicate the feasibility of future work in this area.

D. NUCLEI: DNA AND RNA POLYMERASES

Two of my former students (A. Mastro and E. Augustine) have successfully developed assays for measuring levels of DNA and RNA polymerase in isolated pituitary nuclei. The optimum conditions for each assay have been carefully worked out and the products synthesized shown to be either DNA or RNA. In both studies nuclei were isolated by a non-ionic detergent method (Hymer and Kuff, 1964). This procedure permits recovery of pituitary nuclei in high yield (90%) with RNA/DNA and protein/DNA ratios of 0.21 and 3.0, respectively (Augustine and Hymer, 1973).

Interestingly, enzyme activity in these nuclear preparations appears to reflect their native physiological state. Thus RNA polymerase activity is significantly greater (60%) in pituitary nuclei prepared from 56-day-old male rats than in those from 25-day-old animals (Augustine and Hymer, 1973). Moreover, incorporation of [³H]TMP in the presence of activated primer DNA is greater in nuclei prepared from glands of young animals than from old, or greater in nuclei from estrone-treated or castrated rats than from their respective controls (Mastro and Hymer, 1973).

III. Dissociation and Separation of Cells from the Rat Adenohypophysis

The theoretical advantages of using populations of pure cell types in the study of pituitary gland function are self-evident. Three years ago we set out to obtain such cell preparations; and the progress which has been made to date on this challenging problem is summarized below. This progress could not have been made without the expert help of Drs. Warren Evans of the NIH, and Jack Kraicer and Steve Haskill of Queen's University.

Ideally, pituitary cells in isolation should satisfy the following requirements if they are to be useful tools for adenohypophysial study.

(1) The cells should retain the staining properties characteristic of those in intact tissue. Furthermore, they should also retain their ultrastructural features at the electron microscope level.

(2) The cells should retain their full hormone complement during dissociation and separation.

(3) The cells should pass certain viability tests, e.g., actively exclude dye, respire, synthesize protein (as well as protein hormone).

(4) The cells should respond to physiological secretagogues by release of hormone into the incubation medium; release should be dependent on dose of agent used.

In the following discussion, the extent to which these requirements have been met will be considered.

A. TISSUE DISSOCIATION

A number of methods have been developed to prepare suspensions of single pituitary cells. The list of methods described below is by no means complete, but it is representative of the general approaches used to date. In all cases, glassware is pretreated with silicone to avoid the problem of surface cell adhesion.

1. *Method of Portanova et al. (1970)*

Pituitaries from 6 to 12 adult (250–450 gm) male rats are minced and the tissue fragments dispersed in Krebs–Ringer bicarbonate buffer containing 0.2% glucose (KRBG) and 0.25% trypsin. Tissue fragments are mechanically agitated for 20 min by a glass paddle stirring (500 rpm) in a 50-ml Erlenmeyer flask kept at 37°C with gassing (95% O_2:5% CO_2). After 20 min the supernatant is removed and fresh trypsin solution added. Cell yield is 0.65–1.44 × 10^6 cells/ gland after a 1 hr dissociation period.

The fine structure of cells dissociated by this procedure has been described by Malamed *et al.* (1971). The cells appear intact and essentially undamaged (see chapter by Farquhar).

The capacities of these dissociated cells to release ACTH in response to a crude hypothalamic median eminence extract (HME) has also been studied by this group. When cells are incubated in KRBG containing 3% bovine serum albumin (BSA) and lima bean trypsin inhibitor, 2–6% of the total cellular ACTH (2300–5200 microunits/100,000 cells) is released into the incubation medium.

Addition of HME causes an additional 1.2–3-fold increase in ACTH release. This release is (a) dependent on HME dosage, (b) inhibited when calcium or glucose is omitted from the incubation medium, (c) depressed when physiological concentrations of corticosterone are added to the incubation medium (see Malamed *et al.*, 1971, for references), and (d) increased when the cells are kept at 0–4°C.

2. *Method of Vale et al. (1972)*

Tissue from 15 to 30 male rats (150–200 gm) is dissociated in HEPES buffer containing 3% BSA, 0.1% hyaluronidase, and 0.35% collagenase. Fragments are stirred (100 rpm) by a glass paddle in a flask kept at 31°C. Dispersion rate is increased by "gently drawing the fragments in and out of a siliconized Pasteur pipette every 10 min" throughout the 45-min procedure. Cell yield is approximately 1.0×10^6 cells/gland.

Vale *et al.* (1972) have used these cells for culturing purposes. It is of considerable interest that after 4 days in culture, the cells respond to synthetic TRF or LRF by releasing 5–20 times the TSH or LH levels compared to those seen in control cultures. This release is dose dependent. Moreover, high K^+, prostaglandin E_2, and theophylline all stimulate TSH release from these cells. Surprisingly, the secretory response of freshly dispersed cells to TRF is "slight."

3. *Method of Leavitt et al. (1973)*

In this study, pituitary cell suspensions have been used to measure binding of [³H]estradiol. Pituitaries from ovariectomized rats are digested with 0.05% pronase in calcium–magnesium free Hank's solution. By electron microscopy the cells appear virtually identical with those *in vivo*. The ratios of RNA/DNA and protein/DNA are similar to those of intact tissue.

4. *Method of Hymer et al. (1973)*

Pituitary fragments from 8 to 10 male rats (200–250 gm) are dispersed in 10 ml of Ca-free Spinners minimum essential medium containing 0.1% trypsin and 0.1% BSA buffered to pH 7.3 with $NaHCO_3$. Fragments are stirred by a Teflon impeller (~200 rpm) placed in a 25-ml Spinner suspension flask (Bellco Glass) which is kept at 37°C under constant gassing. To further aid in dispersion, tissue pieces are expelled 20–30 times through a Pasteur pipette every half-hour. No tissue pieces remain after 2 hr. A single cell suspension with a yield of 2.5–3.0×10^6 cells/gland is consistently obtained. This yield accounts for 70–80% of the cells in the gland; an estimate based on counts of nuclei in adenohypophysial homogenates. (We have also tried to dissociate cells in KRBG and find that the yield is in the range that Sayers reports. In Ca-free KRBG, however, this yield increases significantly.)

Cells are prepared for histological study by centrifugation onto a microscope slide (Shandon Cytocentrifuge). Details of the fixation and staining procedures (Herlant's tetrachrome or Herlant's Alcian blue–PAS–Orange G), as well as the morphological criteria used to identify cells, have been described in detail (Hymer *et al.*, 1973). The staining characteristics of the cells after dissociation are essentially identical to those seen in the gland. Distribution of cell types in the suspension is somatotrophs, 35%; chromophobes, 47%; basophils, 4%; lactotrophs, 1%; and unknowns (i.e., cells which are either damaged or classifiable as follicular or endothelial cells), 16%. This distribution is approximately similar to that reported in intact tissue of male rats. In thin section, the ultrastructural morphology of the cells in suspension is virtually indistinguishable from those *in vivo* (Fig. 7) (Hymer *et al.*, 1972b).

We have also studied the STH content of pituitary cells, their secretory responsiveness to GRF and cAMP, and their protein synthesizing capacities. These results will be discussed in the next section when these functions are evaluated in relation to the separated cells.

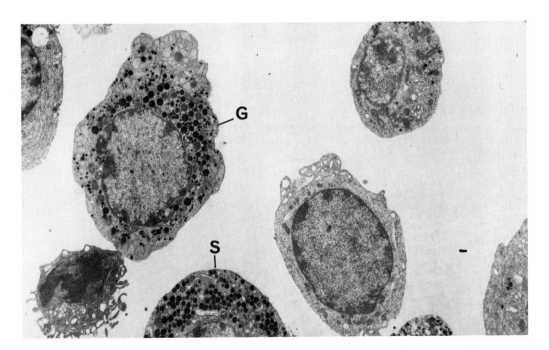

FIG. 7. Electron micrograph of dissociated rat adenohypophysial cells prior to separation by velocity and density gradient centrifugation. Cell types are identified on the basis of secretion granule size. Somatotroph (S), gonadotroph (G). ×4000. Micrograph taken by Dr. S. Bencosme. From Hymer *et al.* (1972b).

B. Separation of Pituitary Cells

A number of procedures have been developed to separate different types of cells contained in tissues of mixed cell populations. Thus far, the most fully developed separation methods appear to be based largely on differences in either cell size, cell density, or cell charge.

1. Sedimentation at Unit Gravity (1G) (Hymer et al., 1973)

We have had some success in separating pituitary somatotrophs by velocity sedimentation using the earth's gravitational field. In this method, cells settle through a shallow stabilizing gradient of BSA. The theoretical basis for the 1G separation method is considered in the original paper of Peterson and Evans (1967) and in the later study of Miller and Phillips (1969); it is generally agreed by these authors that cells separate primarily on the basis of differences in cell size. A brief discussion of the methodology involved and representative results obtained follow (see Hymer et al., 1973, for additional details).

a. *Operation of Sedimentation Chamber.* The chamber and gradient generating devices are shown in the line drawing in Fig. 8. The lucite sedimentation chamber has a 500-ml capacity, is 11 cm o.d. and 11 cm high, and is designed for the separation of relatively small numbers of cells. Experiments are carried

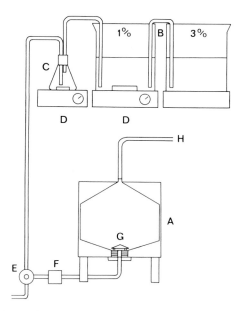

Fig. 8. Apparatus for sedimentation of adenohypophysial cells at unit gravity. See text for details of operation. From Hymer et al. (1973).

out at room temperature under conditions where mechanical vibrations and thermal variations are kept at a minimum to maintain stability of the shallow gradient. The gradient is generated with two beakers (Fig. 9B) containing 300 ml of either 1% or 3% BSA prepared in Medium 199 at pH 7.3. Vessel (C) contains 40 ml of 0.3% BSA in Medium 199. The gradient flows into the chamber (A) through a three-way valve (E). Flow rate is regulated at (F) and gradient solutions are mixed by magnetic stirrers (D). Freshly dispersed pituitary cells are applied on top of the threaded baffle (G) through the opening in the top of the chamber; the cells are then quickly lifted into the chamber by the incoming gradient solution. Cells fall continuously into the gradient during the time it takes to fill the chamber (1 hr). After the chamber has been filled, an additional settling time of 15–85 min is used depending on the experiment. During this time 4–6 discrete cell bands are consistently seen at different levels in the gradient. Cell fractions are collected from the top of the chamber (H) by displacement of the gradient with a 7% sucrose solution introduced from a vessel (not shown in Fig. 9) via (E).

b. *Gradient Characteristics.*   Under the conditions employed above, the BSA under the initial cell layer rises rapidly from 0.3 to 1% in the first 60 ml of the gradient. It then becomes linear with a slope of 0.0037% BSA/ml for the remaining distance. A total gradient of 0.3–2.4% BSA is generated.

Initially, the sample layer is smoothly distributed over the gradient as a thin layer because of the design of the baffle and conical bottom of the chamber. As originally indicated by Peterson and Evans (1967) and later stressed by Miller and Phillips (1969), an important factor in determining the degree of cell enrichment is the concentration of cells in the original suspension. Above a certain critical concentration, rapidly settling cells will cause vertical streams at the bottom of the initial cell band, presumably due to an increased density at that point. This cell streaming phenomenon is undesirable because it increases the width of the initial cell band thereby diminishing resolution. Under the conditions employed above, streaming occurs above a concentration of 1.4×10^6 cells/ml. In the usual experiment 10×10^6 cells are applied, but effective enrichments have been achieved with as many as 20×10^6 cells. Gradients of sucrose, sucrose plus BSA, and fetal calf serum have also been tried, but BSA in Medium 199 appears to give the best results.

c. *Distribution of Cell Types after 1G Sedimentation.*   After 1.25 hr sedimentation, cells in the original suspension are distributed as shown in Fig. 9 (top panel). A major cell peak at fraction 12 (i.e., 120 ml into the gradient), with a progressive decrease in cell numbers to fraction 32 is observed. The positions of these cell peaks are virtually identical from run to run; however, the number of cells in each peak is variable. A majority of the somatotrophs sediment to an area encompassed by fractions 18–32 (Fig. 2, lower panel). Sixty to 70% of the cells in this region of the gradient are somatotrophs, chromophobes, basophils, and

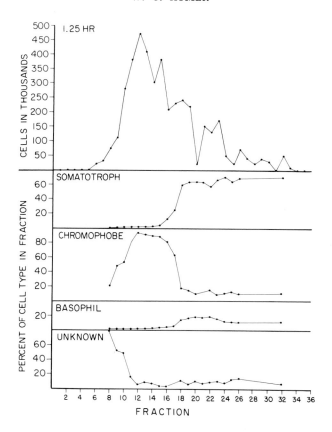

Fig. 9.   Top panel: distribution profile of separated pituitary cells after 1.25 hr sedimentation at unit gravity. Lower panels: localization of specific cell types expressed as percentage of cell type in consecutive 10-ml fractions. From Hymer *et al.* (1972b).

unknowns contaminate this fraction to about an equal extent ($\sim$10%). If the sedimentation time is increased to 2.25 hr, somatotrophs are primarily found between fractions 24–34, but they become considerably more contaminated with chromophobes (20–40%) (Hymer *et al.*, 1972b, 1973). Evidently the somatotrophs are among the most rapidly sedimenting cell type; however, once they reach the higher density regions of the gradient their sedimentation rate is retarded and chromophobes begin to settle into the somatotroph region. Attempts to further purify the somatotrophs by retarding their sedimentation rate with (a) introduction of Ficoll into the gradient or (b) use of shorter sedimentation times have proved unsuccessful. In our experience, cells in the basophil series (i.e., gonadotrophs and corticotrophs) cosediment with the somatotrophs and cannot be further separated by the 1G technique.

It is important to note that the cells retain their characteristic staining proper-

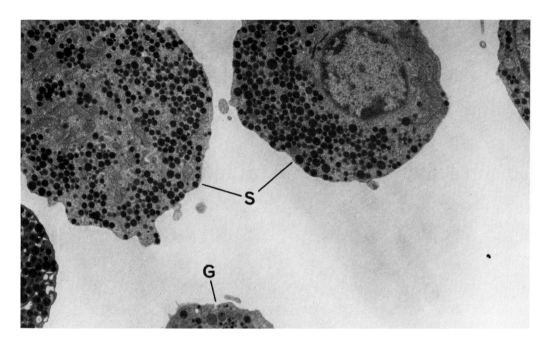

Fig. 10. Electron micrograph of cells recovered from the somatotroph fraction (tubes 18–32, Fig. 10) after unit gravity sedimentation. Four of the five cells are somatotrophs. Gonadotroph (G). ×5500. Micrograph taken by Dr. S. Bencosme. From Hymer *et al.* (1972b).

ties after disassociation and sedimentation. This result would suggest that the cells retain hormone during these separation procedures (confirmed by radioimmuno-assay in the case of GH, see below). Equally important, the staining responses have enabled us to monitor cell purity and recovery under different experimental conditions in at least semi-quantitative fashion.

The ultrastructural appearance of cells in the somatotroph fraction (Fig. 10, from pooled fractions 18–32), compares favorably with those seen in the original cell suspension. Study of the chromophobe fraction (11–14, Fig. 9) by electron microscopy has revealed that it consists of a heterogeneous group of cells which contains variable numbers of granules of varying size. This observation certainly reflects the limit of sensitivity of the histological staining procedure, a finding which has been mentioned by many pituitary cytologists on numerous occasions.

2. *Density Gradient Centrifugation (Hymer et al., 1972b)*

a. *Operation.* We have used the density gradient procedure of Shortman (1968) to purify the somatotrophs collected after velocity sedimentation. Separation by this procedure is based entirely on differences in buoyant densities of the cells. Cells in fractions 18–32 from the 1G step are pooled and concentrated by centrifugation (3.5×10^6 cells); these are dispersed in an isotonic, linear gradient of BSA ($\sim$14–28%) and centrifuged at 2000 g for 45 min. The techniques for gradient generation, centrifugation, and subsequent sampling are all carried out exactly as described in Shortman's report. Recovery of cells from the gradient is 80–100%.

b. *Distribution of All Cell Types in the Density Gradient.* After the cells are centrifuged to isodensity, three cell peaks are consistently recovered (Fig. 11). The first peak is at density 1.0601 ± .0007 gm/cm³; the second at 1.0695 ± .0007 gm/cm³; and the third at 1.0761 ± .0007 gm/cm³ (SEM, 6 experiments). Differential counts of the cells banding between 1.0705 and 1.0850 reveal that 80–92% are somatotrophs. Approximately 75% of the somatotrophs applied to the gradient have a density greater than 1.070 and can therefore be separated in good purity. Those somatotrophs banding at lower density appear partially degranulated by both light and electron microscopy. Interestingly, a majority of the basophils which cosediment with the somatotrophs in the unit gravity system band at densities 1.0575–1.0680. In some experiments, basophils make up 50% of the

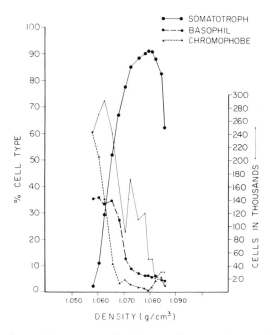

Fig. 11. Distribution and percentage of pituitary cell types after centrifugation in a linear gradient of 14–28% BSA. These cells were obtained by pooling fractions 18–32 (Fig. 10) from the unit gravity system and concentrating them prior to density gradient centrifugation. From Hymer *et al.* (1972b).

cells in this density region. It is tempting to speculate that the apparent density differences between these cell types may be partially explained on the basis of granule densities in their cytoplasms (cf. Table I). Chromophobes appear to be the least dense cell type since they are found at the top of the gradient.

Cells in the purified somatotroph fraction also show good morphological preservation (Fig. 12). Incubation of the cells for 3 hr in Medium 199 does not appreciably alter their ultrastructure (Fig. 13).

It should be pointed out that damaged cells, including some with pycnotic nuclei, are encountered at all stages of the dissociation and separation procedures. We estimate, after examination of cells in 1-μm sections stained with toluidine blue, that a total of 25% of the cells may show some damage after the entire separation procedures. Ten percent show damage after dissociation, 15% after sedimentation at unit gravity, and 20–25% after density gradient centrifugation. Finally, it is worthwhile pointing out that the entire procedure, i.e., dissociation, velocity, and density gradient centrifugation, has the major disadvantage of requiring approximately 5 hr to obtain the somatotrophs. In spite of this relatively long time, however, our data indicate that these cells retain function (see below).

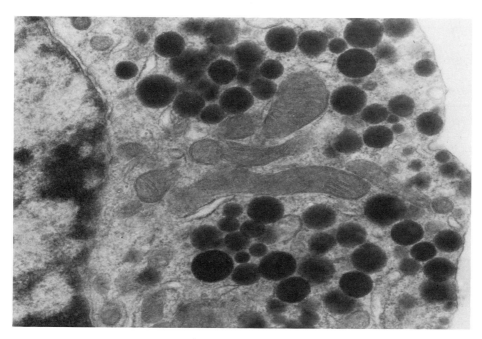

Fig. 12. Electron micrograph of a portion of a somatotroph obtained from the purified somatotroph fraction (densities 1.0687–1.0796, Fig. 12). ×29,000. Micrograph taken by Dr. S. Bencosme. From Hymer et al. (1972b).

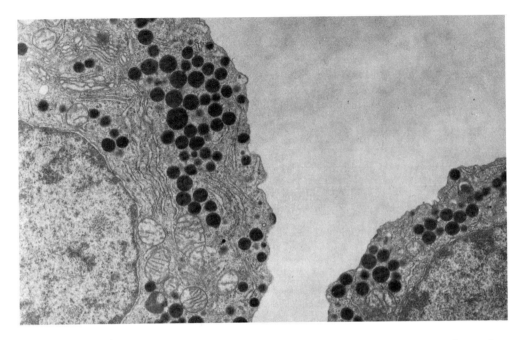

Fig. 13. Electron micrograph of somatotrophs obtained after velocity and density gradient centrifugation. These cells were incubated 3 hr in Medium 199. ×13,000. Micrograph taken by Dr. S. Bencosme. From Hymer *et al.* (1972b).

3. *Functionalities of Separated Pituitary Cells*

a. *Dye Exclusion.* Approximately 85–95% of the cells at the different stages of separation are able to exclude trypan blue. It should be emphasized, however, that this test of cell viability is a relatively insensitive one.

b. *Protein Synthetic Capacity.* Incorporation rates of ^{14}C-amino acids into the TCA precipitable protein fraction of (a) cells in the original suspension, (b) chromophobes from the unit gravity step, and (c) somatotrophs (cells at density 1.070–1.0835) from the density gradient are compared in Table VIII (Hymer *et al.,* 1972b). The data show that the incorporation rates are essentially linear during the 3-hr incubation period for all cell fractions tested; a result which shows that the cells retain viability even after the relatively long time required to obtain them. It is of interest that the relative specific activity (dpm/cell) of the somatotrophs is about twice that observed in the starting cell suspension; while the chromophobes are much less active. In one experiment (unpublished) we found that a major portion of the radioactivity incorporated by cells in the somatotroph fraction was associated with a protein whose electrophoretic mobility on polyacrylamide gels was similar to that of growth hormone.

In another series of experiments (Hymer *et al.,* 1973) we measured the level

TABLE VIII

INCORPORATION OF ^{14}C-AMINO ACIDS INTO TCA PRECIPITABLE PROTEIN OF
DISPERSED AND SEPARATED RAT ADENOHYPOPHYSIAL CELLS

Cell fraction	Incubation time (min)	dpm/cell $\times 10^{-3}$
Dispersed cells[a]	60	2.63 ± 0.15[b]
	120	5.67 ± 0.42
	180	7.29 ± 0.21
Chromophobes[c]	60	0.93 ± 0.12
	120	1.65 ± 0.11
	180	1.85 ± 0.10
Somatotrophs[d]	60	6.75 ± 0.43
	120	9.56 ± 0.60
	180	14.26 ± 0.54

[a] In all cases, total cells, purified chromophobes, or somatotrophs were incubated in Medium 199 containing 0.1% BSA and ^{14}C-amino acid mixture at a concentration of 1 μCi/ml. Cell recoveries after incubation were 70–95%.

[b] Standard error of mean.

[c] Fractions 11–14 from the unit gravity system (Fig. 10).

[d] Cells which banded between 1.070 and 1.0835 gm/cm³ in the 14–28% BSA gradient (Fig. 12). From Hymer et al. (1972b).

of incorporation of ^{14}C-amino acids into TCA-precipitable protein of different cell fractions obtained from either (a) glands incubated in isotope *prior* to dissociation and separation or (b) cells incubated in isotope *after* dissociation and separation. The labeling patterns and level of activity (dpm/cell) in cells after separation was similar to that found in the tissue before separation. This result implies that tissue integrity is not essential for the final result.

c. *Hormone Content.* We have determined, by radioimmunoassay, that the concentration of growth hormone in the initial cell suspension is 30–50 ng/1000 cells (Hymer and Kraicer, 1972). The distribution of hormone in the 0.3–2.4% and 14–28% BSA gradients correlates extremely well with the distribution of somatotrophs in these gradients as judged by light and electron microscopy. In the somatotroph fractions from both gradients hormone levels range from 100 to 200 ng GH/1000 cells. Cells in the other gradient regions contain 5–25 ng GH/1000 cells. Such data not only show that the somatotrophs retain GH during the dissociation and separation procedures, but on the basis of hormone content are concentrated 2–5-fold. Assay of the different cell fractions for the presence of the other adenohypophysial hormones remains to be done.

4. *Effects of Secretagogues on Isolated Somatotrophs*

a. *Dibutyryl Cyclic AMP (DcAMP).* Many workers report that addition of DcAMP to the medium of incubating pituitary tissue stimulates release of immunoassayable GH. We were interested to see if this physiologically important compound had similar effects on the isolated somatotrophs. Thus far our experiments show that (1) the cyclic nucleotide causes degranulation of the somatotrophs (evidenced by a 10–25% loss in stained somatotrophs and a concomitant increase in chromophobe number) after 1 hr incubation and (2) the cyclic nucleotide causes a significant decrease in protein synthetic rates in the somatotrophs while having no effect on synthesis rates in the chromophobes (Hymer *et al.,* 1972a). In a preliminary experiment we also noted that DcAMP causes a 2-fold stimulation in release of immunoassayable GH, with a concomitant depletion of intracellular GH (Table IX). Taken together these results indicate that the somatotrophs in isolation respond in a fashion similar to their *in situ* counterparts when stimulated by DcAMP.

   b. *Growth Hormone Releasing Factor (GRF).*   One of the ultimate tests of the usefulness of isolated cells for adenohypophysial study must be their responsiveness to the natural secretagogue, i.e., releasing factor. In one experiment we examined the ability of a highly purified GRF preparation (kindly furnished by Drs. Fawcett and McCann of the University of Texas Southwestern Medical School) to stimulate release of GH from the isolated somatotrophs. As seen in Table IX, GRF at the low dose had no effect on release, but may have increased the intracellular GH content. At the high dosage, however, GRF did stimulate GH release, but had little apparent effect on intracellular hormone content. Many more experiments will have to be carried out to see if GRF stimulates GH biosynthesis. A more detailed account of these results can be found in the recent report by Kraicer and Hymer (1974).

   On the basis of these results we feel that the isolated pituitary cell preparations provide an ideal model for study of the mechanism(s) by which secretagogues act on the somatotroph to effect secretion of growth hormone.

TABLE IX

EFFECTS OF DcAMP AND GRF ON ISOLATED SOMATOTROPHS[a]

	μg GH/ml incubation medium	Ng GH/1000 cells after incubation
DcAMP		
Control	2.40 ± 0.09 (4)[b]	150.2 ± 3.46 (4)
DcAMP	4.63 ± 0.07 (4)	122.4 ± 3.36 (4)
GRF[c]		
Control	9.48 ± 0.20 (5)	152.5 ± 3.83 (5)
GRF, 50 μl/ml	9.88 ± 0.61 (3)	166.3 ± 6.20 (3)
GRF, 100 μl/ml	13.36 ± 0.14 (3)	141.0 ± 8.61 (3)

   [a] Cells from fractions 18–32 (e.g., Fig. 10) incubated 1 hr in 1 ml Medium 199 containing 0.1% BSA, pH 7.3. Each flask contained 0.2–0.6×10^6 cells.
   [b] Number of flasks. Data represent average $\pm$ SEM.
   [c] Prepared by Fawcett and McCann.

c. *RNA Synthesis.* After incubation of pituitary glands for 2.5 hr in Medium 199 containing [³H]uridine, a majority of the labeled RNA is ribosomal in character (Hymer and Stere, 1967). In a recent experiment we dissociated pituitary tissue after identical incubation conditions and separated the cells by unit gravity sedimentation. A major portion (80%) of the radioactivity in RNA was localized in cells sedimenting in the chromophobe-enriched region of the gradient. Moreover, we determined by radioautography that 70% of the labeled cells were also localized in this same region of the gradient. These cells were also more intensely labeled than the acidophils. These results are similar to those reported by Städtler *et al.* (1970) and suggest that the chromophobes comprise the most active RNA-synthesizing cell compartment in the pituitary.

In a preliminary experiment we have found that dissociated pituitary cells incorporate [³H]uridine into RNA in linear fashion during short-term incubation; a result which provides further evidence for functionality of the dispersed cells.

TABLE X

PERCENTAGE OF ADENOHYPOPHYSIAL CELL TYPES IN CONSECUTIVE
40-ML FRACTIONS AFTER SEDIMENTATION AT UNIT GRAVITY[a]

| Fraction | A. Tissue Source: Thyroidectomized Male Rats | | | | |
	Thyrotroph	Chromophobe	Basophil	Somatotroph	Unknown
1[b]	11.7	3.8	3.1	0.9	80.5
2	17.3	5.3	2.8	0	74.6
3	11.2	4.5	1.4	0	82.9
4	11.3	9.0	1.6	0	78.1
5	41.6	15.0	5.8	0	37.6
6	54.7	10.4	6.2	0.6	28.1
7	59.9	2.8	4.6	0.6	32.1
8	53.8	8.5	5.5	0.8	31.4
Starting material	44	27	1	1	27

| Fraction | B. Tissue Source: Lactating Female Rats | | | | |
	Lactotroph	Chromophobe	Basophil	Somatotroph	Unknown
1	43.5	1.9	0.7	2.6	51.3
2	20.2	0	0.4	2.7	76.7
3	11.9	0	0.5	0.8	86.8
4	20.0	4.1	1.1	1.1	73.7
5	61.4	8.7	6.3	7.0	16.6
6	43.9	2.3	7.9	30.2	15.7
7	17.6	3.1	6.0	49.8	23.5
8	12.4	0.9	6.6	46.5	33.6
Starting material	38	9	2	22	29

[a] Cell numbers counted in A were 3299, 3651, 2933, 7855, 8092, 6205, 3257, and 1704 for fractions 1 to 8, respectively. Cell numbers counted in B were 5926, 1370, 5430, 5782, 13,228, 8019, 1228, and 1691 for fractions 1 to 8, respectively. From Hymer *et al.* (1973).

[b] Represents cells collected in 1–40 ml from the gradient; fraction 2, 41–80 ml, etc.

5. *Separation of Cells from Tissues of Physiologically Altered Animals*

The distribution of cell types within the adenohypophysis is known to reflect the physiological state of the animal. The experiments briefly described in this section are included to demonstrate the kinds of data which can be obtained when the 1G separation technique is applied to tissue from the thyroidectomized or lactating rat.

a. *Thyroidectomized Rats.* Among the more characteristic changes brought about by thyroidectomy are (1) degranulation of the somatotroph to an extent that this cell type becomes unrecognizable by light microscopy and (2) an increase in size and number of basophils identifiable as thyrotrophs. Glands from male rats which had received propylthiouracil in their drinking water for 36 days served as the source of pituitary cells. As expected, essentially none of the cells could be identified as somatotrophs (Table XA). However, a large peak of cells sedimenting 240–280 ml into the gradient (fractions 6 and 7, Table XA) were thyrotrophs. Such cells could prove useful in future studies which require a rich thyrotroph population (Hymer *et al.,* 1973).

b. *Lactating Rats.* Removal of suckling young from lactating female rats results in blockage of prolactin release from the pituitary. It is known that 12–24 hr after removal of the suckling stimulus there is an accumulation of cytoplasmic granules in the lactotrophs. The sedimentation profile of cells from such tissue indicated that a majority of the lactotrophs were found in fraction 5 (i.e., 200 ml into the gradient; Table XB), with a frequency of over 60%. In a related experiment we found cells in this same region of the gradient synthesized a protein with an electrophoretic mobility ($R_f = 0.77$) identical to that of rat prolactin (Hymer *et al.,* 1973). This result thus provides strong support for the morphological evidence of the presence of lactotrophs in this gradient region.

6. *Separation of Prolactin-Producing Cells (Mammotrophs) from the Female Rat Pituitary*

In a recent study (Hymer *et al.,* 1974) we showed that enriched populations of mammotrophs could also be routinely obtained by the velocity sedimentation technique. These cells, which were identified by (a) Herlant's tetrachrome stain or (b) an immunoperoxidase staining technique, were enriched to frequencies of 70% in peak fractions. At the electron microscopic level the separated mammotrophs were virtually indistinguishable from their *in vivo* counterparts. It is of considerable interest that the sedimentation rate of the mammotrophs was found to be dependent upon the physiological status of the donor pituitary tissue. Thus, mammotrophs from estrogen-primed rats sedimented $\sim$100 ml further into the gradient than those prepared from untreated control animals; whereas those from ovariectomized animals were recovered in the upper gradient regions. We feel that these sedimentation patterns reflect differences in size of the pituitary mammotrophs in varying physiological states.

A natural extension of this work has been the cultivation of the separated mammotrophs. Ongoing studies in our laboratory clearly indicate the marked capacity for prolactin synthesis and secretion by these cells in culture. Moreover, they also suggest that not all mammotrophs secrete the same quantity of hormone *in vitro*.

IV. Summary

It is probably fair to generalize that today's biologist has at his disposal the techniques required for the isolation of virtually all the subcellular organelles contained within a pituitary cell. With such isolated preparations one can anticipate future studies aimed at the duplication of intracellular processes involved in the synthesis, packaging, and release of hormones. Of course the major obstacle to such studies will continue to be the isolation of cell type-specific organelles; an obstacle which may be partially overcome when separated cells are used as the tissue source.

A. SECRETORY GRANULES

Highly purified preparations of granules containing either STH, PL, or hormone activity from the "basophilic" cell types can be isolated in routine fashion. Since granules possessing ACTH, TSH, FSH, and LH activity all seem to be of similar size and density, it does not appear likely that they will be further separated by methods used thus far (see Table I). However, methods which utilize potential differences in surface properties of granules in this series remain to be tested.

The provocative paper of LaBella and his colleagues (1971) gives the first real indication of the biochemical makeup of isolated STH and PL granules. This kind of study, which could not have been done without sensitive radioimmunoassay techniques, begins to provide insight as to how the hormone may be packaged within the secretion granule. With the radioimmunoassay procedure, it should also be possible to devise *in vitro* systems to test compounds which might be involved in the granule release process. While addition of TRF to isolated secretion granules possessing TSH activity has no effect on granule stability or hormone release from these particles (McCann and Porter, 1969); addition of cAMP, Ca^{2+}, prostaglandin, ATP, etc., to isolated granules remains to be tested.

B. CELL-FREE PROTEIN-SYNTHESIZING SYSTEMS

The usefulness of these systems (see pp. 155–156) in the study of the pituitary is only now being realized. With it, one can begin to evaluate the way in which physiologically active compounds may act on the hormone synthetic machinery of the pituitary cell. As another example, this system has also been used to evaluate the functionality of pituitary ribosomes in the presence of 100,000 g supernatant fraction prepared from pituitaries of animals bearing pituitary tumors (Reagen *et al.*, 1972).

C. Cell Separation

A list of requirements that isolated pituitary cells should meet if they are to serve as useful tools for adenohypophysial study was given on p. 161. Thus far our results indicate that the isolated somatotrophs satisfy these requirements. The observation that they respond to purified GRF by releasing GH into the surrounding medium is a crucial one since it shows that the receptor sites on the somatotroph membrane remain intact during cell isolation.

In future work, isolated pituitary cell preparations may also prove useful for (1) study as to how releasing factor brings about hormone secretion and/or hormone synthesis; (2) isolation of cell type-specific subcellular organelles, e.g., receptor molecules on plasma membranes, Golgi membranes, nuclei, etc.; (3) study of interaction of target hormones with the pituitary cell, e.g., site of T4 action on the thyrotroph; and (4) culture of different cell types for long-term studies.

Acknowledgments

The studies from the author's laboratory were supported in part by grants from the National Science Foundation (GB-33686), the National Institutes of Health (N01-CB-23863), and a Public Health Service research career development award (1-K04-AM-15808).

REFERENCES

Adiga, P. R., Rao, P., Hussa, R., and Winnick, T. (1966). Biosynthesis of adrenocorticotropin and protein in a cell free system from bovine anterior pituitary tissue. *Biochemistry* **5**, 3850–3855.

Adiga, P. R., Hussa, R., Robertson, M., Hohl, H., and Winnick, T. (1968a). Polysomes of bovine anterior pituitary gland and their role in hormone and protein biosynthesis. *Proc. Nat. Acad. Sci. U. S.* **60**, 606–613.

Adiga, P. R., Hussa, R., and Winnick, T. (1968b). Ribonucleoprotein particles of bovine anterior pituitary gland. Physiochemical and biosynthetic characteristics. *Biochemistry* **7**, 1808–1817.

Adiga, P. R., Murthy, P., and McKenzie, J. (1971). Stimulation by adenosine 3',5'-cyclic monophosphate of protein synthesis by adenohypophysial polyribosomes. *Biochemistry* **10**, 711–715.

Augustine, E. C., and Hymer, W. C. (1973). Characteristics of nuclear RNA polymerase from rat anterior pituitary tissue. *Endocrinology* **92**, 1790–1794.

Baker, B. L. (1970). Studies on hormone localization with emphasis on the hypophysis. *J. Histochem. Cytochem.* **18**, 1–8.

Birge, C. A., Peake, G., Mariz, I., and Daughaday, W. (1967). Radioimmunoassayable growth hormone in the rat pituitary gland: Effects of age, sex and hormonal state. *Endocrinology* **81**, 195–204.

Catchpole, H. (1948). Cell fractionation and gonadotropin assays of anterior pituitary gland. *Fed. Proc., Fed. Amer. Soc. Exp. Biol.* **7**, 19.

Costoff, A., and McShan, W. H. (1969). Isolation and biological properties of secretory granules from rat anterior pituitary glands. *J. Cell Biol.* **43**, 564–574.

DuPont, A., and Chavancy, G. (1973). Prostaglandins and cyclic AMP as mediators of thyrotropin-releasing hormone action. *Proc. Int. Congr. Endocrinol., 4th, 1972* Int. Congr. Ser. No. 256, p. 84.

Ellis, S. (1960). Pituitary proteinase I: Purification and action on growth hormone and prolactin. *J. Biol. Chem.* **235**, 1694–1699.

Grant, G., Vale, W., and Guillemin, R. (1972). Interaction of thyrotropin releasing factor with membrane receptors of pituitary cells. *Biochem. Biophys. Res. Commun.* **46**, 28–34.

Grieshaber, C., and Hymer, W. (1968). Effect of hypothalamic extracts on protein synthesis in rat anterior pituitary tissue. *Proc. Soc. Exp. Biol. Med.* **128**, 459–463.

Hartley, M., McShan, W., and Ris, H. (1960). Isolation of cytoplasmic pituitary granules with gonadotrophic activity. *J. Biophys. Biochem. Cytol.* **7**, 209–218.

Hodges, D. R., and McShan, W. H. (1970). Electrophoretic separation of hormones associated with secretory granules from rat anterior pituitary glands. *Acta Endocrinol. (Copenhagen)* **63**, 378–384.

Hymer, W. C., and Kraicer, J. (1972). Growth hormone content of somatotrophs separated from the rat adenohypophysis. *Physiologist* **15**, 178.

Hymer, W. C., and Kuff, E. L. (1964). Isolation of nuclei from mammalian tissues through the use of Triton X-100. *J. Histochem. Cytochem.* **12**, 359–363.

Hymer, W. C., and McShan, W. H. (1963). Isolation of rat pituitary granules and the study of their biochemical properties and hormonal activities. *J. Cell Biol.* **17**, 67–86.

Hymer, W. C., and Stere, A. (1967). Effect of hypothalamic extracts on RNA synthesis in rat anterior pituitary tissue. *Proc. Soc. Exp. Biol. Med.* **125**, 1143–1149.

Hymer, W. C., Glenn, L., and Kraicer, J. (1972a). Effects of dibutyryl cyclic AMP on dissociated cells from the rat adenohypophysis. *Physiol. Can.* **3**, 19.

Hymer, W. C., Kraicer, J., Bencosme, S., and Haskill, J. (1972b). Separation of somatotrophs from the rat adenohypophysis by velocity and density gradient centrifugation. *Proc. Soc. Exp. Biol. Med.* **141**, 966–973.

Hymer, W. C., Evans, W., Kraicer, J., Mastro, A., Davis, J., and Griswold, E. (1973). Enrichment of cell types from the rat adenohypophysis by sedimentation at unit gravity. *Endocrinology* **92**, 275–287.

Hymer, W. C., Snyder, J., Wilfinger, W., Swanson, N., and Davis, J. (1974). Separation of pituitary mammotrophs from the female rat by velocity sedimentation at unit gravity. *Endocrinology* **95**, 107–122.

Jones, A., Fisher, J., and Vanderlaan, W. (1965). Electrophoretic comparison of pituitary glands from male and female rats. *Endocrinology* **76**, 578–583.

Kobayashi, T., Kobayashi, T., Kigawa, T., Mizuno, M., Amenomori, Y., and Watanabe, T.

(1965). Autoradiographic studies on ³H-leucine uptake by adenohypophysial cells *in vitro*. *Endocrinol. Jap.* **12**, 47–55.

Kohler, P., Bridson, W., and Chrambach, A. (1971). Human growth hormone produced in tissue cultures: Characterization by polyacrylamide gel electrophoresis. *J. Clin. Endocrinol. Metab.* **32**, 70–76.

Kragt, C. L., and Meites, J. (1966). Separation of rat anterior pituitary hormones by polyacrylamide gel electrophoresis. *Proc. Soc. Exp. Biol. Med.* **121**, 805–808.

Kraicer, J., and Hymer, W. C. (1974). Purified somatotrophs from rat adenohypophysis: Response to secretagogues. *Endocrinology* **94**, 1525–1530.

LaBella, F., and Brown, J. (1958). Distribution of hydrolases among anterior pituitary cell fractions. *J. Biophys. Biochem. Cytol.* **4**, 833–835.

LaBella, F., and Brown, J. (1959). Cell fractionation of anterior pituitary glands from beef and pig. *J. Biophys. Biochem. Cytol.* **5**, 17–23.

LaBella, F., Krass, M., Fritz, W., Vivian, S., Shin, S., and Queen, G. (1971). Isolation of cytoplasmic granules containing growth hormone and prolactin from bovine pituitary. *Endocrinology* **89**, 1094–1102.

Labrie, F., Lemaire, S., and Courte, C. (1971a). Adenosine 3',5'-monophosphate-dependent protein kinase from bovine anterior pituitary gland. I. Properties. *J. Biol. Chem.* **246**, 7293–7302.

Labrie, F., Lemaire, S., Poirier, G., Pelletier, G., and Boucher, R. (1971b). Adenohypophysial secretory granules. Their phosphorylation and association with protein kinase. *J. Biol. Chem.* **246**, 7311–7317.

Labrie, F., Barden, N., Poirier, G., and DeLean, A. (1972). Binding of thyrotropin releasing hormone to plasma membranes of bovine anterior pituitary gland. *Proc. Nat. Acad. Sci. U. S.* **69**, 283–287.

Leavitt, W., Kimmel, G., and Friend, J. (1973). Uptake of ³H-estradiol by anterior pituitary cell suspensions. *Proc. Int. Congr. Endocrinol., 4th, 1972* Int. Congr. Ser. No. 256, p. 119.

Lemaire, S., Pelletier, G., and Labrie, F. (1971). Adenosine 3',5'-monophosphate-dependent protein kinase from bovine anterior pituitary gland. II. Subcellular distribution. *J. Biol. Chem.* **246**, 7303–7310.

McCann, S. M., and Porter, J. C. (1969). Hypothalamic pituitary stimulating and inhibiting hormones. *Physiol. Rev.* **49**, 24–284.

McDonald, J., Ellis, S., and Reilly, T. (1966). Properties of dipeptidyl arylamidase I of the pituitary. *J. Biol. Chem.* **241**, 1494–1501.

McDonald, J., Reilly, T., Zeitman, B., and Wellis, S. (1968a). Dipeptidyl arylamidase II of the pituitary. *J. Biol. Chem.* **243**, 2028–2037.

McDonald, J., Leibach, F., Grindeland, R., and Ellis, S. (1968b). Purification of dipeptidyl aminopeptidase II (dipeptidyl arylamidase II) of the anterior pituitary gland. *J. Biol. Chem.* **243**, 4143–4150.

McDonald, J., Callahan, P., Ellis, S., and Smith, R. (1971). Polypeptide degradation by dipeptidyl aminopeptidase I (cathepsin C) and related peptidases. *In* "Tissue Proteinases" (A. J. Barrett and J. T. Dingle, eds.), pp. 69–107. North-Holland Publ., Amsterdam.

McKeel, D., and Jarett, L. (1970). Distribution and characterization of adenylcyclase in subcellular fractions of pig adenohypophysis. *J. Cell Biol.* **47**, 135.

MacLeod, R., and Lehmeyer, J. (1970). Release of pituitary growth hormone by prostaglandins and dibutyryl adenosine cyclic 3',5'-nonophosphate in the absence of protein synthesis. *Proc. Nat. Acad. Sci. U. S.* **67**, 1172–1179.

MacLeod, R., and Lehmeyer, J. (1972). Regulation of the synthesis and release of prolactin. *Lactogenic Horm., Ciba Found. Symp.* pp. 53–82.

MacLeod, R., Abad, A., and Eidson, L. (1969). *In vivo* effect of sex hormones on *in vitro* synthesis of prolactin and growth hormone in normal and tumor bearing rats. *Endocrinology* **84**, 1475–1483.

McShan, W., and Meyer, R. (1949). Gonadotropic activity of granules isolated from rat pituitary glands. *Proc. Soc. Exp. Biol. Med.* **71**, 407–410.

McShan, W., and Meyer, R. (1952). Gonadotropic activity of granule fractions obtained from anterior pituitary glands of castrate rats. *Endocrinology* **50**, 294–303.

McShan, W., Rozich, R., and Meyer, R. (1953). Biochemical properties of fractions obtained from rat anterior pituitary glands by differential centrifugation. *Endocrinology* **52**, 215–222.

Malamed, S., Portanova, R., and Sayers, G. (1971). Fine structure of trypsin-dissociated cells of the rat anterior pituitary gland. *Proc. Soc. Exp. Biol. Med.* **138**, 920–926.

Mastro, A., and Hymer, W. C. (1973). Characteristics of nuclear DNA polymerase from rat anterior pituitary tissue. *J. Endocrinol.* **59**, 107–119.

Meyer, R., and Clifton, K. (1956). Effect of diethylstilbestrol on the quantity and intracellular distribution of pituitary proteinase activity. *Arch. Biochem. Biophys.* **62**, 198–209.

Miller, R., and Phillips, R. (1969). Separation of cells by velocity sedimentation. *J. Cell. Physiol.* **73**, 191–202.

Neville, D. (1960). The isolation of a cell membrane fraction from rat liver. *J. Biophys. Biochem. Cytol.* **8**, 413–422.

Pastan, I., Lefkowitz, R., and Roth, J. (1970). Radioreceptor assay of adrenocorticotropic hormone: New approach to assay of polypeptide hormones in plasma. *Science* **170**, 633–635.

Perdue, J. F., and McShan, W. H. (1962). Isolation and biochemical study of secretory granules from rat pituitary glands. *J. Cell Biol.* **15**, 159–172.

Perdue, J. F., and McShan, W. H. (1966). Association of adrenocorticotropic hormone activity with small secretory granules from rat anterior pituitary glands. *Endocrinology* **78**, 406–408.

Peterson, E., and Evans, W. (1967). Separation of bone marrow cells by sedimentation at unit gravity. *Nature (London)* **214**, 824–825.

Pickup, J., and Hope, D. (1971). Protease and ribonuclease activity in bovine pituitary lobes. *Biochem. J.* **123**, 153–162.

Poirier, G., Barden, N., Labrie, F., Borgeat, P., and DeLean, A. (1973). Purification and some properties of adenyl cyclase and receptor for TRH from anterior pituitary gland. *Proc. Int. Congr. Endocrinol., 4th, 1972* Int. Congr. Ser. No. 256, p. 85.

Pooley, A. (1971). Ultrastructure and size of rat anterior pituitary secretory granules. *Endocrinology* **88**, 400–411.

Portanova, R., Smith, D., and Sayers, G. (1970). A trypsin technique for the preparation of isolated rat anterior pituitary cells. *Proc. Soc. Exp. Biol. Med.* **133**, 573–576.

Racadot, J., Olivier, L., Porcile, E., and Droz, B. (1965). Appareil de golgi et origine des grains de secretion dans les cellules adenohypophysisaires chez le rat. Etude radioautographique en microscopie électronique après injection de leucine tritiée. *C. R. Acad. Sci.* **261**, 2972–2974.

Rao, P., Robertson, M., Winnick, M., and Winnick, T. (1967). Biosynthesis of prolactin and growth hormone in slices of bovine anterior pituitary tissue. *Endocrinology* **80**, 1111–1119.

Reagen, C., Sproul, G., and MacLeod, R. (1972). Cell free protein synthesis by rat anterior pituitary. *Fed. Proc., Fed. Amer. Soc. Exp. Biol.* **31**, 275.

Samli, M., Lai, M., and Barnett, C. (1971). Protein synthesis in the rat anterior pituitary. Interrelationship of amino acid uptake to protein synthesis. *Endocrinology* **88**, 540–547.

Schally, A. V., Arimura, A., Kastin, A. J., Matsuo, H., Baba, Y., Redding, T. W., Nair, R. M. G., Debeljuk, L., and White, W. F. (1971). Gonadotropin-releasing hormone:

One polypeptide regulates secretion of luteinizing and follicle-stimulating hormones. *Science* **173**, 1036–1037.

Shortman, K. (1968). The separation of different cell classes from lymphoid organs. *Aust. J. Exp. Biol. Med. Sci.* **46**, 375–396.

Siperstein, E. (1963). Identification of the adrenocorticotrophin-producing cells in the rat hypophysis by autoradiography. *J. Cell Biol.* **17**, 521–545.

Smith, R. E., and Farquhar, M. G. (1970). Modulation in nucleoside diphosphatase activity of mammotropic cells of the rat adenohypophysis during secretion. *J. Histochem. Cytochem.* **18**, 237–250.

Snyder, G., and Hymer, W. (1971). Divalent cations and growth hormone synthesis. *Amer. Zool.* **11**, 656.

Städtler, F., Stöcker, E., Dohm, G., and Tietze, H. (1970). Autoradiographic studies on nuclear DNA and RNA synthesis in the adenohypophysis of castrated rats. *Acta Endocrinol. (Copenhagen)* **64**, 324–328.

Tesar, J. T., Koenig, H., and Hughes, C. (1969). Hormone storage granules in the beef anterior pituitary. *J. Cell Biol.* **40**, 225–235.

Vale, W., Grant, G., Amoss, M., Blackwell, R., and Guillemin, R. (1972). Culture of enzymatically dispersed anterior pituitary cells: Functional validation of a method. *Endocrinology* **91**, 562–572.

Vanha-Perttula, T. (1969). Aminoacyl and dipeptidyl arylamidases (aminopeptidases) of the pituitary gland as related to function. *Endocrinology* **85**, 1062–1069.

Wilber, J. F., and Seibel, M. J. (1973). TRH interactions with a specific anterior pituitary membrane receptor. *Proc. Int. Congr. Endocrinol., 4th, 1972* Int. Congr. Ser. No. 256, p. 85.

Yamamoto, K., Taylor, L., and Cole, F. (1970). Synthesis and release of GH and prolactin from the rat anterior pituitary *in vitro* as functions of age and sex. *Endocrinology* **87**, 21–26.

Zor, V., Kaneko, T., Schneider, H., McCann, S., and Field, J. (1970). Further studies of stimulation of anterior pituitary cyclic adenosine 3',5'-monophosphate formation by hypothalamic extract and prostaglandins. *J. Biol. Chem.* **245**, 2883–2888.

ULTRASTRUCTURE OF ANTERIOR PITUITARY
CELLS IN CULTURE

A. Tixier-Vidal

GROUPE DE NEUROENDOCRINOLOGIE CELLULAIRE, LABORATOIRE DE PHYSIOLOGIE CELLULAIRE
COLLÈGE DE FRANCE, PARIS, FRANCE

I. Introduction

In studying the biology of anterior pituitary cells, one has to overcome serious difficulties pertaining to the original features of the adenohypophysis: (1) its extreme cellular heterogeneity already mentioned in several chapters of this book, (2) its anatomical situation at one of the less accessible regions of the head, and (3) its complex interactions, generally through the circulatory system, with neuroendocrine areas of the brain and numerous target endocrine glands.

Culture methods offer an elegant approach to the simplification of this system and have been widely used for a long time. Before considering the results obtained in this field it is necessary to define what should be considered as "culture" of the pituitary gland. In this respect we will refer to Willmer (1965) who distinguishes among "tissue culture" proper, in which small fragments of tissue are explanted into a suitable medium and encouraged to grow in isolation to form colonies and perhaps to continue some of their normal functions, "cell culture" in which the cells of a tissue or even individual cells are made to grow in much the same way as bacteria are grown . . . , and "organ culture" where growth is only of minor interest, and embryological development and the maintenance of normal physiological functions are the chief aim and object. We will therefore not take into account the other *in vitro* systems (see chapter by Farquhar *et al.*).

In their review of 1965, Gaillard and Schaberg gave an extensive report of the results obtained with pituitary cultures. On the basis of this review one may distinguish, schematically, two periods in the evolution of research in this field. During the first period, starting in 1935, the authors focused their attention on the maintenance of the morphological and functional differentiation of the pituitary cells *in vitro*. Although the conditions of cultures and the methods for the morphological and functional analysis were widely different and often imprecise, some general features appeared from these studies: (1) the cultivated anterior pituitary maintained several of its hormonal activities in culture, but only for a limited time, and (2) these activities were still present after the light microscopic characteristics of morphological differentiation had disappeared.

The second period was concomitant with the development of research on hypothalamic factors and their role in the regulation of secretion of pituitary hormones. This period started in 1955 with the work of Guillemin and Rosenberg (1955) showing that the production of ACTH by 2–4-week explants disappeared from the fourth day in culture and could be reinitiated by introduction of fragments of hypothalamus and median eminence into the culture tubes. Further demonstration of the importance of tissue culture methods in neuroendocrinology was brought forth by work concerning the hypothalamic control of prolactin; prolactin production increased in culture (Meites *et al.*, 1961, Pasteels, 1961), and the addition of median eminence tissue to the cultures inhibited prolactin production and release (Pasteels, 1961).

In recent years culture methods have been widely used to study the mechanism of action of hypothalamic releasing hormones on pituitary cells. This development has been greatly facilitated by progress in the purification of pituitary and hypothalamic hormones and by significant improvements in assay methods for pituitary hormones, especially radioimmunoassay.

In contrast, studies on the ultrastructural features of cultured pituitary tissue or pituitary cells have been far less numerous, with the main reports on this subject appearing between 1961–1963 (Pasteels, 1961, 1963; Petrovic, 1961, 1963). Because of the importance of ultrastructural data for "structure–function" studies, a reexamination of this question is necessary in the light of more recent findings.

In this chapter we examine, first, recent biological data concerning the *in vitro* secretion of anterior pituitary hormones and modifications in their secretion produced by introduction of hypothalamic hormones. The second part is devoted to ultrastructural data.

II. The Secretion of Pituitary Hormones by the Anterior Pituitary in Culture

This review concerns papers published since 1961. For the publications which appeared before 1961, the reader is referred to the review of Gaillard and Schaberg (1965) already mentioned in the introduction. The results are summarized in Table I for each of the following hormones: melanophorotropin (MSH, Table IA), adrenocorticotropin (ACTH, Table IB), somatotropin or growth hormone (STH, Table IC), prolactin (Pl, Table ID), gonadotropins (Table IE) [follicle stimulating hormone (FSH) and luteinizing hormone (LH)], and thyrotropin (TSH) (Table IF). Experiments in which the secretory activity was estimated by *in vivo* transplantation of the cultured material do not appear in these tables. Although such experiments demonstrate that the tissues *in vitro* retain their secretory potential, they do not allow its exact estimation.

Several general conclusions may be drawn from the examination of Table I.

A. SECRETION OF PITUITARY HORMONES IN NORMAL MEDIA

1. *Organ Culture, Tissue Culture, and Primary Cell Culture*

First of all, although release of each anterior pituitary hormone continues in culture, important differences for the various hormones appear with increasing time. Numerous experiments have shown that the release of prolactin increases during the first hours, days, and weeks; comparison of the total amount released *in vitro* and the initial content of the tissue demonstrates conclusively that there is a continuous prolactin synthesis *in vitro*. Autonomy of prolactin secretion is well established in several mammalian species, as well as in some reptiles and amphibians; in birds, the autonomy is reduced or absent, depending on the species. The secretion of MSH, like that of prolactin persists or increases in culture, but this has been established with only few experiments and in few species (fetal human pituitary and two teleostean species).

For the other pituitary hormones, secretory ability decreases with time in culture until it reaches a low but sustained level. This has been shown for: ACTH, STH, FSH, LH, and TSH. With progress in assay methods, a weak autonomy of secretion has been shown to exist for several pituitary hormones such as STH in fetal human and monkey pituitary, LH and FSH in fetal human and adult rat pituitaries. Secretion of TSH seems to stop *in vitro* sooner than that of the gonado-

TABLE I

Summary of Data Concerning the Hormone Medium Content of Anterior Pituitary Culture

Reference	Species	Culture method	Maximum duration of culture	Assay method	Medium hormonal content in control condition	Effect of hormones
A. MSH						
Hermanus et al., 1964	Human fetus	Tissue culture, hanging drop	8 weeks	Frog skin in vitro	Increasing secretion from 24 days to the end of experiment	—
Baker, 1964, 1965, 1967	Fishes: trout, eel	Organ culture	28 days	Anolis skin in vitro	Continuous secretion	—
B. ACTH						
Stark et al., 1965a	Human fetus	Tissue culture	20–25 days	Combined cultures with adrenal tissue; measure of corticosteroid by chromatography	Continuous production of ACTH with a fall from 6th to 10th day	—
Stark et al., 1965b	Adult rat	Monolayer of trypsinized hypophysial cells	3–4 weeks	Effect of hypophyseal culture medium on secretion of corticosteroids by adrenal tissue cultures	Continuous production of ACTH to a decreasing rate	—
Siler et al., 1972	Human fetus	Tissue culture	11 months	Radioimmunoassay	Decreasing release during the first 20 days to basal or undetectable level	—
Buonassissi et al., 1962 Yasumura et al., 1966	Rat pituitary corticotropic tumor	Cell cultures established after alternate passage from host to culture	Continuous clonal cell lines; several years	Effect of hypophyseal culture medium on secretion of corticosteroid by adrenal cell cultures	Persistent production to a high level	—
C. STH						
Reusser et al., 1962	Human fetus or adult	Cell and tissue culture	1 month	Immunochemical assay (Ouchterlony technique)	No STH	
Pasteels et al., 1963 Pasteels, 1963	Human fetus	Tissue culture (hanging drop)	33 days	Immunoassay (complement deviation)	Decreasing release during the first 10 days to a low constant level	Hypothalamic extract on the 17th day increases STH release

Reference	Material	Culture type	Duration	Assay	Results	Comments
Pasteels, 1969	Human fetus	Long-term organ cultures	30 months	Injection of culture medium to hypophysectomized rats	Weight increase with medium from 18-month cultures	
Gailani et al., 1970	Human fetus	Tissue culture on collagen rocker tray	150 days	Radioimmunoassay	Higher level of STH release with older fetus; reduction in rate of STH production after 1 to 3 weeks; in some cases stable level for 2 to 3 months	Fetal hypothalamic extract increases STH release
Siler et al., 1972	Human fetus	Tissue culture	11 months	Radioimmunoassay	Decreasing release during the first 20 days to basal or undetectable level	
Kohler et al., 1969a	Human normal or adenomas	Tissue culture	Up to 10 months	Radioimmunoassay	*Normal pituitary cultures:* rapid decrease over the first 30 days low level up to 125 days *Adenomas from acromegalic* higher level and for up to one year	
Bridson and Kohler, 1970	Human pituitary adenomas from acromegalic patients	Tissue culture	6–12 weeks or 37 weeks	Radioimmunoassay	Same results as Kohler et al., 1969a	Cortisol (10^{-6} to 10^{-7} M) increases STH content of the medium
Peillon et al., 1972	Human normal and acromegalic adenomas	Organ culture	1 month	Radioimmunoassay	Decreasing release in normal pituitary culture higher level in adenomas cultures	
Batzdorf et al., 1971	Human acromegalic or hypopituitary patients	Suspension culture of trypsinized cells; Rose chambers and pyrex flask	5 weeks	Immunoassay	STH content maximum between the 7th to 14th day in culture undetectable after 5 weeks	
Kohler et al., 1968	Monkey (macaca)	Tissue culture	1 to 6 months	Radioimmunoassay	Slow decline to undetectable level at 4 to 6 months	Cortisol (10^{-6} to 10^{-7} M) increases STH content of medium (172 to 190%)

(Continued)

TABLE I (*Continued*)

Reference	Species	Culture method	Maximum duration of culture	Assay method	Medium hormonal content in control condition	Effect of hormones
Channing *et al.*, 1970	Adult monkeys	Tissue culture	15 to 36 days	Radioimmunoassay	Decreasing release over the first 10 days to a low constant level	
Yasumura *et al.*, 1966 Tashjian *et al.*, 1968	Rat pituitary tumor Rat pituitary tumor	Monolayer cell cultures established by alternate culture and animal passage	Continuous clonal cell line; several years	Effect on rat body weight immunoassay (complement fixation)	Continuous production to high level	
Bancroft *et al.*, 1969	Rat pituitary tumor	Monolayer cell cultures established by alternate culture and animal passage; GH3 strain	Continuous clonal cell line; several years	Immunoassay complement fixation	Continuous production to high level	Hydrocortisone increases STH production
Tashjian *et al.*, 1970	Rat pituitary tumor	Monolayer cell cultures established by alternate culture and animal passage; GH3–GH1	Continuous clonal cell line; several years	Immunoassay complement fixation	Continuous production to high level	Hypothalamic extract and various tissue extracts inhibits STH production
Tashjian *et al.*, 1971	Rat pituitary tumor	Monolayer cell cultures established by alternate culture and animal passage	Continuous clonal cell line; several years	Immunoassay complement fixation	Continuous production to high level	Synthetic TRF inhibits STH production
Kohler *et al.*, 1969	Rat pituitary tumor	Monolayer cell culture; clonal line same origin; GH1 strain	Continuous clonal cell line; several years	Radioimmunoassay	Continuous production of STH	Cortisol (5×10^{-8} to 5×10^{-6} *M*) increases STH production
			D. Prolactin (Pl)			
Pasteels, 1961, 1963	Man and rat	Tissue culture (hanging drop)	6 weeks	Pigeon crop sac assay	Increased secretion with time in culture	Hypothalamic fragments or extracts inhibit prolactin secretion; oxytocin without effect
Pasteels, 1969, 1972a	Human fetus	Long-term organ cultures	30 months	Pigeon crop sac assay	Evidence of prolactin secretion after several months in culture	

Reference	Species	Culture	Duration	Assay	Observation	Comment
Siler et al., 1972	Human fetus	Tissue culture	10 months	Radioimmunoassay LH–FSH	Rapid decrease over the first 20 days followed by a low and sporadic release	
Franchimont and Pasteels, 1972	Human fetus	Long-term "organ culture"	4 or 9 months	Radioimmunoassay LH, FSH, LHα, and LHβ	Very rapid decrease of FSH, LH and LHβ release; maintenance of LHα release	
Kobayashi et al., 1961, 1963, 1971	Rat	Tissue culture and cell culture	3 weeks	Injection of culture media to immature mouse (uterus weight)	After 1 week disappearance of the gonadotropin release	Crude hypothalamic extract reinitiates the release of gonadotropin
Mittler et al., 1970	♂ rat	Organ culture	5 days	Radioimmunoassay LH, FSH	LH and FSH into the medium and into the tissues	Purified LRF depletes LH and FSH tissue content and increases LH and FSH medium content
Tixier-Vidal, et al., 1970	♂ rat	Organ culture	2 weeks	Radioimmunoassay LH	Decrease of LH medium content over the first week, then sustained low level; dramatic decrease of the tissue content	Purified LRF Over the first days depletes the tissue content and increases LH medium content After the first week increases LH medium content without depletion of the tissue content
Redding et al., 1972	♀ rat	"Tissue culture" ?	5 days	Radioimmunoassay LH, FSH	LH and FSH into the medium and the tissues	Purified and synthetic LRF stimulates release and total content of both LH and FSH
Mittler, 1972	♂ rat	Organ culture	5 days	Radioimmunoassay LH–FSH	LH and FSH release	Testosterone or dihydro-testosterone (30 ng/ml) elevate the FSH/LH ratio in medium
Vale et al., 1972a	♂ rat	Cell culture primary culture of enzymatically dispersed anterior pituitary cells	21 days	Radioimmunoassay LH, FSH	Decline in LH and FSH cellular content	Synthetic LRF stimulates the release of LH more than that of FSH; decline of the response to LRF with time in culture

(Continued)

TABLE I (Continued)

Reference	Species	Culture method	Maximum duration of culture	Assay method	Medium hormonal content in control condition	Effect of hormones
Tixier-Vidal et al., 1973	♂ rat	Cell culture primary cultures of enzymatically dispersed anterior pituitary cells	2 months	Radioimmunoassay LH, FSH	Rapid decrease over the first week then irregular low level up to 40 days; increase in culture of the FSH:LH ratio	Chronic administration of synthetic LRF (2 ng/ml); increases the release of both LH and FSH; decrease of the magnitude of stimulation with time in culture
Steinberger and Chowdhury, 1971 Steinberger et al., 1973a,b	♂ rat	Cell culture of enzymatically dispersed cells	Nonpreeised	Radioimmunoassay LH, FSH	Separate release of LH and of FSH in 45% of the clones LH and FSH reached undetectable level after 16 days	Stimulation with acidic rat hypothalamic extract increases both LH and FSH release
Kohler et al., 1969a	Human normal or adenomas	Tissue culture	10 months	Radioimmunoassay	Rapid decrease over the first 30 days; undetectable after 75 days	—
Gailani et al., 1970	Human fetus	Cell culture	5 months	Radioimmunoassay	Rapid decrease over the first 20 days	—
Siler et al., 1972	Human fetus	Tissue culture	10 months	Radioimmunoassay	TSH release in the first 28 days in culture	

F. TSH

Reference	Species	Culture method	Maximum duration of culture	Assay method	Medium hormonal content in control condition	Effect of hormones
Sinha and Meites, 1966	♂ rat	Organ culture synthetic media	9 days	McKenzie method	Dramatic fall of TSH medium content after the first 3 days	Hypothalamic extract increases the TSH release
Vale et al., 1972a	♂ rat	Cell culture primary culture of enzymatically dispersed anterior pituitary cells	21 days	McKenzie method	Decrease release with time in culture; TSH undetectable in the media at 21 days	Synthetic TRF stimulates TSH release; decline of the response to TRF with time in culture; TRF ineffective at 21 days

tropins. Secretion of the LHα subunit is considerably higher than that of LH and FSH in long-term organ culture of human fetal pituitaries. This is a very important finding because of the similarity of the α subunit of the three glycoprotein hormones, LH, FSH, and TSH. Does this release reflect a physiological or pathological phenomenon, and does it reflect an *in vitro* hormonal synthesis? Morphological study of cultured tissue brings some evidence in favor of its being a physiological phenomenon (see Section II of this chapter). Moreover, the incorporation of radioactive tracer into newly synthesized hormone has been shown for *STH* in tissue culture (Kohler *et al.,* 1971) and organ culture (Peillon *et al.,* 1972) of normal and adenomatous human pituitaries; for STH and prolactin in cell cultures of rat pituitary (Vale *et al.,* 1972a,b); and for LH in organ cultures of rat pituitary (B. Kerdelhué, M. Jutisz, and A. Tixier-Vidal, unpublished observation). Nothing is known about the half-life of the RNA messenger of the pituitary hormones.

Depending on the authors, the experimental conditions, and the hormones, secretory autonomy lasts from 1 week to 4 months. The empiricism of culture methods does not allow precise interpretation of this cessation. The influence of culture conditions seems nevertheless predominant. The organ culture method is less favorable than the tissue or cell culture methods, which allow keeping the cells in a healthy state for a longer time. The long-term organ cultures described by Pasteels (1969) are the only exceptions, but here one may question whether one deals with true organ culture, since the author describes an outgrowth of the explants. Results concerning FSH and LH release do not differ whether one uses tissue cultures or cell cultures. However, Kobayashi *et al.* (1961, 1963), using a low sensitivity assay, pointed out that gonadotropic secretion disappeared sooner in tissue culture than in cell culture. Composition of the culture medium also seems important. As shown by Gala (1971) and Gala and Kuo (1972), rat pituitary organ cultures on synthetic media undergo progressive necrosis and lose their ability to secrete prolactin between 21 and 33 days *in vitro*. In general for long-term cultures synthetic media are supplemented with biological fluids, mainly fetal or adult mammalian sera. However, the necessary presence of these sera introduces several unknown parameters in the culture conditions. One may presume that the establishment of specific synthetic media for the secretion of each of the pituitary hormones would provide more rigor in culture experiments.

2. *Continuous Cell Lines*

In contrast with these features of hormonal secretion in organ, tissue, or cell cultures, continuous secretion of several pituitary hormones has been obtained with pituitary clonal cell lines. These serial strains have been obtained by Sato and his colleagues from mammotropic rat pituitary tumors by alternate culture and animal passages. One clone producing ACTH and several clones producing both STH and prolactin have thus been isolated and are still hormone secreting

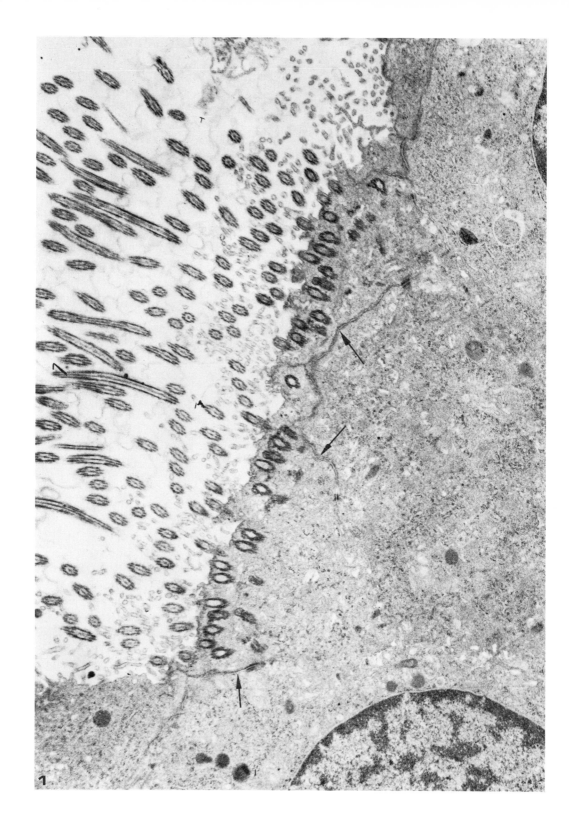

after several years. These cultures differ in some ways from tissue or cell cultures of normal pituitary. First of all they are continuously dividing (generation time: 20 to 30 hr), while the latter have a low mitotic ability, except for their fibroblastic components. They are heteroploid (Sonnenshein *et al.*, 1970). The hormonal content of the cell is very low and is rapidly renewed: 15 min for STH (Bancroft *et al.*, 1969) and 90 to 120 min for prolactin (Tashjian *et al.*, 1970). These values are much lower than those obtained from *in vivo* data: for prolactin 91 hr in male rat, 21 hr in estrogenized female rat, and 10 hr in grafted rat pituitaries (Nicoll and Swearingen, 1971). This ability to secrete ACTH and STH continuously may be related to the tumoral nature of the donor tissues. Indeed organ cultures and tissue cultures of human pituitary adenomas produce STH at a higher level and for a longer time than culture of normal human pituitary.

Finally, to our knowledge, no continuous cell line able to secrete any pituitary glycoprotein hormone has been isolated since the work of Thompson *et al.* (1959). These authors started from tissue cultures of human adult and fetal pituitaries. After serial propagation of heterogeneous cell cultures, they selected two cell lines which were propagated in a suspension system and which produced both FSH and LH. This work needs confirmation. More recently, Steinberger *et al.* (1971, 1973b) obtained in secondary culture of normal rat pituitary cells, small clones secreting either FSH alone, LH alone, or both FSH and LH. However, no precise information was given about their survival time.

B. Influence of Hormones and Drugs on Hormonal Secretion by the Anterior Pituitary Gland in Culture

Examination of Table I shows that cultured pituitaries retain the ability to respond to factors which are known to regulate the secretion of several pituitary hormones *in vivo*. In this respect, culture methods offer a valuable tool to analyze the action of these substances at the cell level. We will mainly discuss the effect of crude hypothalamic extracts or of purified and synthetic hypothalamic hormones, in view of their major importance in the regulation of the pituitary hormones.

The discrepancy between the release of prolactin and the other pituitary hormones in culture was very early attributed to differences in the nature of hypothalamic control of the hormones (see reviews by Pasteels, 1963, 1969; Meites *et al.*, 1963; Nicoll *et al.*, 1970a). Hypothalamic extract or fragments introduced into the medium inhibits prolactin release from mammalian pituitary cultures and demonstrates the existence of a prolactin inhibiting factor in that species. These

---

Fig. 1. Organ culture of rat anterior pituitary. Transverse section of an explant showing the layer of cells at the surface. The cells which form this "membrane" are joined by tight junctions (arrows) and display numerous microvilli and cilia. They do not have the features of glandular cells. ×12,500.

results were obtained in organ culture as well as in tissue culture and are in perfect agreement with *in vivo* data. On the other hand, with a clonal strain of rat pituitary cells, hypothalamic extracts stimulate prolactin production, but various other tissue extracts exert the same effect. In similar clonal strains of rat pituitary cells, a stimulation of prolactin production was also observed with synthetic thyrotropin releasing factor (TRF). This finding is in agreement with *in vivo* results obtained with TRF in man (Bowers *et al.*, 1971; Jacobs *et al.*, 1971), in cow (Convey *et al.*, 1973), in sheep (Kann *et al.*, 1973), and more recently in rats (Meites, 1973).

A stimulatory effect of hypothalamic extract on the secretion of the other anterior pituitary hormones has been observed frequently in culture. This was obtained in organ culture as well as in tissue and cell culture, with crude hypothalamic extracts for STH and TSH and later with purified or synthetic hypothalamic hormones (LRF, TRF) for FSH, LH, and TSH. In short-term cultures the major effect of LRF was hormonal release from a previous tissue store and an

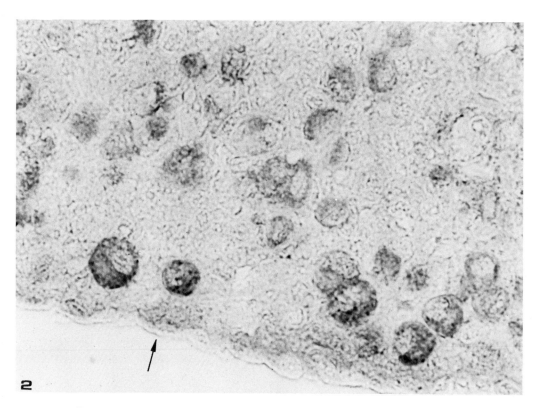

FIG. 2. Fourteen-day organ culture of rat anterior pituitary. Section treated for the immunocytochemical localization of LH. The gonadotropic immunoreactive LH cells are dark. Notice that the limiting cells are all negative (arrow). ×750.

increase of the medium + tissue content. This last finding suggests a stimulation of hormonal synthesis which was also confirmed by an increased incorporation of labelled glucosamine into LH (Redding *et al.,* 1972). Recently, both a purified and a synthetic hypothalamic "somatotropin-release inhibiting factor" (SRIF) have been shown to inhibit the release and the synthesis of STH from 4-day monolayers of dispersed rat anterior pituitary cells (Vale *et al.,* 1972b).

In contrast, very few experiments deal with the ability of a purified or synthetic hypothalamic factor to maintain secretion of the corresponding pituitary hormone in long-term cultures. In 2-week organ culture the repeated introduction of LRF into the medium induces a higher level of LH release but does not prevent its decrease with time (Tixier-Vidal *et al.,* 1970). An analogous result was observed for both LH and FSH in cell cultures maintained for 2 months on an LRF-enriched medium (Tixier-Vidal *et al.,* 1973). Similarly, the secretory response of anterior pituitary cell culture to synthetic LRF and TRF sharply decreases between 4 and 21 days in culture, together with the tissue LH, FSH, and TSH content (Vale *et al.,* 1972a). Such results suggest that unknown factors in addition to the specific hypothalamic factor are required in the medium for the continuous production of a given hormone.

III. Ultrastructural Data

Although hormonal secretion appears to follow a comparable pattern in organ culture, tissue culture, and cell culture, one might expect that from the morphological point of view the situation would be different.

A. ORGAN CULTURES

The ultrastructure of pituitary explants maintained in organ culture for several weeks was first described by Petrovic (1959, 1961, 1963) in the guinea pig, the rat, and the hamster, using semisynthetic medium with roller tubes or agar semisolid medium and pituitaries. Similar studies were reported later for several vertebrate classes: birds [duck (Tixier-Vidal and Gourdji, 1965, 1972; Tixier-Vidal and Picart, 1967), pigeon and quail (Tixier-Vidal and Gourdji, 1972)], amphibian [rana (Doerr-Schott, 1967)], man [human fetus (Pasteels, 1969, 1972a), and pituitary adenomas (Peillon *et al.,* 1972)]. Effects of estrone, thyroxine (Cohere *et al.,* 1964), ergocornine (Ectors *et al.,* 1972), and of purified LRF (Tixier-Vidal, 1972a) have also been described.

1. *Tissue Organization of the Explant*

As usually observed in organ cultures, the pituitary explants are rapidly surrounded by a unicellular "membrane" which may be related to a healing process. In some cases, this "membrane" is restricted to the superior face of the explant in

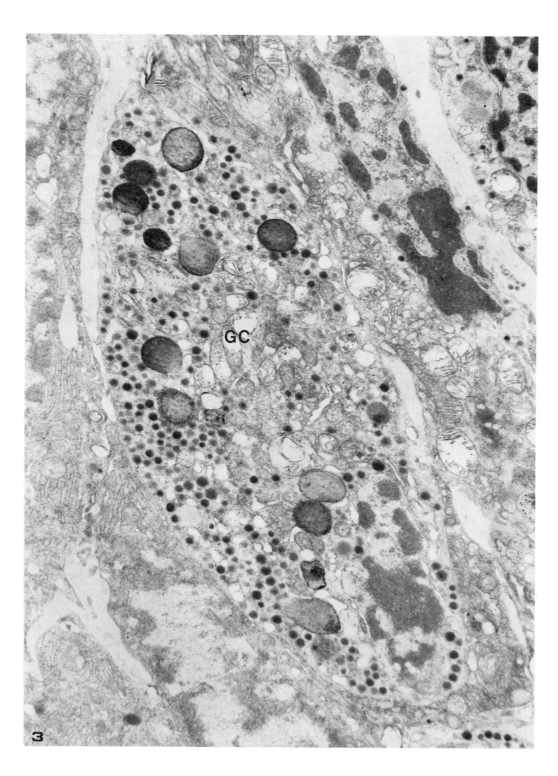

contact with the atmosphere. The exact nature of the cells which form the "membrane" is not well established; they could be glandular (Petrovic, 1961) or "fibroblastic" (Doerr-Schott, 1967). In our own experiments there was no evidence of secretory granules in these limiting cells, but there were microvilli and in some cases numerous typical cilia (Fig. 1). Such features do not suggest a fibroblastic origin of the cells but would be consistent with possible migration of nonsecreting pituitary elements such as follicular cells or stellate cells which represent a non-negligible fraction of the pituitary cell population. The formation of a unicellular membrane can coexist with an outgrowth of cells in the case of mixed "organotypic–histiotypic" culture [Petrovic, 1961 (roller tube); Pasteels, 1963 (hanging drop)].

Inside the explant, the lobular structure of the pituitary tissue remains unchanged, but the interspace between lobules becomes larger and is occupied by abundant collagen fibrils. A basement membrane lines the glandular cells as in the intact pituitary tissue. In other lacunar spaces which are of different origin, the basement membrane is absent. Some authors (Peillon *et al.*, 1972) have described a central necrosis of the explant, but according to the available data and to our own experience, this is not a constant feature of organ cultures.

Finally, a reduction in size and weight of the explants has been consistently observed. Mitoses were rarely seen [Pasteels, 1969, 1972a (human fetal pituitaries); Pawlikowski and Kunert-Radek, 1973 (rat)].

2. *The Glandular Cells*

The glandular cells undergo a reduction in size and contain fewer secretory granules. Besides these common features, important differences exist between the diverse cell types which may be related to differences in the secretion of the various pituitary hormones.

Some cell types display ultrastructural features characteristic of intense secretory activity, that is, important development of the ergastoplasm (ER), enlargement of the Golgi zone and paucity of secretory granules. This is particularly true for the "epsilon" cells identified as *prolactin cells* by Pasteels (1963, 1972b) in rat and human pituitary. These cells are recognized by their large, irregularly shaped secretory granules. The ultrastructural modifications of prolactin cells appear shortly after explantation, concurrent with an increase in prolactin secretion. Eighteen hours after explantation, Pasteels (1972b) described numerous small secretory granules which coexist with an older population of large secretory

FIG. 3. Organ culture of rat anterior pituitary maintained on control medium for 14 days. Section treated for the localization of acid phosphatase activity. A gonadotropic cell (GC) contains one class of round secretory granules and large rounded dense bodies. A positive reaction is seen on the large dense bodies. ×18,000.

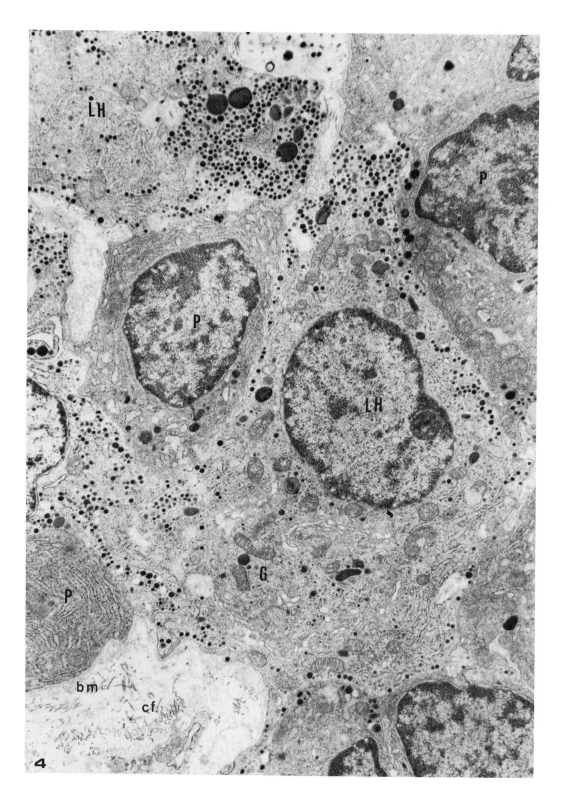

granules. In older cultures, the explants contain a heterogeneous population of prolactin cells. They show either a well-developed ergastoplasm and Golgi zone with few secretory granules or many secretory granules. The differences result presumably from an asynchrony of their secretory cycle. Similar findings are also observed in 2-week organ cultures of duck pituitaries which produce a constant level of prolactin. These cells are able to incorporate [³H]leucine into proteins which partly migrate into secretory granules. The more degranulated these cells, the higher the level of radioactivity present (Tixier-Vidal and Picart, 1967; Tixier-Vidal and Gourdji, 1970; Tixier-Vidal, 1972b).

Very few other cell types display such a high level of secretory activity in pituitary organ culture. In human fetal pituitaries, the MSH cells retain the same aspect for 2½ years, but their ultrastructure has not been described (Pasteels, 1969). In duck pituitary culture, besides the prolactin cell, another cell type is able to incorporate [³H]leucine into proteins which follow the same intracellular course as in prolactin cells. It has been proposed that MSH is secreted by these cells (Tixier-Vidal and Picart, 1967; Tougard, 1971). In an amphibian pituitary (*Rana temporaria*), Doerr-Schott (1967) described chromophobic cells which show after 1 month in culture a high degree of development of the Golgi zone and of the ergastoplasm; their function was not studied. In teleost pituitaries (trout, eel), besides the erythrosinophile cells (presumed prolactin cells), the MSH cells and the TSH cells undergo degranulation and hypertrophy. Their ultrastructure was not described (Baker, 1969).

The majority of the other pituitary cell types in organ culture undergo an opposite evolution characterized by a reduction of the Golgi zone and of the ergastoplasm and by a decrease in the number of secretory granules. These features agree well with the decreasing level of several pituitary hormones in the medium (see Section I). In some conditions, or with some species, the cells retain numerous secretory granules for several weeks, and the various pituitary cell types can be identified for example in the trout (Baker, 1963), the frog (Doerr-Schott, 1967), and the pigeon (Tixier-Vidal and Gourdji, 1972). In mammals nevertheless, the progressive disappearance of the secretory granules with the time in culture appears to be a general feature. Under such conditions, identification of the pituitary cell types becomes difficult in the absence of specific methods such as immunocytology. This last method allowed the identification of LH cells and STH cells in 1½-year organ cultures of human fetal pituitaries with fluorescein-labeled antibodies (Pasteels, 1969). In one- and two-week organ cultures of adult rat anterior pituitaries strongly positive LH cells were found with peroxidase-labeled antibodies

FIG. 4. Organ culture of rat anterior pituitary maintained on LRF (2 μg/ml) enriched medium for 14 days. Two LH cells show increased secretory activity (G, Golgi zone) (compare with Fig. 3). P, prolactin cells with ergastoplasmic cisternae. One notices the basement membrane (bm) lining the glandular cells and the numerous collagen fibrils (cf) within the interspace between the cell cords. ×12,000.

and an anti-ovine LH (Fig. 2). These cells are less numerous than *in vivo* and localized in the superior part of the explants (Tixier-Vidal, 1972a).

The processes involved in the progressive disappearance of the secretory granules *in vitro* may be followed by comparing the ultrastructure of explants of increasing age. Such work has been done for the gonadotropic cells in organ cultures of adult rat pituitaries (Tixier-Vidal, 1972a), the evolution of tissue and medium LH content being followed at the same time by radioimmunoassay (Tixier-Vidal *et al.*, 1970). After 3 and 5 days in culture the ultrastructure of the gonadotropic cells was still similar to that of gonadotropic cells identified *in vivo* by electron microscopic immunocytochemistry (Tougard *et al.*, 1973a,b). Cells with two types of round secretory granules (mean diameter: 500 and 200 mμ) and rounded ergastoplasmic cisternae were found, generally grouped as *in vivo*. After 2 weeks such cells were no longer seen, but small oval-shaped cells were occasionally encountered. These latter contained one class of small round secretory granules more or less numerous, and several rounded large dense bodies (700 mμ diam.) which react positively with the acid phosphatase method and are considered to be lysosomes. The ergastoplasmic cisternae were almost absent and the Golgi zone

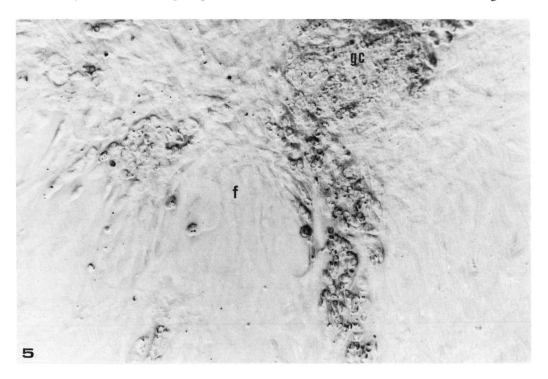

FIG. 5. Phase contrast microscopy of a monolayer of rat anterior pituitary cell showing a glandular cell colony (gc) surrounded by fibroblasts (f). ×240.

was reduced (Fig. 3). These modifications are accompanied by a decrease of the LH content of the tissue (from 838 to 127 ng/mg of fresh tissue).

In what measure do these ultrastructural observations bring some light to the problem of the physiological origin of hormones found in the culture medium? The rapid decrease of hormonal secretion which follows explantation seems to correspond both to the destruction of some cells and the ultrastructural reorganization of some other cells, the last being accompanied by the formation of abundant lysosomes related to intracellular degradation of the hormonal content. Later, the low store of secretory granules and the maintenance of the organization of the Golgi zone which occasionally contains some condensing granules agree with the hypothesis of a low level of autonomy of hormonal synthesis.

At present, the available data are in agreement with the suggestion of Petrovic (1961, 1963) concerning the maintenance *in vitro* of the respective differentiation of the pituitary cell types.

3. *Effect of Hormones and Drugs on the Ultrastructure of Pituitary Cell Types in Organ Culture*

a. *Estron and Thyroxine.* In 8-day organ cultures of rat pituitaries maintained from the first days on an estron-enriched medium, Cohere *et al.* (1964) described a conspicuous development of the ergastoplasmic membranes and of the Golgi zone and the appearance of large secretory granules. In the presence of thyroxine the same development of the ergastoplasm and of the Golgi zone occurred, but without the appearance of large secretory granules. These results are in agreement with those which reported an increased production of prolactin in the presence of estradiol and thyroxine (Table ID).

b. *Ergocornine.* Eighteen hours after introducing bromoergocryptine (10 μg/ml) into the medium of rat pituitary organ cultures, Ectors *et al.* (1972) observed a selective inhibition of granule exocytosis without any action on the RER and the Golgi zone. The authors postulate a selective action of the drug on the plasma membrane which leads to a partial inhibition of prolactin release as determined by bioassay of the culture medium on the pigeon crop sac (Pasteels *et al.,* 1971).

c. *Purified LRF (Luteinizing Hormone Releasing Factor) (Tixier-Vidal, 1972a).* The effect of a purified LRF (prepared by Dr. Jutisz, Paris) on the ultrastructure of rat pituitary organ culture was followed in correlation with its effects on tissue and medium LH content (Tixier-Vidal *et al.,* 1970). The most striking changes were found in cultures maintained in the presence of LRF renewed each 2 days from the second to the fourteenth day. The LH cells enlarged and took on an irregular shape between the unmodified prolactin cells. Their secretory granules were smaller (100 mμ) and localized in the vicinity of the cell membrane (Fig. 4). The ergastoplasmic cisternae were numerous with an irregular outline and sometimes bore a filigree aspect as described by Farquhar and Rinehart in pituitaries of castrated animals (1954). The most striking modifications occurred

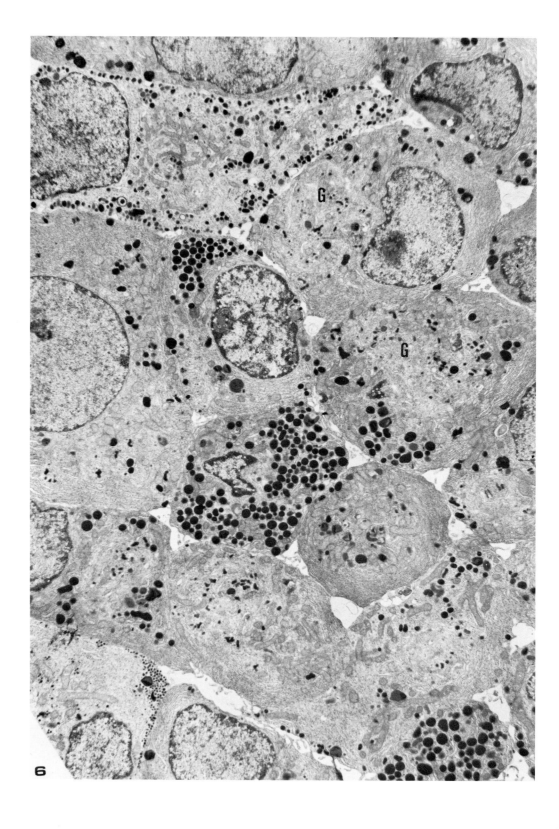

6

in the Golgi zone which was considerably enlarged with typical features: numerous stacks of thin saccules dispersed among a great number of small vesicles sometimes filled with dense material. These features of the Golgi zone facilitated the recognition of the gonadotropic cells among the surrounding degranulated prolactin cells. One must notice, however, that some gonadotropic cells were not modified, which suggests that they had become unreactive.

These results show that LRF has a selective and stimulating action on the gonadotropic cells, although it is not essential to their maintenance *in vitro*. We found no evidence of a transformation of prolactin cells into gonadotropic cells in the presence of LRF as well as no significant effect on prolactin secretion (unpublished results). Finally, the hypothesis of the irreversibility of the differentiation of the respective pituitary cell types in organ cultures remains valid.

B. TISSUE CULTURES

The ultrastructure of pituitary tissue cultures was described by Petrovic (1961) and by Pasteels (1963) who used different culture methods: roller tubes and serum supplemented synthetic media (Petrovic) or hanging drop on a plasma clot with serum (25%) and embryo juice (5%) supplemented Tyrode's solution (Pasteels).

1. *Morphology of Outgrowth Cells*

In both cases, these authors observed a cell outgrowth from the explants, as already seen by previous workers (see review from Gaillard and Schaberg, 1965). The exact nature of the cells has unfortunately not been studied with the electron microscope. By light microscopy, the first outgrowing cells have a fibroblastic aspect. Later polyhedric cells with an epithelial aspect grew out to form epithelial membranes. These cells were all identical and according to Pasteels contained erythrosinophilic granules, which led him to consider them as prolactin cells since at the same time he found an increased secretion of prolactin into the medium. According to Pasteels only the prolactin cells were able to grow out of the explants and divide *in vitro*. With human pituitaries, adult as well as fetal, even the fibroblastic cells disappeared when proliferation of the prolactin cells increased.

The presence in the medium of an increased proportion of chick embryo extract selectively stimulated fibroblastic proliferation (Gaillard, 1942; Pasteels, 1963). The introduction of chick embryo brain extract on the contrary stimulated the migration of authentic glandular cells characterized by their specific granulations (Petrovic, 1961).

FIG. 6. Two-month monolayer of rat anterior pituitary cells. *In situ* embedding. Horizontal section of a glandular cell colony. The glandular cells are arranged as in a parenchyme. Most are prolactin cells. One notices the striking extension of the Golgi zone (G), the development of parallel ergastoplasmic cisternae. The size and number of secretory granules greatly vary depending on the cells. ×6000.

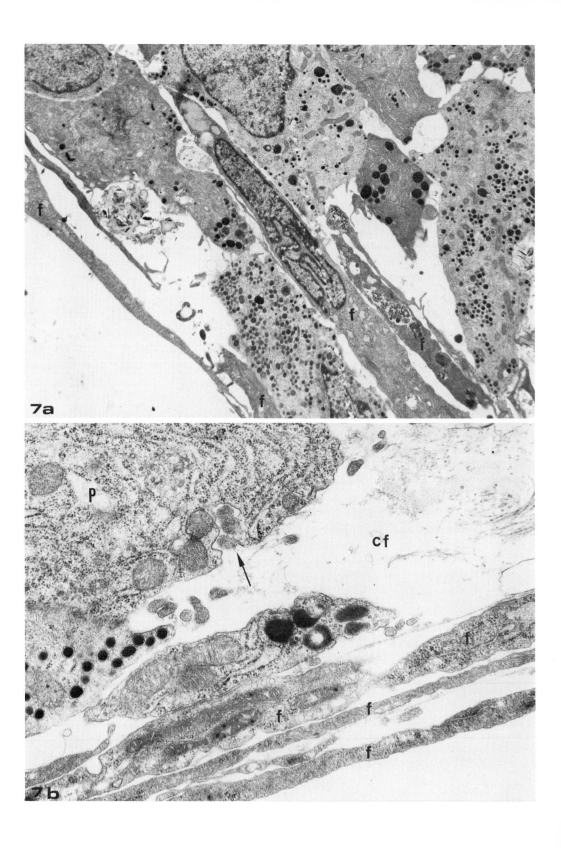

2. *Ultrastructure of the Explants*

Pasteels (1972a,b) used mixed organotypic–histiotypic cultures and tried to limit the cell outgrowth. In these conditions, there was an inverse relationship between the outgrowth and the maintenance of tissue organization.

As previously seen by Gaillard (1942) with the light microscope, the explants were rapidly destroyed when the culture medium contained the juice of young chick embryo. Similarly Petrovic observed ultrastructural signs of tissue degeneration within explants which were cultured in the presence of chick embryo brain extract: chromatin margination in the nucleus, mitochondrial alterations, rapid disappearance of secretory granules, and formation of cysts with empty cavities. One cell type, the epsilon cells, remained recognizable by the development of its ergastoplasmic cisternae.

Even under conditions of limited outgrowth, Pasteels (1963) described a central necrosis of the tissue explants, with a rapid involution of the majority of the pituitary cell types within the first 2 weeks. He pointed out, moreover, the ultrastructural signs of increased secretory activity of the prolactin cells which are identical to those described for organ culture. This together with the outgrowth and proliferation of numerous prolactin cells leads "to an apparently pure culture of prolactin cells." One may nevertheless suppose that this proliferation is limited in time since Pasteels (1969) could not keep such cultures more than 6 weeks.

In summary, tissue cultures, contrary to pituitary organ cultures, are mainly characterized by a selective proliferation of prolactin cells and a more rapid disappearance of the other glandular cell types. From the functional point of view, there is in the two cases an increased secretion of prolactin which might be caused in organ cultures by an increased activity of preexisting cells and in tissue culture by both proliferation and a high secretory activity. As concerns the other anterior pituitary hormones there is no data that might suggest their more rapid disappearance in tissue culture than in organ culture.

3. *Effects of Hypothalamic Extracts*

As already reported above, Pasteels was the first to show that a crude saline hypothalamic extract introduced into the medium of 2-week tissue cultures of rat

---

Fig. 7a. Same culture as in Fig. 6. *In situ* embedding. Perpendicular section of another glandular cell colony. The bottom of the plastic bottle was on the lower left-hand corner. A sheet of fibroblasts (f) is seen underneath the glandular cells; other parallel fibroblasts are mixed with the glandular cells. ×4000.

Fig. 7b. Same section as in Fig. 7a showing at a larger magnification the relation between glandular cells and fibroblasts (f). The bottom of the plastic bottle was on lower right-hand corner. A prolactin cell (P) displays granule exocytosis (arrow). One notices the absence of a basement membrane along the glandular cells and the presence of collagen fibrils in the interspaces (cf). ×9000.

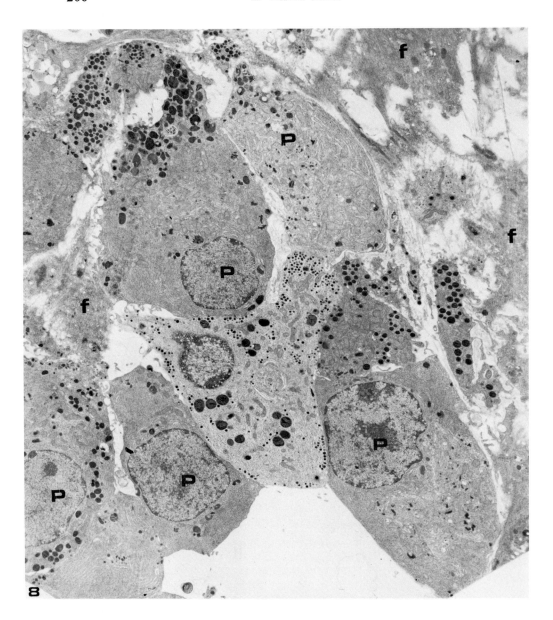

FIG. 8. Same culture as in Fig. 6. *In situ* embedding. Horizontal section of glandular cell colony showing the surrounding fibroblasts (f). P, prolactin cell. One notices the striking differences between fibroblasts and glandular cells. ×4000.

or human pituitary inhibits prolactin secretion (1961). He followed (1963) the cytological modifications induced by such extracts in human pituitary cultures treated for 4 or 9 days and described in the light microscope the appearance of numerous PAS positive cells. Since he found before the treatment neither glyco-protein cells nor undifferentiated cells, he postulated the transformation of pro-lactin cells into gonadotropic cells and confirmed this interpretation by electron microscopic examination. At the same time, he found somatotropic cells whose presence corresponded with the reappearance of somatotropic hormone within the medium. He suggested also that these cells were derived from transformation of prolactin cells.

Such a postulation of the possible transformation of prolactin cells into other pituitary cell types would be of considerable importance for the understanding of the unsolved problem of pituitary stem cells. It cannot however be definitely accepted since Pasteels himself has recently (1972a) questioned his own con-clusion: "We were unable to confirm this in other experimental conditions, so we cannot even ascertain whether a specific pituitary cell can change its function."

4. Effect of 3′,5′-Cyclic Adenosine Monophosphate

In connection with studies on the mechanism of action of hypothalamic releasing hormones, the effect of cyclic AMP on 10-day rat anterior pituitary tissue culture has been described recently (Pelletier et al., 1972). Twenty-four and 36 hr after the addition of cyclic AMP the secretory cells were degranulated. The ergasto-plasm was very well developed and the Golgi apparatus was prominent with an increased number of smooth vesicles and abundant smooth membranes in its vicinity. It is of interest to note that such prominent development of the rough and smooth reticulum was also observed in LH cells in rat pituitary organ cultures treated with purified LRF (Tixier-Vidal, 1972a, see above).

According to Pawlikowski and Kunert-Radek (1973), dibutyryl cyclic AMP would increase the mitotic index in 48-hr rat pituitary organ cultures.

C. Cell Cultures

1. Monolayer of Dispersed Anterior Pituitary Cells

In contrast with the increasing use of monolayers of dispersed anterior pituitary cells to study anterior pituitary hormone secretion, only few data have until now been published on the ultrastructural aspect of pituitary cells in this condition (Rappay et al., 1973). Ours were the first attempt to correlate ultrastructure with LH, FSH, and prolactin content of the medium (Tixier-Vidal et al., 1973a, 1974) and with immunocytochemical detection of LH and FSH cells (Tougard et al., 1974).

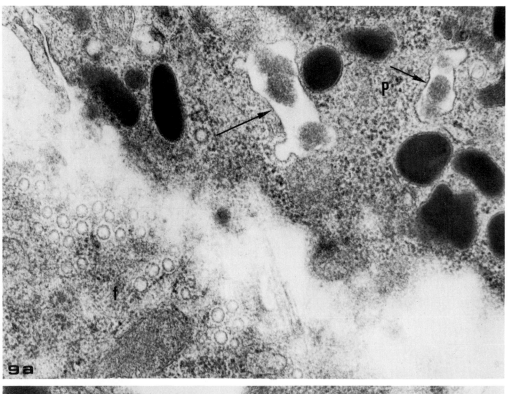

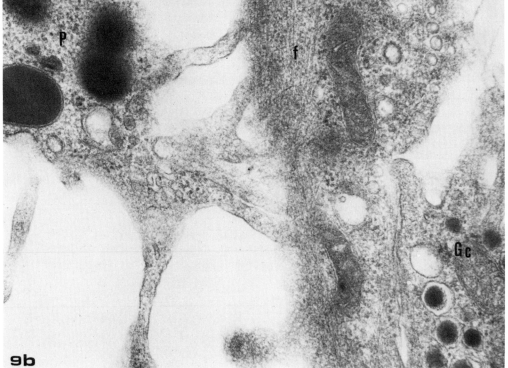

a. *Morphological Evolution of the Monolayers: Evidence for Two Cell Populations.* After dispersion by enzymatic treatment, pituitary cells had a rounded shape. During the first 3 days in culture an important proportion of them were tightly attached to the dishes. They were grouped in small irregular clumps which flattened and seemed to stretch or to grow. After the seventh day, their appearance remained almost unchanged.

After the fifth day, another cell population appeared and thereafter became more and more numerous. These cells had a fibroblastic aspect, elongated and transparent (Fig. 5) and differed from the previously attached cells which were more refractile with very distinct outlines. They divided quickly and formed a continuous monolayer which surrounded the cell clumps. During the second month several sheets of fibroblastic cells were superimposed onto the dish, which led to the progressive regression of the cell clumps. Such invasion by actively dividing fibroblasts is considered a general feature of primary cultures of normal cells (Sato *et al.,* 1960).

The nature of these two cell populations and the relation between them have been determined by studies in the electron microscope of monolayers embedded *in situ* and sectioned in two perpendicular planes, horizontal or vertical (Tixier-Vidal *et al.,* 1974).

The cell clumps are composed of typical glandular cells characterized by the presence of secretory granules and a very well defined cell membrane. In horizontal as well as vertical sections the cells appear associated as in a parenchyme (Figs. 6 and 7). Specialized junctions between them were, however, never observed and the presence of a basement membrane was hardly seen. The spatial distribution of the cytoplasmic organelles, particularly as observed in secretory granule exocytosis (Figs. 7b and 9a), was the same in the two planes which suggests that in such conditions the glandular cells have no morphological polarity.

The other cell population cannot be mistaken for glandular cells. In perpendicular sections these cells are elongated and narrow with a dark cytoplasm. They form a continuous sheet over the plastic and under the glandular cells. Their nuclei are flattened, often lobulated, and the chromatin forms a dense layer along the nuclear membrane. In horizontal sections they display other features (Fig. 8). The cytoplasm is more extended and often contains stacks of cytofilaments mainly at the periphery of the cells. The cell membrane is ill defined with abundant pinocytosis vesicles (Fig. 9a). Fibrillar structures appear to come out of the cell membrane and to establish connections with the neighboring glandular cells some-

FIGS. 9a and b. Same culture as in Fig. 6. Horizontal section showing details of the spatial relations between glandular cells and fibroblasts. Fig. 9a shows numerous pinocytic vesicles at the periphery of a fibroblast (f); compare with the periphery of the prolactin cell (P) which moreover displays a horizontal view of granule exocytosis (arrows). ×50,000. Fig. 9b. Another type of arrangement between a fibroblast (f) and a prolactin cell (P) (on the left) and a gonadotropic cell (Gc) (on the right). ×50,000.

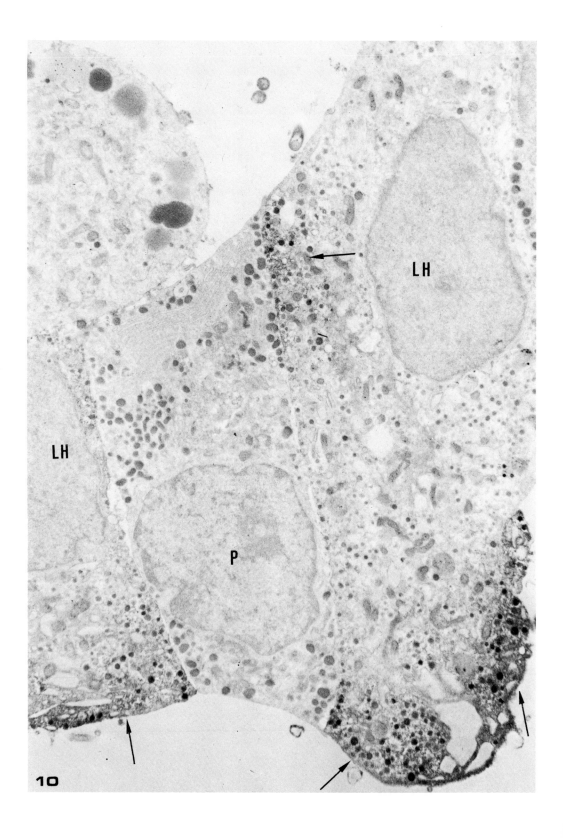

10

times through microvilli (Fig. 9b). The nucleus appears rounded or oval. The distribution of the chromatin is often assymetrical and abundant nuclear pores are seen over the large zones of chromatin. These cells sometimes contain dense material surrounded by a smooth membrane which might, at first sight, be taken for secretory granules, but they have a heterogeneous structure and correspond to lysosomes. All the different features enumerated allow these cells to be identified as fibroblasts. Since no intermediate cells between glandular ones and fibroblasts were observed, it may be suggested that they represent two independent populations each following *in vitro* a different numerical evolution. Rappay *et al.* (1973) have found in 5- and 6-day monolayers, after a short pulse of [³H]thymidine, that 19 and 28% of the nuclei were labeled. They concluded that the pituitary cells divided although they did not determine the exact nature of the cells with labeled nuclei. Using colchicine blockage and staining of the monolayer with Herlant's tetrachrome we found at least 50% of the mitoses within typical fibroblasts; mitosis of typical glandular cells were infrequent.

Our ultrastructural observations therefore are not in favor of a dedifferentiation of the pituitary cells leading to undifferentiated "fibroblastic" cells. Moreover they reveal the existence of interesting cellular interrelations in pituitary cell cultures. The glandular cells are able to recognize each other and to reassociate. They seem to need contact with fibroblasts which could initially play the role of a natural feeder layer but which later limit the survival of the glandular cells, because of their high mitotic rate.

b. *Evolution of the Glandular Cell Types.* Immediately after dissociation the pituitary cell types can be recognized in the electron microscope by their ultrastructural features which are unmodified (Ishikawa, 1969; Malamed *et al.,* 1970; see chapters by Farquhar *et al.* and Hymer).

In 5-day monolayers, the different pituitary cell types retain their typical ultrastructural features and may therefore be easily identified: somatotropic cells, corticotropic cells, and thyrotropic cells were found. As concerns prolactin cells and gonadotropic cells findings are similar to those described above for 5-day organ cultures. The ultrastructural identification of the LH cells has been confirmed by direct application of the immunocytochemical technique to the monolayer (Fig. 10) (Tougard *et al.,* 1974).

In 15-, 18-, 22-, and 27-day monolayers there was a progressive modification of the glandular cell population. The prolactin cells increased in number with the

---

FIG. 10. Five-day monolayer of anterior pituitary cells. Horizontal section of a clump of glandular cells. The immunocytochemical detection of the binding sites of an anti-ovine LHβ has been made within the culture dishes. The positive cells were recognized by the light microscope and then sectioned for electron microscopy. The positive reaction is restricted to small areas of the cytoplasm (arrows). LH, LH cells; P, prolactin cells. ×10,500.

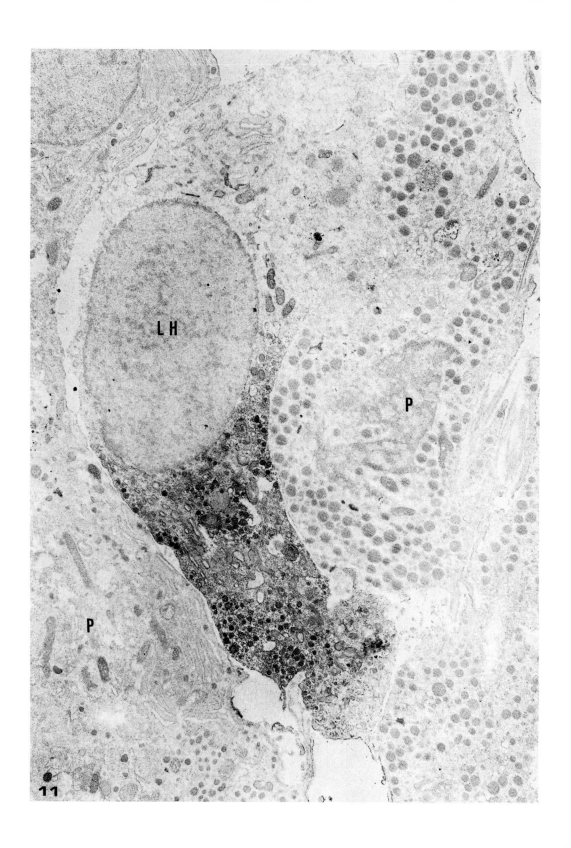

same development of the ergastoplasm and of the Golgi zone. This is in agreement with the increasing production of prolactin in the medium (Tixier-Vidal *et al.,* 1974). The other cell types seemed less abundant and difficult to identify without the use of specific methods. Thus immunocytochemical staining enabled us to recognize gonadotropic cells. These underwent modifications (Figs. 11 and 12) that are similar to those observed for organ culture (see above). Finally a single gonadotropic cell type remained. It was characterized by the low electron density of the cytoplasm, a single class of small round secretory granules (125–150 mμ diameter), linear ergastoplasmic cisternae, and large round or oval dense bodies (600 mμ diam.). With antisera against ovine LH and ovine LHβ, the ground cytoplasm and the small secretory granules were always positive (Fig. 12). The

TABLE II

NUMBER OF CELLS IMMUNOREACTIVE WITH ANTISERA AGAINST OVINE LH, OVINE LHβ, AND OVINE LHα AS COMPARED TO THE LH AND FSH MEDIUM CONTENT[a] WITHIN THE SAME CULTURE DISH, IN MONOLAYERS OF ANTERIOR RAT PITUITARY CELLS OF INCREASING AGE

Age of the monolayer	SD 11 series				SD 16 series		
	% of positive cells	Medium LH content into the same dish (ng/ml)	Medium FSH content into the same dish (ng/ml)		% of positive cells	Medium LH content into the same dish (ng/ml)	Medium FSH content into the same dish (ng/ml)
5 days	LH 2	185	45		LH 5	37.5	82
	LHβ 7.5	129	—		LHβ 5	28.5	82
	LHα 13	107	48				
12 days	LH 5.5	7.75	7.3		LH 2.3	5.5	123
	LHβ 12	6.5	11.8		LHβ 4	5.2	44.8
	LHα 7	7.3	6.8				
18 days	LH 3.5	7.4	11		LH 2.2	4.5	32.2
	LHβ 6.5	9.3	13		LHβ 9	4.1	30.8
	LHα 13.5	7.1	10.5				
26 days					LH 3.7	2.6	33.8
					LHβ 4.5	2.4	27.2
35 days	LHβ 0	4.5	9.9		LHα 12	2.2	30.8

[a] For the LH and FSH assays see Table III.

FIG. 11. Twenty-seven-day monolayer of rat anterior pituitary cells. Immunocytochemical detection of the binding sites of LHβ. A positive gonadotropic cell (LH). The positive reaction is concentrated in one area of the cytoplasm. P, prolactin cells. ×9000.

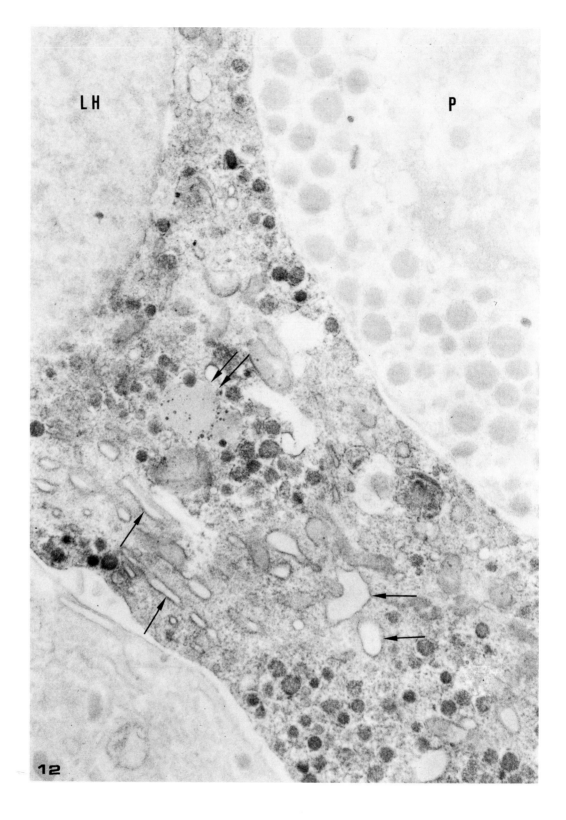

LH

P

12

positive reaction was often strikingly confined to a large area of the cell cytoplasm (Fig. 12).

In 2-month monolayers the prolactin cells were the major component of the glandular cell population and few gonadotropic cells were dispersed among them. The latter have essentially the same features as in 27-day monolayers. It is interesting to note that at this stage generally only FSH is to be found in the medium, LH having disappeared sooner (Tixier-Vidal *et al.,* 1973).

During the same period the relative number of gonadotropic cells immunoreactive with antisera against ovine LH, ovine LHβ, and ovine LHα was followed by direct count in the culture dishes (Tougard *et al.,* 1974) and compared with the level of LH and FSH in the medium of the same dish (Table II). The number of immunoreactive cells for A-LH, A-LHβ, and A-LHα did not decrease dramatically between 5 and 26 days in culture. This contrasts with the sharp decrease of the LH and FSH medium content and indicates that the decreasing ability of the cultures to secrete LH and FSH results more from a modification of the secretory activity of the gonadotropic cells than from their disappearance either by necrosis or dedifferentiation. The persistence of a well-defined Golgi zone in these cells together with the development of large dense bodies suggests that there is both continuous synthesis and intracellular degradation of the hormone.

These studies of the ultrastructural evolution of the glandular cells in monolayers corroborate and extend the conclusion drawn from the examination of organ cultures: (1) The increase in number and in secretory activity of the prolactin cell, (2) the maintenance *in vitro* of the specific differentiation of the gonadotropic cells. The important difference in cell interrelations in cell cultures and in organ cultures, does not therefore interfere with the secretory activity of the glandular cells.

  c. *Effect of Hypothalamic Factor on the Ultrastructure of the Glandular Cells.*

  i. *Corticotropic releasing factor (CRF).* Ohtsuka *et al.* (1971, 1972) have studied the effect of purified CRF on the morphological and functional differentiation of cultured chromophobe cells isolated from rat anterior pituitaries. In fact, these authors did not use monolayers although they started from dispersed pituitary cells. They isolated a "pure chromophobe" pellet which was cut into small pieces 2 × 2 × 2 mm, the pieces then being cultured as "explants" in roller tubes, for up to 1 month. In "control" culture (50 μg/liter L-thyroxine), chromophobes divided during the first 6 days and partly differentiated into "immature ambiguous cells." Such explants contained small amounts of ACTH and GH, but a large amount of prolactin. In presence of CRF plus thyroxine combined with

---

FIG. 12. Same section as in Fig. 15. *In situ* embedding. Detail of the immunocytochemical staining of the LH cell. A positive reaction is found within the small secretory granules and the cytoplasm. The content of the ergastoplasmic cisternae (arrows) and the dense bodies (double arrow) are negative. P, prolactin cell. ×27,000.

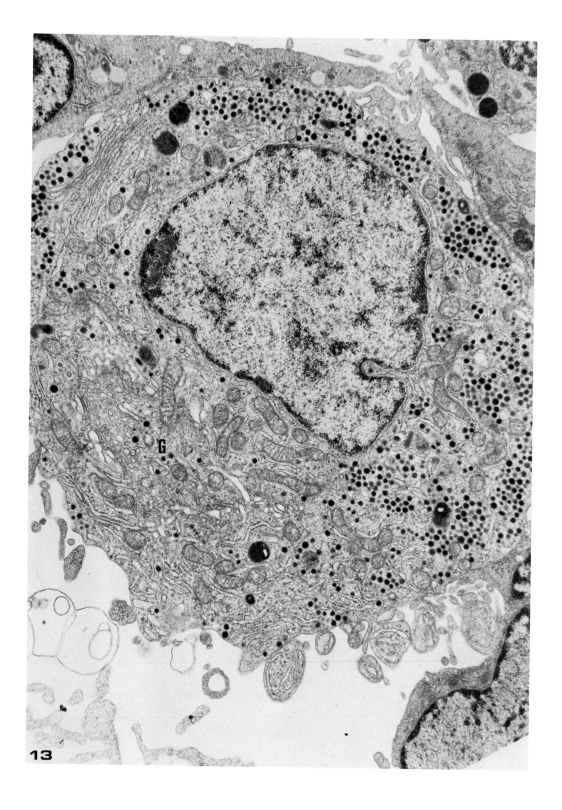

13

colchicine, the chromophobe cell division was arrested and many cells became "acidophils." The explants contained a larger amount of ACTH but neither GH nor prolactin. The authors concluded that CRF induced the transformation of the chromophobes into acidophils which produce ACTH. Surprisingly enough, the ACTH activity was found associated with small (100 to 200 mμ in diam.) as well as large (300–500 mμ) and "huge" irregularly shaped granules. These last results contrast with *in vivo* finding concerning the hormonal content of these three types of granules (see chapter by Nakane).

ii. *Synthetic LRF*. In 5-day monolayers treated for 4 hr with synthetic LRF (5 ng/ml), the gonadotropic cells were already clearly modified. The secretory granules decreased in size and migrated to the plasma membrane which presented small invaginations. Pictures of granule exocytosis were however not frequent. The Golgi zone was strikingly enlarged with several stacks of saccules and abundant condensing secretory material. Twenty-seven day monolayers chronically treated from the beginning with LRF (2 ng/ml) showed no modifications when observed in the living state. In the electron microscope, findings are similar to those observed in 2-week organ culture chronically treated with purified LRF (see above). The prolactin cells were unchanged which agrees with an absence of modification of the medium prolactin content (Table III). In contrast the gonadotropic cells were strikingly modified (Fig. 13) although not increased in number. This stimulation of the gonadotropic cells agrees with the increase of medium LH and FSH content (Table III).

In monolayer as in organ culture, therefore LRF appears able to stimulate the activity of preexisting gonadotropic cells but not to induce the differentiation of new gonadotropic cells. Moreover although both FSH and LH release were stimulated, only a single gonadotropic cell type was observed as time increased. Development of the smooth reticulum together with the decreasing size of the secretory granules were the most striking modifications in the stimulated cells.

2. Monolayers of a Continuous Cell Line: GH3 Strain

As reported above (see Table IC and D and Section I of this chapter), the GH3 strain is a clonal line of pituitary cells isolated by Yasumura *et al.* (1966) (as in Table I origin is from a rat pituitary tumor). Although continuously dividing, the cells have retained both specific differentiation (secretion of growth hormone and of prolactin) and reactivity to one of the physiological stimulating agents of prolactin secretion, the hypothalamic thyrotropic stimulating factor (TRF) (Tashjian *et al.*, 1971). The ultrastructure of the cells has been studied by Gourdji *et al.* (1972) in both control and TRF-enriched medium.

FIG. 13. Twenty-seven-day monolayer of rat anterior pituitary cells maintained on LRF (2 ng/ml) enriched medium. Section through a pellet of trypsinized cells. A stimulated gonadotropic cell showing numerous small secretory granules, development of the Golgi zone (G) and ergastoplasmic cisternae. Compare with Fig. 4 (organ culture). ×13,000.

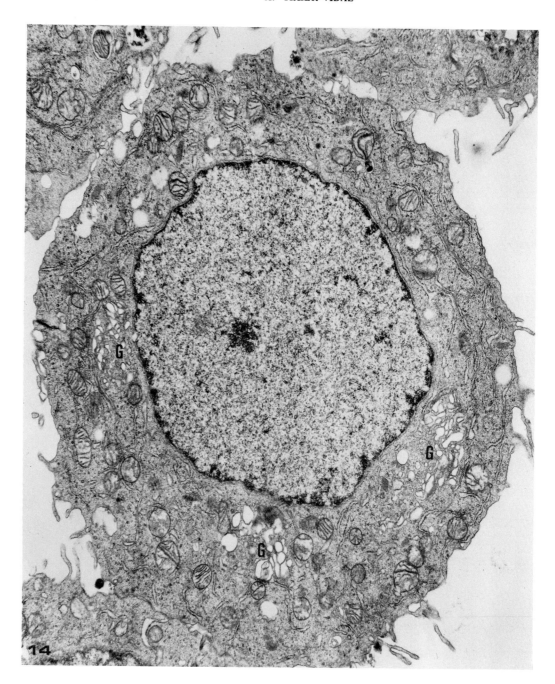

TABLE III

PROLACTIN, LH, AND FSH CONTENT (ng/ml)[a] OF THE MEDIUM IN PRIMARY CULTURES OF
DISPERSED ANTERIOR PITUITARY CELLS. CHRONIC EFFECT OF LH-RH[b]

Culture	Age of culture		
	0 → 3 days	5 → 7 days	12 → 14 days
Prolactin			
Control	148	262	325
	90	135	312
	107	150	
LH-RH	160	255	325
	90	135	300
	107	175	
LH			
Control	1320	31.5	12.2
	780	43	13.4
	800	46.2	12.8
LH-RH	1040	108.3	51.2
	700	116.6	50
	900	126.2	50
FSH			
Control	191	38.6	8.6
	155	56.5	16.6
	158	72.5	16.6
LH-RH	208	85.3	40.6
	135	68.8	37.4
	165	78.9	34

[a] Samples were assayed for rat LH and rat FSH using radioimmunoassay methods as described in Tixier-Vidal et al., 1973a.

Rat LH laboratory preparation (0.95 × NIH-LH-SI) and rat FSH (S1-1, 100 × NIH-FSH-S1) kindly provided by NIAMDD program were used for labeling and as standards. The radioimmunoassay of prolactin was performed by B. Kerdelhué and D. Grouselle using rat prolactin and corresponding antiserum kindly provided by the NIAMDD program.

[b] Individual values of 6 culture bottles (2 ng/ml).

a. *Control Medium.* In control medium, the GH3 cells appear as small cells (10 μm mean diameter) with a high nucleus/cytoplasm ratio (Fig. 14). The secretory granules are sparse or absent. Their diameter varies from 50 to 250 mμ and they are localized near the cell membrane. Their shape is polymorphic in some

FIG. 14. GH3 cells grown in control medium. Notice the extreme paucity of secretory granules and the several units of the Golgi zone (G) with dilated cisternae. ×12,000. Courtesy of D. Gourdji.

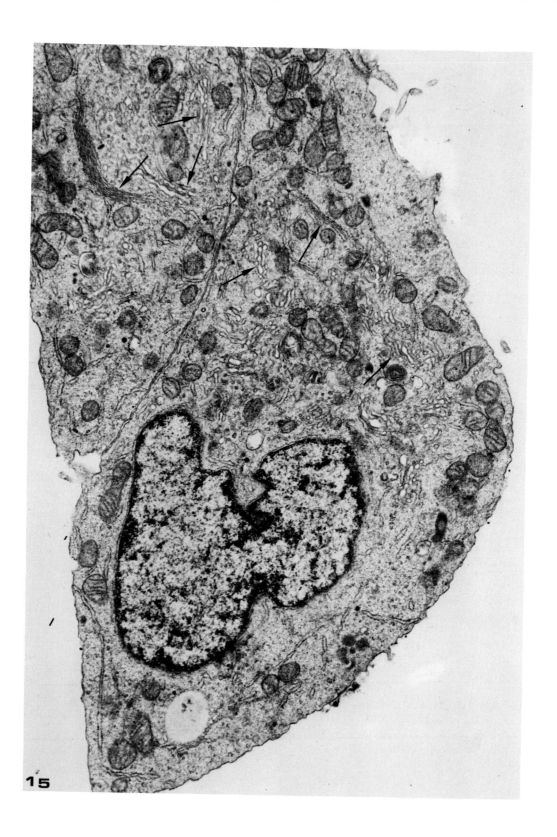

15

cells, but more often rounded. The Golgi zone is made of several units scattered within the cytoplasm. Each unit consists of dilated saccules and several smooth vesicles. Condensing secretory granules are unfrequent. Bundles of cytofilaments are also observed. The ergastoplasmic cisternae are extremely flattened and numerous free ribosomes and polysomes are seen within the cytoplasm. The cells are often joined together by tight junctions.

Such ultrastructural organization differs markedly from that of prolactin cells in monolayers of primary cultures. The difference is to be related to the tumoral origin of this strain. Indeed similar modifications in the ultrastructure of prolactin cell have been described in estrogen-induced pituitary tumors (Waelbroeck-Van Gaver and Potvliège, 1969; Pantic et al., 1971) and in human pituitary adenomas (see chapter by Olivier et al.). Moreover, by budding from the plasma membrane the cells release viral particles morphologically similar to those observed in murine and avian leukemias and sarcomas. The loss of typical features of the normal morphological phenotype of prolactin or growth hormone cells contrasts with the maintenance of their biochemical differentiation: the synthesis of both prolactin and growth hormone. As a consequence the ultrastructural study does not permit deciding if both hormones are produced simultaneously within the same cell or independently within different cells which are derived from a single stem cell, since they have been obtained by cloning (see Tashjian et al., 1970). The extreme paucity of secretory granules is in agreement with the low hormonal content of the cells (see Section I of this chapter). Nothing is known about their secretory cycle.

b. *TRF-Supplemented Medium.* A few hours after the addition of synthetic TRF (10 and 100 ng/ml), the cells spread and take on an angular shape. In the electron microscope, important ultrastructural modifications are seen (Fig. 15). The secretory granules augment in number and remain small. The Golgi zone is hypertrophied and occupies a large part of the cytoplasm. It consists of several groups of flattened saccules with numerous smooth vesicles and maturing secretory granules. Dense bodies were more frequent than in control cells. In contrast, the ergastoplasm was not very modified. These ultrastructural modifications were already seen after 30 min of contact with TRF but after a longer treatment a larger proportion of the cells was modified.

Several inferences are to be drawn from this study. Synthetic TRF seems to act preferentially on the smooth reticulum of the Golgi zone. Whether this effect is direct or not remains to be demonstrated. A similar effect was observed on gonadotropic cells treated with purified or synthetic LRF. The hypothesis of a

---

FIG. 15. GH3 cells maintained for 30 min on TRF (10 ng/ml) enriched medium. Notice the striking modification of the Golgi zone with abundant stacks of flattened saccules and vesicles (arrows). ×15,000. Courtesy of D. Gourdji.

specific action of hypothalamic hormones on the smooth endoplasmic reticulum may therefore be advanced. Since the smooth reticulum of the Golgi zone is more involved in the segregation of the secretory products than in their synthesis, this hypothesis would agree with the present concept of an indirect rather than a direct effect of the hypothalamic hormones on the synthesis of the pituitary hormones.

Considering that the GH3 cells have a low prolactin store, Tashjian et al. (1971) claim that in this system, TRF acts mainly by stimulating prolactin synthesis. This does not agree in our opinion with the weak effect of TRF on the ergastoplasm. In regards to monolayers of dispersed pituitary cells the GH3 cell strain offers a homogeneous population of pituitary glandular cells and a promising system to study the mechanism of action of a synthetic hypothalamic factor. Recent results on the interaction between tritiated TRF (Pradelles et al., 1972) and GH3 cells (Gourdji et al., 1973; Hinkle and Tashjian, 1973) open the perspective of a molecular approach to this problem.

IV. Conclusion

This review shows that there is good correlation between ultrastructural data and hormonal secretory abilities of pituitary cultures using either organ, tissue or cell culture methods. Moreover, interesting perspectives appear concerning the main problems of pituitary cell biology: differentiation of the pituitary cell types and regulation of secretory activity.

The increased secretion of prolactin together with ultrastructural features of a hyperactivity of the prolactin cells have been widely confirmed in mammals since the initial work of Pasteels. Several cellular mechanisms could be involved in this increase of prolactin secretion: (1) stimulation of preexisting prolactin cells (predominant in organ culture), (2) differentiation of chromophobe cells into prolactin cells, and (3) selective proliferation of prolactin cells (mainly in tissue culture). The current physiological explanation for this rapid increase of prolactin secretion in vitro implies the suppression of an inhibiting hypothalamic factor. Such high level of prolactin secretion, because of its maintenance for several months in vitro, could also reveal a genetic feature of these cells, at least in mammals.

For the majority of the other pituitary hormones, secretory ability decreases with time in culture, but increase in sensitivity of the pituitary hormone assays allows the existence of a low autonomy of secretion to be detected. In this respect, ultrastructural study brings definite evidence for the maintenance, in vitro, of several pituitary cell types. Such studies allow the conclusion to be drawn that the hormonal content of the medium is mainly due to a physiological release by living cells. Besides, in the special case of the gonadotropic cells, it was also shown that the cells undergo morphological modifications leading to a single cell type still present after 2 months. It would be of interest to study in a similar way the other pituitary cell types displaying a low autonomy of secretion such as TSH, STH, and ACTH. This could be a possible approach to the problem of the pituitary stem cell.

Autonomy of secretion might therefore be considered an intrinsic feature of pituitary cells. Among the cell types the differences would be quantitative only. From a phylogenetic point of view, in mammals the prolactin cells and perhaps the melanotropic cells display the maximum level of autonomy, but in lower vertebrates other cell types, such as TSH cells, seem to share this high level of autonomy. Malignant transformation is also known to increase the autonomy of secretion of several pituitary hormones (see chapter by Olivier *et al.*).

Pituitary culture studies also show that the function of hypothalamic releasing hormones consists mainly in a quantitative modulation of secretion. The study of ultrastructural modifications induced by purified or synthetic hypothalamic hormones as well as by cyclic AMP suggests at least in culture a selective action on the smooth reticulum including the Golgi zone. An increase in secretory granule exocytosis was not particularly evident in culture. It appears that the hypothalamic hormones would act mainly at the site of segregation and migration of the secretory product. Another aspect of the function of hypothalamic hormones concerns their possible role in the differentiation of pituitary cells. The available data do not allow a positive conclusion on this point. No true evidence of transformation of one pituitary cell type into another has yet been obtained in culture. The absence of any effect of hypothalamic factors on the differentiation of pituitary cells is consistent with the concept of an autonomy of pituitary cells. Nevertheless, further experimental modalities must be explored before reaching a final conclusion.

Another interesting point brought out in this review is the similarity of the biological and morphological results obtained with organ cultures and with cell cultures. These two methods differ in their results by one factor: the duration of culture which is much greater in cell cultures. In this last condition, ultrastructural study reveals the neoformation of a tissuelike structure, which could explain the similarity with organ culture. On the contrary, the available data in tissue culture indicates a possible selection in favor of prolactin cells. More work on the ultrastructure of the outgrowing cells would be needed to verify this conclusion.

In spite of its usefulness the cell culture method has limitations due to the overgrowth of fibroblasts. Until now, the only possibility to escape this difficulty consisted in starting from pituitary tumors in which the dividing ability of the glandular cell could compete with that of the fibroblasts. This enabled Sato and his colleagues to isolate pure pituitary cell lines. The ultrastructure of such cells, however, is modified with respect to normal prolactin and somatotropic cells. This fact must be kept in mind in using such cell lines for the study of pituitary cell regulation for which they offer a promising approach.

ACKNOWLEDGMENTS

This work was supported by grants from the CNRS (ERA 89) and the DGRST (Biologie de la Reproduction). We express our thanks to Mrs. R. Picart and Mr. C. Pennarun for their excellent technical assistance.

We are indebted to Dr. F. Haguenau (from the Laboratoire de Medecine Experimentale, Paris) and to Pr. L. L. Rosenberg (from the University of Berkeley, California) for their help and advice in the preparation of the English text.

REFERENCES

Baker, B. I. (1963). Comportement en culture organotypique des cellules de l'hypophyse de la Truite. *C. R. Acad. Sci.* **256**, 3356–3358.

Baker, B. I. (1964). Synthèse de l'hormone mélanophorotrope par l'hypophyse de la Truite *in vitro*. *C. R. Acad. Sci.* **258**, 5082–5085.

Baker, B. I. (1965). The site of synthesis of the melanophore stimulating hormone in the trout pituitary. *J. Endocrinol.* **32**, 397.

Baker, B. I. (1967). Sécrétion de l'hormone mélano-stimulante par l'hypophyse de Poisson "in vitro." *Biol. Med. (Paris)* **56**, 351–358.

Baker, B. I. (1969). The response of teleost pituitary thyrotrophs to thyroxin *in vitro*. A histological study. *Gen. Comp. Endocrinol.* **12**, 427–437.

Bancroft, F. C., Levine, L., and Tashjian, A. H. (1969). Control of growth hormone production by a clonal strain of rat pituitary cells: Stimulation by hydrocortisone. *J. Cell Biol.* **43**, 432–441.

Batzdorf, U., Gold, V., Matthews, N., and Brown, J. (1971). Human growth hormones in cultures of human pituitary tumors. *J. Neurosurg.* **34**, 741–748.

Bowers, C. Y., Friesen, H. G., Hwang, P., Guyda, H. J., and Folkers, K. (1971). Prolactin and thyrotropin release in man by synthetic pyroglutamyl-histidyl-prolinamide. *Biochem. Biophys. Res. Commun.* **45**, 1033–1041.

Bridson, W. E., and Kohler, P. O. (1970). Cortisol stimulation of growth hormone production by human pituitary tissue in culture. *J. Clin. Endocrinol. Metab.* **30**, 538–540.

Buonassisi, V., Sato, G., and Cohen, A. (1962). Hormone producing cultures of adrenal and pituitary tumor origin. *Proc. Nat. Acad. Sci. U. S.* **48**, 1184.

Channing, C. P., Taylor, M., Knobil, E., Nicoll, C. S., and Nichols, C. W., Jr. (1970). Secretion of prolactin and growth hormone by cultures of adult simian pituitaries. *Proc. Soc. Exp. Biol. Med.* **135**, 540.

Cohere, G., Bousquet, J., and Meunier, J. M. (1964). Ultrastructure of explants of adult female rat's pituitary gland cultivated on artificial media: Action of hormonal variations. *C. R. Soc. Biol.* **158**, 1056.

Convey, E. M., Tucker, H. A., Smith, V. G., and Zolman, Y. (1973). Bovine prolactin, growth hormone, thyroxine and corticoid response to thyrotropin releasing hormone. *Endocrinology* **92**, 471.

Danon, A., Dikstein, S., and Sulman, F. G. (1963). Stimulation of prolactin secretion by Perphenazine in pituitary–hypothalamus organ culture. *Proc. Soc. Exp. Biol. Med.* **114**, 366–368.

Doerr-Schott, J. (1967). Cytologie de l'hypophyse distale d'un Amphibien (*Rana temporaria*) après greffe homéotypique et culture *in vitro* prolongée. Analyse de la fonction gonadotrope. *Arch. Anat., Histol. Embryol.* **60**, 92–130.

Ectors, F., Danguy, A., and Pasteels, J. L. (1972). Ultrastructure of organ cultures of rat hypophyses exposed to ergocornine. *J. Endocrinol.* **52**, 211–212.

Farquhar, M., and Rinehart, Y. (1954). Electron microscopic studies of the anterior pituitary gland of castrate rats. *Endocrinology* **54**, 516–541.

Franchimont, P., and Pasteels, J. L. (1972). Sécrétion indépendante des hormones gonadotropes et de leurs sous-unités. *C. R. Acad. Sci., Ser. D* **275**, 1799–1802.

Gailani, S., Nussbaum, A., McDongall, J., and McLimans, W. (1970). Studies on hormone production by human foetal pituitary cell cultures. *Proc. Soc. Exp. Biol. Med.* **134**, 27–32.

Gaillard, P. P. (1942). Hormones regulating growth and differentiation in embryonic explants. *Actual. Sci. Ind.* **923**, 1–82.

Gaillard, P. J., and Schaberg, A. (1965). Endocrine glands in "Cells and Tissues in Culture" (E. N. Willmer, ed.), Vol. 2, Chapter 16, pp. 631–643. Academic Press, New York.

Gala, R. R. (1970). Optimum culture conditions for the production of prolactin from rat anterior pituitary organ cultures. *Acta Endocrinol. (Copenhagen)* **65**, 466–476.

Gala, R. R. (1971). Prolactin production by the human anterior pituitary cultured *in vitro*. *J. Endocrinol.* **50**, 637.

Gala, R. R., and Kuo, E. Y. H. (1972). Prolactin production by rat anterior pituitaries cultured *in vitro* as estimated by crop gland assay, densitometric analysis and radioimmunoassay. *Proc. Soc. Exp. Biol. Med.* **139**, 1349.

Gala, R. R., and Reece, R. P. (1964). Influence of hypothalamic fragments and extracts on lactogen. Production *in vitro*. *Proc. Soc. Exp. Biol. Med.* **117**, 833–836.

Gourdji, D., and Tixier-Vidal, A. (1966). Mise en évidence d'un contrôle hypothalamique stimulant de la prolactine hypophysaire chez le Canard. *C. R. Acad. Sci., Ser. D* **263**, 162–165.

Gourdji, D., Kerdelhué, B., and Tixier-Vidal, A. (1972). Ultrastructure d'un clone de cellules hypophysaires sécrétant de la prolactine (clone GH3). Modifications induites par l'hormone hypothalamique de libération de l'hormone thyréotrope (TRF). *C. R. Acad. Sci., Ser. D* **274**, 437–440.

Gourdji, D., Tixier-Vidal, A., Morin, A., Pradelles, P., Morgat, J. L., Fromageot, P., and Kerdelhué, B. (1973). Binding of a tritiated thyrotropin releasing factor (TRF) to a prolactin secreting clonal cell line (GH3). *Exp. Cell Res.* **82**, 39–46.

Groom, G. V., Groom, M. A., Cooke, I. D., and Boyns, A. R. (1971). The secretion of immunoreactive luteinizing hormone and follicle stimulating hormone by the human foetal pituitary in organ culture. *J. Endocrinol.* **49**, 335–344.

Guillemin, R., and Rosenberg, B. (1955). Humoral control of anterior pituitary: A study with combined tissue cultures. *Endocrinology* **57**, 599–607.

Hermanus, J. P., Pasteels, J. L., and Herlant, M. (1964). Localisation cellulaire de la sécrétion d'intermédine dans l'hypophyse du foetus humain en culture de tissus. *C. R. Acad. Sci.* **258**, 6530–6532.

Hinkle, P. M., Tashjian, A. H., Jr. (1973). Receptors for thyrotropin-releasing hormone in prolactin-producing rat pituitary cells in culture. *J. Biol. Chem.* **248**, 6174–6179.

Ishikawa, H. (1969). Isolation of different types of anterior pituitary cells in rats. *Endocrinol. Jap.* **16**, 517.

Jacobs, L. S., Snyder, P. J., Wilber, J. F., Otiger, R. D., and Daughaday, W. H. (1972). Increased serum prolactin after administration of synthetic thyrotropin releasing hormone (TRH) in man. *J. Clin. Endocrinol. Metab.* **33**, 996–1005.

Kann, G., Habert, R., and Denamur, R. (1973). Concentrations plasmatiques de la prolactine et de l'hormone tyro-stimulante au cours de la traite des brebis. Comparaison avec les effets du TRF. *C. R. Acad. Sci.* **276,** (D), 1321–1324.

Kobayashi, T., Kobayashi, T., Kigawa, T., Mizuno, M., and Amenomori, Y. (1961). Influence of the crude hypothalamic extract upon the cultivated rat anterior pituitary cells. *Endocrinol. Jap.* **8,** 223–226.

Kobayashi, T., Kobayashi, T., Kigawa, T., Mizuno, M., and Amenomori, Y. (1963). Influence of rat hypothalamic extract on gonadotropic activity of cultivated anterior pituitary cells. *Endocrinol. Jap.* **10,** 16–24.

Kobayashi, T., Kigawa, T., Mizuno, M., and Watanabe, T. (1971). *In vitro* methods for the study of the adenohypophysial functions to secrete gonadotrophin. *Karolinska Symp. Res. Methods Reprod. Endocrinol., 3rd Symp., 1971* pp. 27–40.

Kohler, P. O., Bridson, W. E., and Rayford, P. L. (1968). Cortisol stimulation of growth hormone production by monkey adenohypophysis in tissue culture. *Biochem. Biophys. Res. Commun.* **33,** 834–840.

Kohler, P. O., Bridson, E., Rayford, P. L., and Kohler, E. (1969a). Adenomas in culture. *Metab. Clin. Exp.* **18,** 782–788.

Kohler, P. O., Frohman, L. A., Bridson, W. E., Vanha-Perttula, T., and Hammond, J. M. (1969b). Cortisol induction of growth hormone synthesis in a clonal line of rat pituitary tumor cells in culture. *Science* **166,** 633–634.

Kohler, P. O., Bridson, W. E., and Chrambach, A. (1971). Human growth hormone produced in tissue culture: Characterization by polyacrylamide gel electrophoresis. *J. Clin. Endocrinol. Metab.* **32,** 70–76.

Malamed, S., Portanova, R., and Sayers, G. (1970). Fine structure of trypsin dissociated cells from the anterior pituitary and adrenal glands. *J. Cell Biol.* **47,** 128a (336).

Meites, J. (1973). Control of prolactin secretion in animals. *In* "Human Prolactin" (J. L. Pasteels and C. Robyn, eds.), pp. 52–64. Excerpta Med. Found., Amsterdam.

Meites, J., Kahn, R. H., and Nicoll, C. S. (1961). Prolactin production by rat pituitary *in vitro*. *Proc. Soc. Exp. Biol. Med.* **108,** 440–443.

Meites, J., Nicoll, C. S., and Talwalker, P. K. (1963). The central nervous system and the secretion and release of prolactin. *Advan. Neuroendocrinol., Proc. Symp., 1961* pp. 238–277.

Mittler, J. C. (1972). Androgen effect on gonadotropin secretion by organ cultured anterior pituitary. *Proc. Soc. Exp. Biol. Med.* **140,** 1140–1142.

Mittler, J. C., Arimura, A., and Schally, A. V. (1970). Release and synthesis of luteinizing hormone and follicle-stimulating hormone in pituitary cultures in response to hypothalamic preparations. *Proc. Soc. Exp. Biol. Med.* **133,** 1321–1325.

Nicoll, C. S. (1965). Neural regulation of adenohypophysial secretion in tetrapods. Indications from *in vitro* studies. *J. Exp. Zool.* **158,** 203–210.

Nicoll, C. S. (1972). Secretion of prolactin and growth hormone by adenohypophyses of rhesus monkeys *"in vitro."* *Lactogenic Horm., Ciba Found. Symp.* pp. 257–268.

Nicoll, C. S., and Meites, J. (1963). Prolactin secretion *in vitro*: Effects of thyroid hormone and insulin. *Endocrinology* **72,** 544–551.

Nicoll, C. S., and Swearingen, K. (1971). Preliminary observations on prolactin and growth hormone turnover in rat adenohypophyses *in vitro*. *In* "The Hypothalamus" (L. Martini, M. Motta, and F. Fraschini, eds.), pp. 1–14. Academic Press, New York.

Nicoll, C. S., Fiorindo, R. P., McKennee, C. T., and Parsons, J. A. (1970a). Assay of hypothalamic factors which regulate prolactin secretion. *In* "Hypophysiotropic Hormones of the Hypothalamus: Assay and Chemistry," (J. Meites, ed.), pp. 115–150. Williams & Wilkins, Baltimore, Maryland.

Nicoll, C. S., Parsons, J. A., Fiorindo, R. P., Nichols, C. W., Jr., and Sakuma, M. (1970b).

Evidence of independent secretion of prolactin and growth hormone *in vitro* by adeno-hypophyses of rhesus monkeys. *J. Clin. Endocrinol. Metab.* 30, 512.

Ohtsuka, Y., Tshikawa, H., and Omoto, T. (1971). Effect of CRF on the morphological and functional differentiation of the cultured chromophobes isolated from rat anterior pituitaries. *Endocrinol. Jap.* 18, 133.

Ohtsuka, Y., Ishikawa, H., Watanabe, T., and Yoshimura, F. (1972). ACTH synthesizing and releasing activities of adenohypophyseal acidophils differentiating from the isolated chromophobes in a chemically defined medium supplemented with TRF. *Endocrinol. Jap.* 19, 237–249.

Pantic, V., Ozegovic, B., Genbacev, B., and Milkovic, S. (1971). Ultrastructure of trans-plantable pituitary tumor cells producing luteotropic and adrenocorticotropic hormones. *J. Microsc. (Paris)* 12, 225–232.

Pasteels, J. L. (1961). Premiers résultats de culture combinée *in vitro* d'hypophyse et d'hypo-thalamus dans le but d'en apprécier la sécrétion de prolactine. *C. R. Acad. Sci.* 253, 3074–3075.

Pasteels, J. L. (1963). Recherches morphologiques et expérimentales sur la sécrétion de prolactine. *Arch. Biol.* 74, 439–553.

Pasteels, J. L. (1969). Nouvelles recherches sur la structure et le comportement des cellules hypophysaires en culture. *Mem. Acad. Roy. Med. Belg.* 7, 1–45.

Pasteels, J. L. (1972a). Tissue culture of human hypophyses: Evidence of a specific prolactin in man. *Lactogenic Horm., Ciba Found. Symp.* pp. 269–277.

Pasteels, J. L. (1972b). Morphology of prolactin secretion. *Lactogenic Horm., Ciba Found. Symp.* pp. 241–256.

Pasteels, J. L., Brauman, H., and Brauman, J. (1963). Etude comparée de la sécrétion d'hormone somatotrope par l'hypophyse humaine *in vitro* et de son activité lacto-génique. *C. R. Acad. Sci.* 256, 2031–2033.

Pasteels, J. L., Danguy, A., Frerotte, M., and Ectors, F. (1971). Inhibition de la sécrétion de prolactine par l'ergocornine et la 2-Br-α-ergocryptine: Action directe sur l'hypophyse en culture. *Ann. Endocrinol. (Paris)* 32, 188–192.

Pawlikowski, M., and Kunert-Radek, J. (1973). The effect of dibutyryl cyclic AMP on cell proliferation of adenohypophysis in organ culture. *Endokrynol. Pol.* 24, 307–310.

Peillon, F., Gourmelin, M., Brandi, A. M., and Donnadieu, M. (1972). Adénomes somato-tropes humains en culture organotypique, ultrastructure et sécrétion étudiées à l'aide de leucine tritiée. *C. R. Acad. Sci.* 275, 2251–2254.

Pelletier, G., Bornstein, M., Stern, J., and Puviani, R. (1972). Etude morphologique des effets du 3'5' adenosine mono-phosphate cyclique (AMP cyclique) sur l'adénohypophyse du rat en culture de tissu. *J. Microsc. (Paris)* 13, 383–390.

Petrovic, A. (1959). Etude cytologique au microscope électronique d'explants préhypophys-aires cultivés *in vitro*. *C. R. Acad. Sci.* 249, 1270–1271.

Petrovic, A. (1961). Recherches sur la préhypophyse en culture organotypique: Evolution structurale et action sur les organes effecteurs associés *in vitro*. Thesis, Imprimerie Alsatia, Colmar, Strasbourg.

Petrovic, A. (1963). Cytophysiologie de l'adénohypophyse des Mammifères en culture organotypique: Fonction gonadotrope, thyréotrope et corticotrope. In "Cytologie de l'adénohypophyse" (J. Benoit and C. Da Lage, eds.), pp. 121–136. CNRS, Paris.

Pradelles, P., Morgat, J. L., Fromageot, P., Oliver, C., Jacquet, P., Gourdji, D., and Tixier-Vidal, A. (1972). Preparation of highly labelled ³H-Thyreotropin releasing hormone (PG2-HIS-PRO (NH2) by catalytic hydrogenolysis. *FEBS Lett.* 22, 19–22.

Rappay, G., Gyevai, A., Kondics, L., and Stark, E. (1973). Growth and fine structure of monolayers derived from adult rat adenohypophyseal cell suspensions. *In Vitro* 4, 301–306.

Redding, T. W., Schally, A. V., Arimura, A., and Matsuo, H. (1972). Stimulation of release and synthesis of luteinizing hormone (LH) and follicle stimulating hormone (FSH) in tissue cultures of rat pituitaries in response to natural and synthetic LH and FSH releasing hormone. *Endocrinology* **90,** 764–771.

Reusser, F., Smith, C. G., and Smith, C. L. (1962). Investigations on somatotropin production of human anterior pituitary cells in tissue culture. *Proc. Soc. Exp. Biol. Med.* **109,** 375–378.

Sato, G., Zaroff, L., and Mills, S. E. (1960). Tissue culture populations and their relation to the tissue of origin. *Proc. Nat. Acad. Sci. U. S.* **46,** 963.

Siler, T. M., Morgenstern, L. L., and Greenwood, F. C. (1972). The release of prolactin and other peptides hormones from human anterior tissue cultures. *Lactogenic Horm., Ciba Found. Symp.* pp. 207–217.

Sinha, D. K., and Meites, J. (1966). Stimulation of pituitary thyrotropin synthesis and release by hypothalamic extract. *Endocrinology* **78,** 1002–1006.

Sonnenschein, C., Richardson, U. I., and Tashjian, A. H., Jr. (1970). Chromosomal analysis, organ specific function and appearance of six clonal strains of rat pituitary tumor cells. *Exp. Cell Res.* **61,** 121–128.

Stark, E., Gyevai, A., Szalay, K., and Acs, Zs. (1965a). Hypophyseal–adrenal activity in combined human foetal tissue cultures. *Can. J. Physiol. Pharmacol.* **43,** 1–7.

Stark, E., Gyevai, A., Szalay, K., and Posolaky, Z. (1965b). Secretion of adrenocorticotropic hormone by hypophysial cells grown in monolayer culture. *J. Endocrinol.* **31,** 291–292.

Steinberger, A., and Chowdhury, M. (1971). LH and FSH production in clone cultures of rat anterior pituitary cells. *Endocrinology* **88,** Suppl., Abstr. No. 64, A-74.

Steinberger, A., Chowdhury, M., and Steinberger, E. (1973a). Effect of repeated replenishment of hypothalamic extract on LH and FSH secretion in monolayer cultures of rat anterior pituitary cells. *Endocrinology* **92,** 12–17.

Steinberger, A., Chowdhury, M., and Steinberger, E. (1973b). Cultures of rat anterior pituitary cells with differential gonadotrophin secretion and tinctorial properties. *Endocrinology* **92,** 18–21.

Tashjian, A. H., Jr., Yasumura, Y., Levine, L., Sato, G. H., and Parker, M. L. (1968). Establishment of clonal strains of rat pituitary tumor cells that secrete growth hormone. *Endocrinology* **82,** 342–352.

Tashjian, A. H., Jr., Bancroft, F. C., and Levine, L. (1970). Production of both prolactin and growth hormone by clonal strains of rat pituitary tumor cells. Differential effects of hydrocortisone and tissue extracts. *J. Cell Biol.* **47,** 61–71.

Tashjian, A. H., Jr., Barowsky, N. J., and Jensen, D. K. (1971). Thyrotropin releasing hormone: Direct evidence for stimulation of prolactin production by pituitary cells in culture. *Biochem. Biophys. Res. Commun.* **43,** 516–523.

Thompson, K. W., Vincent, M. M., Jensey, F. C., Price, R. T., and Schapiro, E. (1959). Production hormones by human anterior pituitary cells in serial culture. *Proc. Soc. Exp. Biol. Med.* **403,** 102.

Tixier-Vidal, A. (1972a). Ultrastructure et immunocytologie des cellules gonadotropes en culture. *In* "Hormones glycoprotéiques hypophysaires" (M. Jutisz, ed.), pp. 49–68. Colloque INSERM., Paris.

Tixier-Vidal, A. (1972b). Ultrastructural features of gonadotropic cells in cultures. *Symp. Biol. Hung.* **14,** 43–58.

Tixier-Vidal, A., and Gourdji, D. (1965). Evolution cytologique ultrastructurale de l'hypophyse du Canard en culture organotypique. Elaboration autonome de prolactine. *C. R. Acad. Sci.* **261,** 805.

Tixier-Vidal, A., and Gourdji, D. (1970). Synthesis and renewal of proteins in duck anterior hypophysis in organ culture. *J. Cell Biol.* **46**, 130–136.

Tixier-Vidal, A., and Gourdji, D. (1972). Cellular aspects of the control of prolactin secretion in birds. *Gen. Comp. Endocrinol., Suppl.* **3**, 51–64.

Tixier-Vidal, A., and Picart, R. (1967). Etude quantitative par radioautographie au microscope électronique de l'utilisation de la DL-leucine ³H par les cellules de l'hypophyse du Canard en culture organotypique. *J. Cell Biol.* **35**, 501–519.

Tixier-Vidal, A., Kerdelhué, B., Bérault, A., and Jutisz, M. (1970). Cinétique de la sécrétion de l'hormone lutéinisante (LH) par l'antéhypophyse de Rat en culture organotypique: Influence d'une préparation purifiée du facteur hypothalamique de libération de LH (LRF). *C. R. Acad. Sci., Ser. D* **271**, 523–526.

Tixier-Vidal, A., Kerdelhué, B., and Jutisz, M. (1973). Kinetics of release of luteinizing hormone (LH) and follicle stimulating hormone (FSH) by primary cultures of dispersed rat anterior pituitary cells. Chronic effect of synthetic LH and FSH releasing hormone, *Life Sci.* **12**, Part I, 499–509.

Tixier-Vidal, A., Gourdji, D., and Tougard, C. (1974). A cell culture approach to the study of anterior pituitary cells. *Int. Rev. Cytol.*, in press.

Tougard, C. (1971). Recherches sur l'origine cytologique de l'hormone mélanophorotrope chez les Oiseaux. *Z. Zellforsch. Mikrosk. Anat.* **116**, 375–390.

Tougard, C., Kerdelhué, B., Tixier-Vidal, A., and Jutisz, M. (1973a). Light and electron microscopic localization of the binding sites of antibodies against ovine luteinizing hormones and its two subunits as revealed by the peroxidase labelled antibody technique. *J. Cell Biol.* **58**, 503–521.

Tougard, C., Tixier-Vidal, A., Kerdelhué, B., and Jutisz, M. (1973b). Cellular and intracellular localization of the luteinizing hormone and its two subunits in the male rat adenohypophysis as revealed by the peroxydase-labelled antibody technique. *Proc. Int. Congr. Endocrinol. 4th, 1972* Int. Congr. Ser. No. 256, Abstr. No. 619.

Tougard, C., Picart, R., Tixier-Vidal, A., Kerdelhué, B., and Jutisz, M. (1974). *In situ* immunochemical staining of gonadotropic cells in primary cultures of rat anterior pituitary cells with the peroxydase labeled antibody technique. A light and electron microscope study. *Int. Symp. Electron Microsc. Cytochem., 2nd, 1973* p. 163.

Vale, W., Grant, G., Amoss, M., Blackwell, R., and Guillemin, R. (1972a). Culture of enzymatically dispersed anterior pituitary cells: Functional validity of a method. *Endocrinology* **91**, 562–572.

Vale, W., Brazeau, P., Grant, G., Nussey, A., Burgus, R., Rivier, J., Ling, N., and Guillemin, R. (1972b). Premières observations sur le mode d'action de la somatostatine, un facteur hypothalamique qui inhibe la sécrétion de l'hormone de croissance. *C. R. Acad. Sci., Ser. D* **275**, 2913–2916.

Waelbroeck-Van Gaver, C., and Potvliège, P. (1969). Tumeurs hypophysaires induites par les oestrogènes chez le Rat. I. Activité fonctionnelle, Histologie et Ultrastructure. *Eur. J. Cancer* **5**, 99–117.

Willmer, E. N. (1965). Cell and tissues in culture introduction. *In* "Cell and Tissues in Culture" (E. N. Willmer, ed.), Vol. 1, pp. 1–18. Academic Press, New York.

Yasumura, Y., Tashjian, A. H., and Sato, G. H. (1966). Establishment of four functional clonal strains of animal cells in culture. *Science* **154**, 1186.

ULTRASTRUCTURE OF PITUITARY TUMOR CELLS: A CRITICAL STUDY

L. Olivier, E. Vila-Porcile, O. Racadot, F. Peillon, and J. Racadot

LABORATOIRE D'HISTOLOGIE-EMBRYOLOGIE (CNRS ERA 42), FACULTÉ DE MÉDECINE
PITIÉ-SALPÊTRIÈRE, PARIS, FRANCE

I. Introduction

Cytopathological and physiopathological problems appeared simple during the first three decades of the present century as pituitary tumors had just been defined in man (cf. Cushing, 1933). Physiopathologically speaking, three important types of tumors seemed to be differentiated: acidophil adenomas, basophil adenomas, and chromophobe adenomas. They were considered as deriving from the three cellular types which could then be identified in the pituitary by trichrome methods: chromophil cells with acidophil granules, chromophil cells with basophil granules, and chromophobe cells. Three clinical syndromes were related to these three physiopathological types: acidophil adenomas were responsible for acromegaly through a hypersecretion of growth hormone, basophil adenomas responsible for hypercorticism through hypersecretion of a corticotroph factor, and, finally, chromophobe adenomas were supposed to be without endocrine secretory activity. More or less implicitly, a correspondence was established between a normal cell and a tumor cell, on the one hand, and between the morphological aspects of cells and their functional aspects on the other.

Later this correspondence no longer appeared as strict. In fact, the increasing number of endocrinological observations in man as well as the systematic exploration of animal experimental tumors were to demonstrate a great variety of secreting tumors which did not fit into the previous classification and that many chromophobe tumors were actually secreting. It became "increasingly clear that conventional cellular morphology is often of little help in predicting physiological activities" (Russfield, 1966).

The study of pituitary tumors was therefore carried on with modern cytological methods of light and electron microscopy which lead to the subdivision of normal pituitary cells in a more refined way (see chapters by Herlant and by Farquhar et al.). Thus new data were gathered. However, for the time being, they do not permit the replacement of the former classification of tumors by a new one expressing both the stem cell line involved in tumorigenesis and its secretory capacities.

Several reasons account for this impossibility. First of all, studies on tumors using modern methods remain until now rather few. Besides, the identification of some cells is still difficult in tumors as well as in the normal pituitary, particularly as regards chromophobe cells which represent an important component of pituitary tumors. Finally, mechanisms involved in pituitary cell cytogenesis remain unknown both in fetal and in adult pituitaries.

Identification criteria and classification of normal adult pituitary cell types have been carefully reviewed in other chapters of this book (see chapters by Herlant and by Farquhar et al.). As concerned pituitary cytogenesis, we will briefly review the present state of this field. We will later emphasize the general features of pituitary tumors, both spontaneous and experimental, before describing their morphology.

II. Cytogenesis in Normal Pituitary

A. Fetal Pituitary

Classically, the embryonic pituitary consists of chromophobe cells, one part of which differentiates into acidophil and basophil cells, the second part remaining chromophobe cells (stem cells or classically "chief-cells," cf. below).

Recent studies show that cytogenesis is more complex and that before typical granular cells can be identified (Falin, 1961; Dubois and Dumont, 1966; Dubois,

1967; Conklin, 1968; Dupouy and Magre, 1973), different actively dividing embryonic cell types appear.

According to their chronological manifestation, embryonic cells are described as: (1) "marginal" cells (Murakami et al., 1968; Sano and Sasaki, 1969) lining the Rathke pouch as well as the "canaliculi" located in the core of the first cords (Dubois, 1968; Andersen et al., 1971); (2) small cells, completely agranular or "primordial" cells (Yoshimura et al., 1970; Daikoku et al., 1973) or "undifferentiated" cells (Yoshida, 1966); (3) cells with sparse granules of very small size (Fink and Smith, 1971), difficult to classify and therefore called "ambiguous" cells (Yoshimura and Harumiya, 1965; Yoshimura et al., 1970); (4) cells with larger granules that are still difficult to classify—"intermediate" cells (Yoshimura and Harumiya, 1965); and (5) cells with low granule content but exhibiting morphological features of functional types in the light and the electron microscope. They seem to correspond to the "Ubergangszellen" of Kraus (1926) or "transitional" cells of Severinghaus (1933).

Very early, some marginal cells of the first cords display features of follicular cells (see chapters by Herlant and by Farquhar et al.) which rapidly seem to constitute a definite stemline (Yoshida, 1966). However, one cannot exclude the possible role of the follicular cells as progenitor toward the other types (Andersen et al., 1971).

As concerns glandular cells, it has not been established whether unipotent functionally determined cells already exist among morphologically undifferentiated primordial cells, or if, on the contrary, these primordial cells are totally multipotent. In this last eventuality, it is not known what morphological stage corresponds to a functionally determined cell. One should note that recent studies showed that embryonic cells might display an early positive immunochemical reaction, although granules are not yet observable under the light microscope (Stokes and Boda, 1968; Dubois, 1971; Dubois et al., 1973; Setaló and Nakane, 1972).

B. Adult Pituitary

In adult pituitary, according to the cellular cycle theory (cf. Severinghaus, 1933, 1937; Romeis, 1940; Racadot, 1949), persistent chromophobe cells are mother cells which turn into acidophil or basophil cells by passing through a transition phase ("Ubergangszelle" of Kraus 1914, cited in Kraus, 1926). At the opposite, chromophil cells can revert to a chromophobe state.

At present, it is clear that formation of new cells can take place in adult pituitary. Although this has been previously questioned, it is demonstrated by the presence of mitoses (Hunt, 1942; Nouët and Kujas, 1973) as well as by the results obtained with labeled thymidine (Leblond and Walker, 1956; Hunt and Hunt, 1966; Mastro et al., 1969a,b; Goluboff et al., 1970; Stratmann et al., 1972).

Whenever an endocrine function is triggered by a physiological hormone imbalance (gestation) or through experimentation (ablation of target organs), there is a considerable increase in the number of corresponding functional cells which causes cord and gland hyperplasia. During the last few years three answers, at least, have been given to the question of the origin of these new cells.

(a) Some authors suppose that the chromophobe cells of light microscopy and the poorly granulated cells of electron microscopy are a reserve of mother cells (or stem cells). These elements completely undetermined or partially determined might be able to divide and differentiate into diverse functional cells (Ishikawa *et al.,* 1971; Yoshimura *et al.,* 1969; Ohtsuka *et al.,* 1971).

(b) Other investigators, questioning the existence of undifferentiated cells, believe that all pituitary cells can be listed under a morphofunctional category (Stratmann *et al.,* 1972). Hyperplasia of cells in either category can only be attributed to the division of cells belonging to the same category.

(c) Finally, other authors assume that in a given functional category differentiation is not definitive. In spite of their mature aspect, the cells from either functional category might be able to change into cells of another category acquiring at the same time the morphological aspect and the secretory capacity of the latter (Kwa and Feltkamp, 1965; Feltkamp and Kwa, 1965; Dingemans, 1969).

Such differences in opinion, often based on comparable experimental methods, underline the present difficulties in classifying cells whenever they diverge from a functional morphotype, the only present reference. This also explains the problems encountered in the interpretation of tumoral pituitary cells.

III. General Features of Pituitary Tumors

The pituitary tumor material described in the literature is greatly diverse in origin. It includes:

Numerous varieties of *spontaneous tumors* observed in different species (cf. Russfield, 1966). In mammals, the most detailed studies have been done in man (important literature, see Section IV,A). These are also studies on the dog, showing similarities with man (Capen and Koestner, 1967; Capen *et al.,* 1967a,b).

Primary experimental tumors, induced by ionizing irradiation and mostly by endocrine homeostasis disturbance (pharmacological or surgical removal of target organs, estrogen administration, or pituitary transplantation). These last tumors develop exclusively when the inducing endocrine disturbance is permanent (host defined as "conditioned," stage defined as "hormonodependent") (cf. review in Gardner *et al.,* 1953; Furth and Clifton, 1966; Russfield, 1966).

Autonomous experimental tumors, obtained through successive transplantations (cf. Gardner *et al.,* 1953; Furth and Clifton, 1966). After the first transplantations, the tumor growth still remains fully dependent on the hormonal inducing disturbance ("conditioned host"). After successive transplantations, the tumors acquire ability to develop in normal hosts: they are "autonomous."

In some cases, the growth of autonomous tumors is still responsive to the hormonal disturbance ("responsive autonomous tumors"). In the other cases, the hormonal imbalance does not influence the tumor growth ("nonresponsive tumors"), and in the last case, the opposite, the initially inhibiting hormonal conditions now stimulate the tumor growth ("reversely responsive tumors") (cf. Furth, 1968, 1969).

Whatever the tumors they may be considered under three different headings: proliferation, endocrinology, and cytology.

A. Proliferative Aspects of Tumors

Cellular proliferation can take on two different anatomical aspects involving different tumorigenesis phenomena as well.

The first, frequently observed in man, consists of neoformed cells developing into a limited tumor, differing from the neighboring pituitary parenchyma (Roussy and Clunet, 1911). Such tumors have variable sizes. Some are nodules seen only under microscopic examination (Costello, 1935). Others can distort the gland or even extend out of it, thus compressing the neighboring areas (optic nerve, chiasma, and encephalon).

In the second form, essentially observed during experimental induction, proliferation of stimulated cells takes place through the whole gland among residual cells. A hyperplasia thus develops, described as "diffuse adenomatous hyperplasia" (Gardner et al., 1953), or "pituitary enlarged," or "pituitary adenomatously enlarged" (Kwa, 1961) when the pituitary weight is above the upper limit of the "normal" weight in the studied animal strain. Within this hyperplasia, independently proliferating cellular islets may trigger tumor nodules (Wolfe and Wright, 1938; Lacour, 1950; Halmi and Gude, 1954; Lacassagne, 1959). Whenever the pituitary weight is multiplied by 3 or 4 the expression "tumor" is arbitrarily used by the authors (Gardner, 1941; Dent et al., 1955; Clifton and Meyer, 1956).

The proliferating potential of these neoplastic cells is more or less easy to predict according to the tumor strain. Generally, histological examination of sections yields but approximate indications. In fact, the presence of mitoses either normal or abnormal is a good indication of rapid growth (Dott and Bailey, 1925). Their scarcity or their absence, however, does not allow for any conclusion as the sampling might have been done between scarce nuclear divisions. Likewise, the morphological "anomalies" of interkinetic nuclei (Kernohan and Sayre, 1956; Lewis and Van Noorden, 1972) do not allow definitive conclusions on their proliferative potential, for polyploidy is frequent among pituitary cells, even normal (Deminatti and Vendrely, 1959; Fand, 1973). Moreover, nuclear and principally nucleolar hypertrophy observed in tumors may be related to hormonal hypersecretion and, as well as, to protoplasmic synthesis phenomena related to tumor growth.

1. *In Man*

It is sometimes possible to evaluate tumor growth by comparing the duration of symptoms and the tumor volume. These measurements show different growth rates within a given tumor type. This may apply to tumors developing over a few months as well as tumors evolving over several years and even several decades. As a whole and excluding rare metastasizing adenocarcinoma (cf. Kernohan and Sayre, 1956; Russfield, 1968; d'Abrera *et al.*, 1973), human pituitary tumors justify the expression "adenoma." In spite of their benignity, the human tumors show frequent karyotypic anomalies. These deviations, different from one tumor type to the other, suggest that clonal populations arise during tumor proliferation (Mark, 1971).

2. *In Animals*

Measurement of tumor growth seems easier than in man, for it is possible to know the tumor weight as a function of time (Kwa, 1961). In fact there are individual variations or sources of errors owing to such contingent phenomena as vascular congestion or intratumoral hemorrhagea, frequently occurring after some time of the evolution (Cramer and Horning, 1936; McEuen *et al.*, 1936; Zondek, 1936; Wolfe and Wright, 1938).

The growth rhythm of the hormone-dependent primary tumors depends on the animal species and on its genetic constitution (cf. Gardner *et al.*, 1953; Russfield, 1966). Measurement has been partly achieved with macroscopic methods (weight × time factor, Kwa, 1961) and partly with tritiated thymidine labeling methods (Messier, 1966; Clifton, 1966). Contradictory proliferation models had been proposed. The first models suggested tumor doubling times from 20 to 50 days (Clifton and Meyer, 1956; Kwa, 1961). But more recent studies indicate that the proliferative process is very complex, certainly involving several diversely growing cellular populations in the same experimental pituitary tumor (Clifton, 1966).

Recent cytogenetic studies, at that hormone dependence stage, show variable chromosomic formulas. Some tumors seem to retain an euploid formula (Bayreuther, 1960), others become rapidly aneuploid but in a variable way (Di Paolo *et al.*, 1964; Waelbroeck-Van Gaver, 1969). This suggests that there are early chromosomal modifications at the beginning of a cellular transformation (Waelbroeck-Van Gaver, 1969).

After autonomization by successive transplantations, the tumor grows faster. This corresponds to an actual malignant transformation of the tumors, demonstrated by their infiltrating capacity and their metastatic possibilities (Bates *et al.*, 1957; Waelbroeck-Van Gaver and Potvliège, 1969). Simultaneously, cytogenetic anomalies are always present. In each tumor variety, aberrant strains arise which differ from each other (Di Paolo *et al.*, 1964; Waelbroeck-Van Gaver, 1969).

B. Endocrine Aspects of Tumors

From this point of view, pituitary tumors can be divided into two categories: functional and nonfunctional, according to the presence or the absence in the host of hormonal hypersecretion signs.

Evaluation of functional activity is accomplished through testing in the host both the tumor and the residual hypophysis. The latter retains only a very low secretory activity except in the case of an oversecretion of prolactin due to a mechanical disconnection of the hypothalamus (Herlant *et al.,* 1966; Young *et al.,* 1967).

The evaluation of the functional activity rests on different criteria:

(a) Morphological modifications of target-tissues stimulated by the overdose of hormones secreted by these tumors, such as somatic deformities and hyperplasia of thyroid, adrenals and mammary glands.

(b) Morphological and pathological modifications indirectly induced by the hypersecretion of a target gland stimulated by hyperpituitarism, e.g., Cushing's syndrome.

(c) Assay of hormone rates secreted by target organs stimulated by hyperpituitarism.

(d) Direct assays of the plasma level of hypophysial hormones. Results are different depending on whether evaluated by bioassay or radioimmunoassay as these methods do not necessarily investigate the same sequence in the hormone molecule (Furth, 1968).

The two first criteria have been, and still are, the most used as they are easy to apply and usually highly positive. On the other hand, either their negativity or the inability to apply them do not exclude any functional activity.

Tumor functional activity is the final step of intracellular synthesis and of a hormonal release process. Between these two stages, a storage phase is intercalated. The intratumoral hormonal storage process has been studied in a small number of tumors (Peake *et al.,* 1969). It differs from one variety to another and in a given variety it is sometimes influenced by exogenous factors (Buonassisi *et al.,* 1962; Sinha and Meites, 1967; Tashjian *et al.,* 1970, 1971; Hagen *et al.,* 1971). In some cases, the hormonal concentration related to the tumor weight is in fact either equal (Steelman *et al.,* 1956; Bates *et al.,* 1957, 1962) or clearly above the normal pituitary concentration (Bates *et al.,* 1957; Cohen and Furth, 1959; Furth and Clifton, 1966). In other cases, it is clearly inferior (Steelman *et al.,* 1956; Bates *et al.,* 1957, 1962; Buonassisi *et al.,* 1962; Pelletier *et al.,* 1971) in spite of the plasma hormonal excess. This discrepancy between tumor concentration and plasma concentration can be explained either by the mass effect of the tumor (the great number of secreting cells compensating their individual low activity) or by a shortening of the intermediate storage phase.

The study of hormone elaboration by functional tumor cells has recently been indirectly approached through cultures either organotypic (Siler *et al.,* 1972) or in suspension (Tashjian *et al.,* 1968; Kohler *et al.,* 1969). These techniques have shown that tumor cells are able to elaborate and to secrete hormones, without any stimulating factors (Peillon *et al.,* 1972; Guyda *et al.,* 1973). In some cases, these factors remain able to modulate the activity of cultured cells even when they come from autonomous tumors (Buonassisi *et al.,* 1962; Tashjian *et al.,* 1968, 1970, 1971; Bancroft *et al.,* 1969; Bancroft and Tashjian, 1971) (see chapter by Tixier-Vidal).

Finally, if one considers the tumors from a functional point of view, numerous varieties may be described according to the type of hormone secreted. It appears that any type of pituitary hormone may be secreted by tumors; however, gonadotropic hormones seem to be rarely secreted (cf. Furth and Clifton, 1966; Russfield, 1966). Furthermore, in some cases the tumor secretes only one hormone and in other cases several hormones (see below).

C. Cytological and Histological Aspects of Tumors

The microscopic organization of tumors (and to a lesser degree that of hyperplasia) is quite different from that of the normal pituitary. The organization into cords is replaced by a massive, sinusoidal or papillary one (Kernohan and Sayre, 1956) with modifications of the connective tissue–vascular stroma explaining the frequent hemorrhagic or cystic degenerative phenomena observed. It can be noticed that neocapillaries of pituitary adenomas may however retain the endothelial fenestrations characteristic of normal pituitary capillaries. They are only decreased in number (Schechter, 1972; Hirano *et al.,* 1972).

The study of the glandular cells of the new parenchyma raises several basic cytological problems. Do the cells in each type of tumor reproduce the normal pituitary cells either adult or fetal, or could they be new abnormal forms? Do the cells in each tumor belong to one or several defined cellular stemlines? What morphological features allow one to conclude whether a tumor cell is still secreting or not? Answers to these questions remain uncertain because of difficulties in identifying cell types, especially when fixation has been inadequate (autopsy material). Since it now appears impossible to describe the tumors only on the basis of their cytology, their study on the basis of a functional classification seems to be more appropriate.

IV. Morphological Features of Pituitary Tumors

A. Description of Spontaneous Tumors (Human Tumors)

There are four general types of functional tumors in man: adenomas in acromegaly, in Cushing's disease, in amenorrhea–galactorrhea and thyrotropic adenomas. In most of the tumor cases published in the literature, hyperpituitarism has been estimated only in reference to clinical data and to exploration of target organs. It concerned but one hormone at a time: respectively, STH, ACTH-MSH, LTH, and TSH. But, we presently know, according to more developed endocrinological studies (often including pituitary hormone assays in plasma and sometimes even in the tumor), that within these four general types there are variants where two hormones can be simultaneously secreted: STH + LTH (Guyda *et al.,* 1973), STH + ACTH (McCormick *et al.,* 1951; Mautalen and Mellinger, 1965), LTH + ACTH (Young *et al.,* 1967; Mahesh *et al.,* 1969), and STH + TSH (Lamberg *et al.,* 1969).

In the following descriptions we are comparing tumor cells to normal cellular types previously described in man under the light microscope by Romeis (1940), Pearse (1952, 1962), Foster (1956), Ezrin and Murray (1963), Pearse and Van Noorden (1963), Purves (1966), Herlant and Pasteels (1967), Russfield (1968), Leleux and Robyn (1971), Robyn *et al.* (1973), and Phifer *et al.* (1973) and in the electron microscope by Foncin and Le Beau (1963, 1964), Foncin (1966, 1971), Paiz and Hennigar (1970), von Lawzewitsch *et al.* (1972), and Deaton and Dugger (1972). Figure 1 shows a general aspect of normal human pituitary in the electron microscope.

Our results are based on the study of 96 cases of acromegaly (of which 24 by electron microscopy), 25 cases of amenorrhea–galactorrhea (12 by electron microscopy), 25 cases of Cushing's disease (16 by electron microscopy), and 150 cases of chromophobe adenomas (20 by electron microscopy). We will more particularly study adenomas in acromegaly as they are the most documented and they actually summarize all problems raised by pituitary cytopathology.

1. *Adenomas in Acromegaly (Somatotroph Adenomas)*

The following types have been described:

(a) Adenomas almost exclusively composed of granular cells, with few chromophobe cells: "pure" acidophil adenomas of Dott and Bailey (1925), typical chromophil adenomas of Young *et al.* (1965) (Fig. 2).

(b) Adenomas composed of mixture in variable proportions of granular cells, poorly granulated, and chromophobe cells: transitional adenomas (Kraus, 1914, cited in Kraus, 1926; Biggart and Dott, 1936), mixed adenomas (Dott and Bailey, 1925), intermediary adenomas (Roussy and Oberling, 1933), oligochrome adenomas (Costero and Berdet, 1939) (Fig. 4).

(c) Adenomas exclusively constituted of chromophobe cells (Fig. 3) recently described (Schelin, 1962; Lewis and Van Noorden, 1972).

By light microscopy, *heavily granulated cells* are large sized elements. Their nucleus with a polyploid aspect is frequently notched, and includes a prominent nucleolus. The granules are orange-stained with Herlant's tetrachrome method.

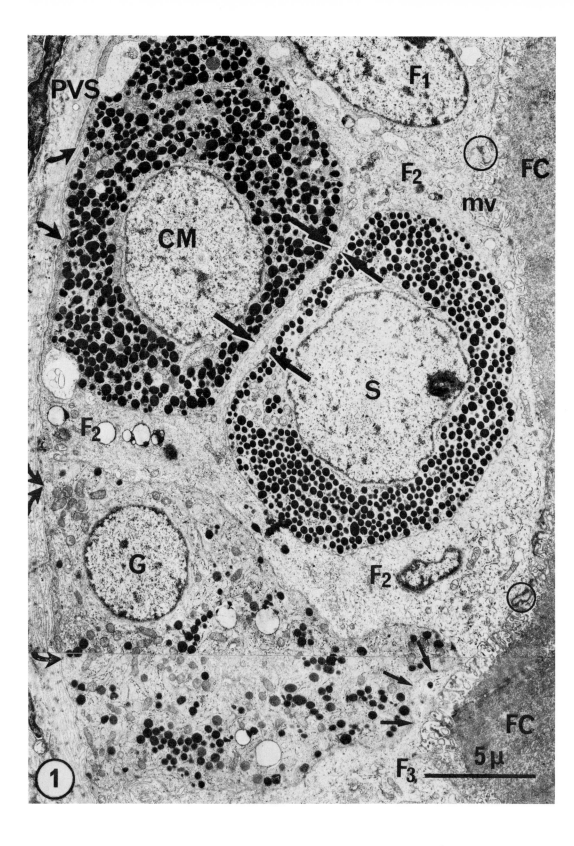

PVS, F₁, F₂, FC, mv, CM, S, G, F₃, 5 μ

In the electron microscope, the range of size of the granular diameters is narrow (average size 350 nm). The intracellular organization is of a "normal" type with, however, frequent hypertrophy of mitochondria (Fig. 6), of the rough endoplasmic reticulum, and of the Golgi apparatus. These cells certainly reproduce and retain the main characteristics of the somatotroph cells (Figs. 5 and 12).

Poorly granulated cells have a more variable aspect compared to the former (see sampling Fig. 4), as the variations concern both the amount of granules and the intracellular organization. Granules remain orange-stained on paraffin sections with Herlant's method.

In the electron microscope, the granules can be separated into two categories (Schelin, 1962). The first category is made of granules scattered in the cytoplasm (Fig. 4) showing rather important size variations in a given cell and also from one cell to another. However, their average sizes (350 nm) as well as their internal structure studied by Rambourg's method for glycoproteins (1967, 1969) are still compatible with that of a somatotroph granule (Porcile *et al.,* 1968) (Figs. 9 to 11).

The second category of granules is much smaller, also showing significant intra- and intercellular variations. Their average diameter is approximately 150 nm. They have a tendency to be distributed along the cellular membrane (Fig. 10). After a glutaraldehyde fixation and uranyl-lead staining, the core of the granule looks separated from the surrounding membrane by a narrow light space. Studied with the Rambourg's method, this granular core seems limited by a glycoproteic shell (Fig. 10).

In these cells the ergastoplasm is scattered and may be replaced by smooth reticulum dilated into large cisternae (Fig. 3), or at the opposite forms masses of narrow and intermingled tubes (Fig. 13). The Golgi apparatus is often hypertrophied and shows many vesicles (Fig. 14). Granules in formation are often visible in this area. Primary and secondary lysosomes are frequently located in its neighborhood (Figs. 13 and 14).

Mitochondria are often more numerous and may conglomerate. Their sizes

---

FIG. 1. Nontumoral human pituitary (electron micrograph montage). Global view of a cellular cord with a perivascular connective space (PVS) and a follicular cavity (FC). Three granular glandular cells: S (somatotroph), CM (corticomelanotroph), and G (gonadotroph), separated by the folliculostellate cells F₁, F₂, F₃. Along the follicular cavity, the folliculostellate cells show microvilli (mv) and are bound by junctional complexes (circles). They completely isolate granular cells from the follicular cavity (cf. arrows at the border of cell G). Within the cord, processes from F₂ cell separate the glandular cells from each other (note the thin process marked by opposite arrows). Alongside the perivascular connective tissue space, "footlike" processes of folliculostellate origin lie against the pericordonal basal lamina (curved arrows). They isolate the granular cells from the connective tissue area. The multiplication of basal lamina within the connective perivascular space is noteworthy.

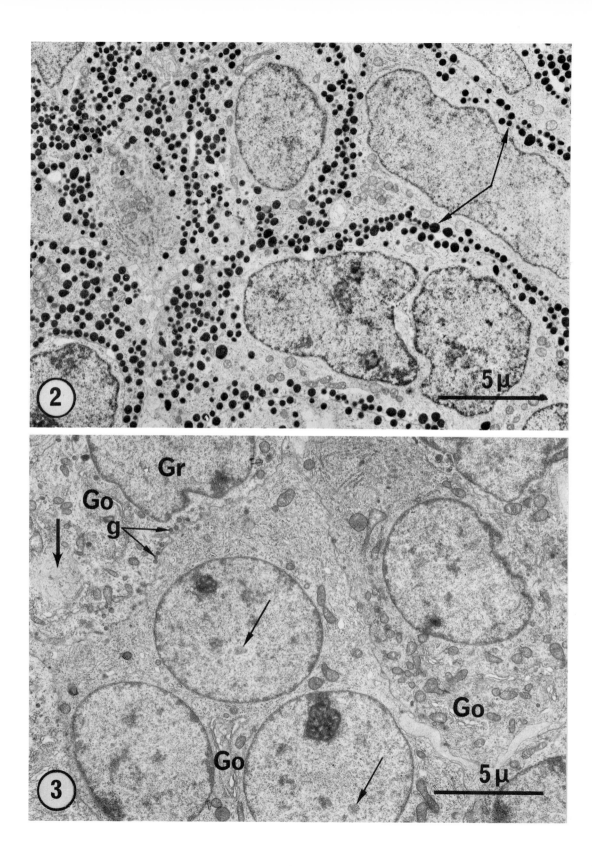

may be modified as well as the density of their matrix and the disposition of their cristae, leading to "inversed" pictures (Fig. 7).

Spheroid filamentous aggregates are frequently observed anomalies of poorly granulated cells (Racadot *et al.,* 1964; Olivier *et al.,* 1965; Cardell and Knighton, 1966; Schochet *et al.,* 1972). These formations can reach up to 5 μm in diameter and are easily seen under the light microscope; they appear like paranuclear "balls" with a hyalin aspect (Müller, 1954; Brilmayer *et al.,* 1957). They are made of intermingled submicroscopic fibrils, enclosing in their network a small number of cellular organelles: tubules of the smooth reticulum, polysomes, secretory granules, and mitochondria isolated or packed together. Golgi elements are often located at their periphery (Fig. 15).

Finally the poorly granulated cells retain some characteristics of the somatotroph cell type. Indeed adenomas with a majority of poorly granulated cells are associated with clinical acromegaly, a strong plasma STH activity, and are able to secrete STH *in vitro* when cultivated (Peillon *et al.,* 1972). But on the other hand the cells differ from the normal somatotroph type in that anomalies in internal organization and smaller size of the granules are observed (compare Figs. 9 and 10). The presence of such small granules might be related to a high level of secretory activity as seen in normal pituitary tissue (see chapter by Herlant). They might also contain another hormone such as prolactin since 20 to 40% of acromegalic patients show prolactin hypersecretion (cf. Jacobs and Daughaday, 1973). However, in some observations on acromegaly with galactorrhea, the prolactin secretion seems to be related to a second cell population, probably prolactin cells, scattered among the poorly granulated somatotroph cells (Peillon *et al.,* 1970; Racadot *et al.,* 1971; Guyda *et al.,* 1973). In these observations, the presence of both morphofunctional cell types suggests that two independent secretory cellular lines are involved in the tumor process.

In the electron microscope, *chromophobe cells* are very polymorphous, either in size or internal organization. Their only common point is a great scarcity of small granules (less than 150 nm), usually lying against the plasma membrane (Fig. 3). After the application of Rambourg's method, these generally appear like the granules of the poorly granulated cells described above. In some cells however the glycoprotein reaction involves the whole granule (Fig. 11).

Figs. 2 and 3. Pituitary adenomas in acromegaly. Fig. 2. Adenoma with heavily granulated cells. Compact parenchyma made up of granular cells. The granules are either scattered in the whole cytoplasm or set along the plasma membrane (double arrow). Nuclei are irregular. Fig. 3. Adenoma with almost agranular cells ("chromophobe" cells of light microscopy). Compact parenchyma made of large, apparently multinucleated agranular cells. Their dark cytoplasm is invaded by cisternae of the ergastoplasm and free ribosomes. Nuclei are regular and include a prominent nucleolus and nuclear bodies (transverse arrows). One cell (Gr) shows some peripheral small granules (double arrows) and a "spheroid filamentous aggregate" (vertical arrow; cf. Fig. 15). The small size of the Golgi areas is to be noted (Go).

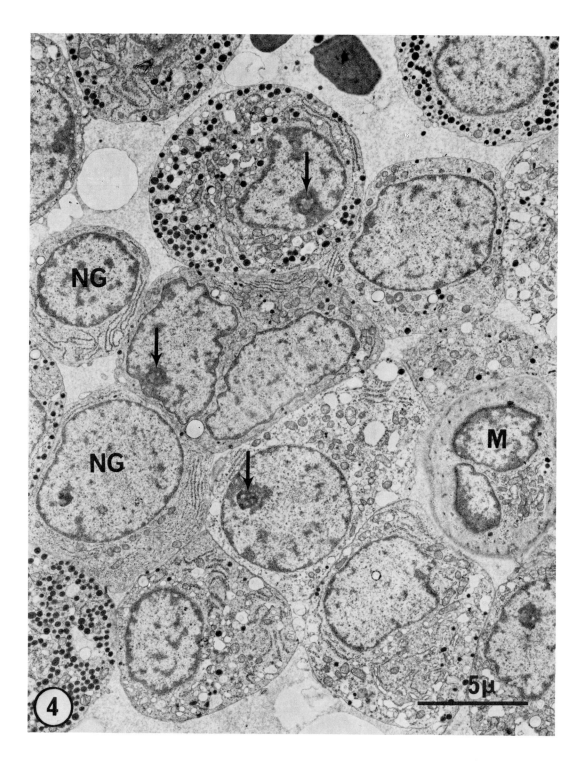

4

The cell nuclei have a variable size with a large nucleolus and important pleo-morphism (Fig. 3).

In a few cells the cytoplasm is dark due to a high concentration of free ribo-somes (Figs. 3 and 16). In other cells, it is light with a vacuolar aspect and the Golgi apparatus is no longer recognizable (Fig. 16).

The mitochondria vary in number. They are either very few or numerous lead-ing to an oncocyte-like aspect. Very often, their structure is abnormal, sometimes with concentric cristae and even inclusions (Fig. 8).

Finally, one sometimes observes in "chromophobe" cells spheroid filamentous aggregates resembling those described in poorly granulated cells (Fig. 3). In addi-tion to other features, these formations possibly indicate a parenthood between almost agranular cells and poorly granulated cells.

Since chromophobe tumors in acromegaly are always accompanied by a high plasma level of growth hormone, even higher than that observed in granular cell adenomas (Lewis and Van Noorden, 1972), we are allowed to interpret chromo-phobe cells as secreting cells. The small number and size of their secretory granules may be related to an immediate release with a short storage phase (Klotz *et al.,* 1964). Important modifications in the cavitary systems (endoplasmic reticulum and Golgi apparatus) associated with mitochondrial anomalies in chromophobe cells could also be related to modifications of the usual process of hormone syn-thesis and secretion, the latter even taking place without any granular stage.

2. *Adenomas of the Amenorrhea–Galactorrhea Syndrome*

The presence of a pituitary tumor associated with an amenorrhea–galactorrhea syndrome in women or with a gynecomasty syndrome in men has been known for a long time. But the hypersecretion of prolactin responsible for galactorrhea has been only recently related to the tumor and not to the residual normal pituitary (Herlant *et al.,* 1965; Linquette *et al.,* 1967; Peake *et al.,* 1969; Peillon *et al.,* 1970; Racadot *et al.,* 1971; Herlant and Pasteels, 1971; Le Beau and Foncin, 1972).

These tumors contain both granular erythrosinophilic cells and chromophobe cells with a large nucleus and prominent nucleolus. Scarce granules are generally found as a thin marginal line along the cell membrane. In the electron microscope, they vary in size between 500 and 300 nm (Peake *et al.,* 1969). Some classic features of prolactin cells remain, particularly a typical lamellar ergastoplasm,

FIG. 4. Pituitary adenoma of "mixed type" in acromegaly. Granular, poorly granulated and nongranulated (NG) cells, making up a nonorganized tumoral parenchyma, dissociated in some places by an amorphous material. The nongranulated cells have a well-developed ergastoplasm. All the nuclei present a segregation of nucleolar components (arrows). M, cell with a myoid peripheral differentiation, normally belonging to the pericapillary space, found here in the epithelial compartment after the disruption of the pericordonal basal lamina.

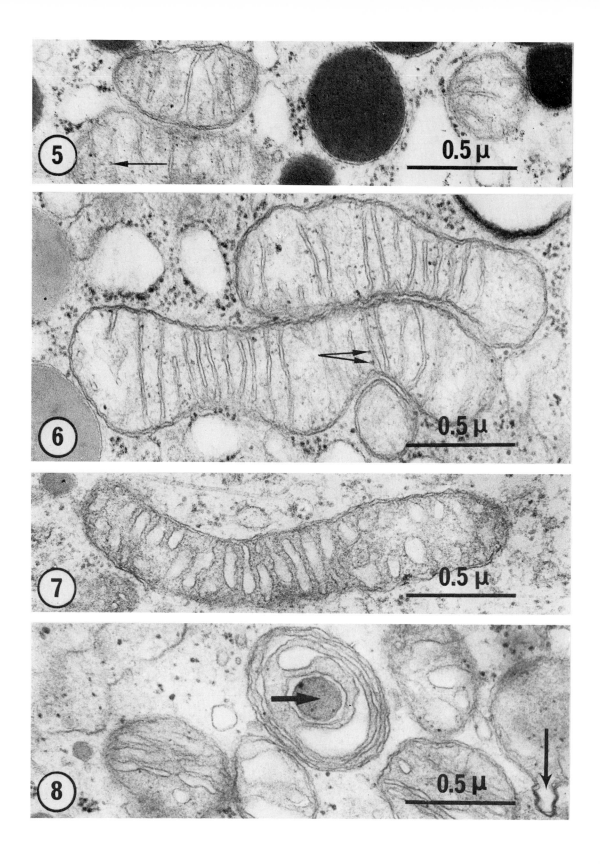

but anomalies of the other organelles and numerous secondary lysosomes are frequent.

Chromophobe cells retain a lamellar ergastoplasm and as in acromegaly exhibit frequent cytological anomalies.

3. Adenomas and Tumors in Cushing's Disease

Cushing's disease is a hypercorticism due to adrenal hyperplasia of pituitary origin (Cushing, 1932). Cushing thought that the pituitary lesion was an adenoma. Later this notion was modified. At present, from systematic autopsies it is known that the frequency of the tumor is approximately 50% in these patients. 40% are adenomas which, due to their small size, can only be identified in pituitary sections. The other 10% are macroscopic tumors that can even lead to compression symptoms (cf. collected statistics in Racadot *et al.,* 1967; Bricaire *et al.,* 1971).

These tumors, studied by light microscopy, have been differently interpreted. Some authors like Cushing identified them as basophil tumors, others as chromophobe tumors and even some others, as a variety of tumors with acidophil granules. (See bibliographic review in Racadot *et al.,* 1966a,b, 1967.) They have been rarely explored with electron microscopy (Foncin and Le Beau, 1964; Bergland and Torack, 1969; Olivier *et al.,* 1972).

Our personal experience covering 25 cases, showed that the cells which greatly vary in their size, their shape, and their nuclear anomalies, could be classified into heavily granulated cells, poorly granulated cells (marginal granule line), and chromophobe cells, just as in acromegaly. When visible, all the granules have staining characteristics in common: cyanophilia, positive reaction with PAS, i.e., they are basophil. Diversity, however, is found in the electron microscope. On the whole, granules differ in their shape (rounded or slightly polyhedral) but in a given cell they form a rather homogeneous population (Figs. 20 to 22).

Two types of data allow one to consider these granular cells as secreting ACTH or MSH or both hormones. Hormonal assays on some tumors led to the conclusion that they contain ACTH (Bahn *et al.,* 1960; Montgomery, 1963) and

---

Figs. 5, 6, 7, and 8. Mitochondria. Fig. 5. Nontumoral pituitary. Small mitochondria in a somatotroph cell, with fine cristae and clear matrix. Intramitochondrial ribosomes (arrow), smaller and less dense than cytoplasmic ribosomes. Fig. 6. Pituitary adenoma in acromegaly (granular cell). Same aspect as on Fig. 5 (fine cristae, clear matrix), but larger size (2 μm). Mitochondrial ribosomes (double arrow). Fig. 7. Pituitary adenoma in acromegaly (poorly granulated cell). Different aspect of a large mitochondrion: dense matrix, enlarged cristae. "Inversed" picture compared to previous figures. Fig. 8. Pituitary adenoma in acromegaly. Almost agranular cell ("chromophobe"). Highly modified mitochondria: cristae have either a longitudinal or a concentric disposition and are even surrounding a more dense inclusion (horizontal arrow). "Bleb" on the periphery of a mitochondrion without cristae (vertical arrow).

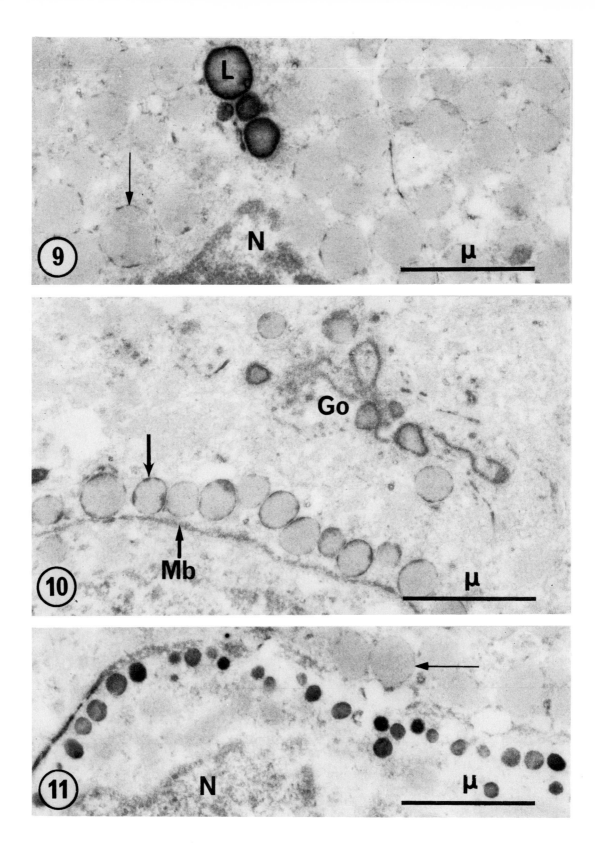

MSH (Bahn *et al.,* 1960; Nelson and Sprunt, 1965; Abe *et al.,* 1967). On the other hand, the observed granular cells correspond to the corticomelanotroph or melanotroph cells described in normal human pituitary (Leznoff *et al.,* 1962; Pearse and Van Noorden, 1963; Brozman, 1967; Breustedt, 1971; Phifer *et al.,* 1970; Dubois, 1973). Cytoimmunological explorations of our personal cases (M. P. Dubois, personal communication, 1973) favor the notion that they contain either ACTH or MSH or both hormones simultaneously.

An original feature of the granulated cell in Cushing's disease adenomas is the frequent occurrence of bundles of filaments (Fig. 19) associated with secondary lysosomes (Olivier *et al.,* 1973a). With the exception of nuclear and mito-chondrial anomalies, such cells are indistinguishable from Crooke's cells observed in nontumoral pituitary of patients with endogenous or exogenous hypercorticism (Fig. 18). Crooke cells are considered as modified melanotrope or corticomelano-troph cells in response to a high plasma level of corticoids (cf. Porcile and Raca-dot, 1966; Kracht and Hachmeister, 1967; Wagermark and Wersall, 1968; Dourov and Herlant, 1970; De Cicco *et al.,* 1972). It is interesting to notice that such a reactivity to glucocorticoids is retained in tumor cells.

4. *Thyrotroph Adenomas*

The thyrotroph tumors have not been studied with the electron microscope, but only with the light microscope.

Most cases surveyed are seen in relation to hypothyroidism. They can be com-pared to animal observations after suppression of the thyroid function (see p. 256) (cf. literature cited in Linquette *et al.,* 1971). Hyperplasia or possibly adenoma have been described, with chromophobe, basophil, or amphophil cells considered as thyrotroph cells (Russfield, 1958; Herlant *et al.,* 1966).

Functional adenomas causing hyperthyroidism seem to be rare. In the cases observed, the adenoma is chromophobe (Werner and Steward, 1958; Nyhan and Green, 1964; Jackson, 1965) or composed of more or less abnormal thyrotroph cells (Linquette *et al.,* 1969).

Finally, if one considers the diverse types of adenomas from the point of view of the origin of the tumor cells, one is forced to admit that the electron microscope is of little help in solving such a problem. Classic authors considered the chromo-phobe cells in functional tumors as stem cells, without any secretory activity. At

---

FIGS. 9, 10, and 11. Pituitary adenoma in acromegaly (Rambourg's method: phospho-tungstic acid at low pH after glycolmethacrylate embedding). Fig. 9. "Typical" somatotroph granule, showing only a thin dark line at its periphery (arrow). L, lysosomes strongly reactive. N, nucleus. Fig. 10. Smaller granules than those of Fig. 9, lying along the plasma membrane (Mb). They are surrounded by a reactive shell (arrow). The saccules of Golgi apparatus (Go) are markedly stained. Fig. 11. Whole stained small sized granules, at the margin of an uni-dentified cell. Upper right, "typical" somatotroph granules with dark reactive line (arrow).

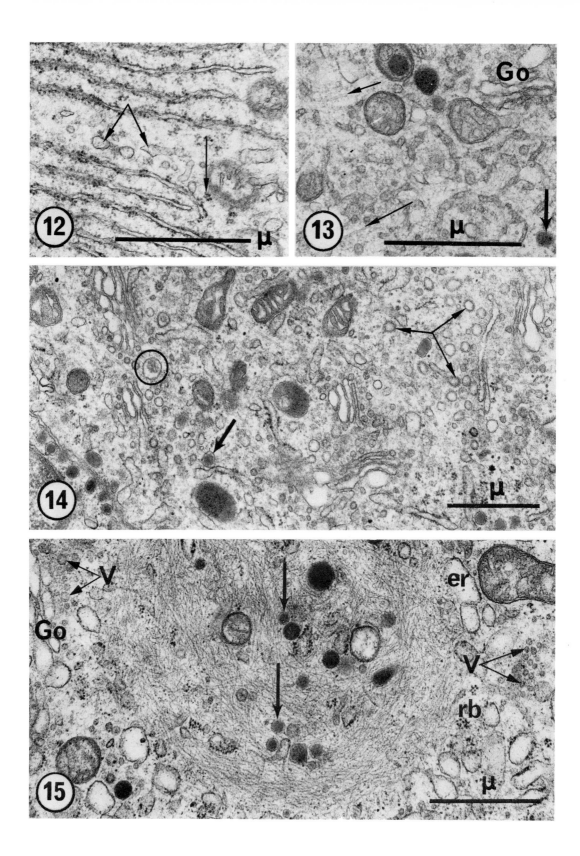

present, such a hypothesis cannot be completely admitted since (1) it is known that chromophobe cells might be secretory in the normal pituitary as well as in the tumors (see above) and (2) it is also known that in adenomas, granular cells as well as chromophobe cells can divide (Cushing, 1933).

Finally, anomalies observed in poorly granulated cells or in almost agranular cells raise the question whether these cells still correspond to normal cells, or whether they are "transformed" cells, i.e., anaplastic (cf. Bernhard, 1969). Tissue anarchy regularly associated with tumors containing these poorly granulated or "almost agranular cells" favors this hypothesis. Usually, a fragmentation or even a lack of the pericordonal basal lamina separating epithelial cords from the pericapillary connective space is observed (Gusek, 1962; Hirano *et al.*, 1972; Schechter, 1972). Modifications in intercellular relationships can also be involved: the intercellular spaces can enlarge and be filled with an amorphous material leading to a partial parenchymal dissociation (Fig. 4) or can be narrowed leading to the formation of a compact parenchyma (Figs. 2, 3, and 16).

These data could explain the apparent contradiction in the first results given by karyotypic analysis of human tumors. Although they exhibit a chromosomal aneuploidy as usually observed in malignant tumors (Mark, 1971), their evolution is benign (cf. Levan, 1969).

5. Nonfunctional Tumors

Tumors are said to be nonfunctional when there is neither clinical nor biological signs of secretion. The validity of this diagnosis is, however, related to the accuracy of the tests used. Some tumors held to be nonfunctional a few years ago are at present considered as prolactin secreting (Jacobs and Daughaday, 1973). It is beyond doubt, however, that some adenomas have no secretory activity as shown by the pituitary defect in these patients which reaches a complete hypopituitarism.

Most nonfunctional tumors *are chromophobe adenomas* with the exception of some granular adenomas too small to be secretory. McCormick and Halmi (1971)

FIGS. 12–15. Ergastoplasm. Fig. 12. Acromegaly (heavily granulated cell). Numerous flattened and parallel cisternae of the ergastoplasm. Between them, smooth cisternae are intermingled (double arrow). Rosettes of ribosomes (vertical arrow). Fig. 13. Acromegaly (poorly granulated cell). Smooth endoplasmic reticulum only. Lysosomal formations of various types, small granulations (vertical arrow) and microtubules (transverse arrows). Fig. 14. Acromegaly. Golgi apparatus (granular cell). The Golgi apparatus of this cell is hypertrophied, characterized by a proliferation of smooth and coated vesicles (triple arrow). Some vacuoles contain condensing material (circle). Small granules (arrow), lysosomes and tubules are visible. Fig. 15. Acromegaly: "spheroid filamentous aggregate" (poorly granulated cell). Detail of filamentous aggregate with intermingled filaments and included mitochondria, lysosomes, and small granules (arrows). On the periphery of the aggregate, numerous enlarged ergastoplasm cisternae (er), rosettes of ribosomes (rb), and stacked up vesicles (V). The Golgi apparatus (Go), close to the filamentous aggregate, is accompanied by groups of vesicles.

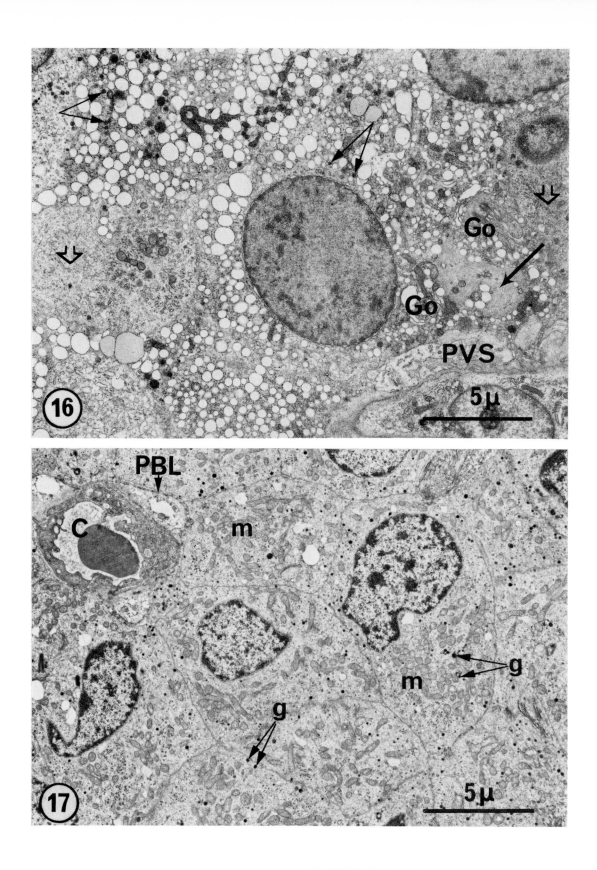

criticized the term "chromophobe" according to observations with the light micro-scope on semithin sections, because under those conditions it is possible to find some granules in these tumor cells. However, it seems possible to retain the classic notion of chromophobe adenomas, i.e., adenomas with cells where it is impossible to identify a definite granular content in usual light microscopic techniques.

In the light microscope, numerous aspects have been described according to their general organization (Dott and Bailey, 1925; Kernohan and Sayre, 1956), their cell size, or their clear or dark aspect (Roussy and Oberling, 1933). Usually, they are believed to correspond to the "chief cells" of the adult pituitary. When they are polyhedral, and set up in regular rows separated by blood capillaries, these are considered as an embryonic form (adenomas with fetal cells, Kraus, 1926).

In the electron microscope, several surveys have been made on chromophobe adenomas (Schelin, 1962; Luse, 1962; Meneghelli et al., 1964; von Weschler et al., 1965; Oliva et al., 1966; Kuromatsu, 1968; Zambrano et al., 1968; Tomiyasu et al., 1973; Kovacs and Horvath, 1973; Landolt and Oswald, 1973; Schechter, 1973).

Usually, cells in nonfunctional adenomas are not characteristic enough to be related to any cellular stemline, either adult or embryonic, and they often give the impression that they are anaplastic cells. None of the observed features or anoma-lies is specific, and a comparison between the cells of functional chromophobe adenomas and cells from nonfunctional chromophobe adenomas sometimes show nearly similar aspects, particularly in regard to the presence of small secretory granules (Fig. 17).

Finally the problem raised by the nonfunctional character of the chromophobe tumors is opposite to that raised by functional chromophobe tumors. Why do these tumors although containing some granules have no apparent endocrine activity? Several hypotheses may be put forward:

(a) The endocrine activity may be too weak to be biologically significant. This might be supported by the fact that these cells present certain slight secretory signs. The volume of the nucleolus is sometimes small compared to that of functional cells (Racadot et al., 1974).

FIG. 16. Pituitary adenoma in acromegaly. Cell with vacuolar aspect due to the scatter-ing of enlarged cisternae of the ergastoplasm. Few Golgi saccules (Go) near a filamentous aggregate (transverse arrow). Some small granules (double arrows) are visible. On the oppo-site, neighboring cells have a dense cytoplasm due to a high ribosome concentration (clear arrow). PVS, perivascular space, the pericordonal basal lamina is lacking.

FIG. 17. Nonfunctional adenoma. Compact parenchyma made of cells with irregular nuclei and numerous clear mitochondria (m). The granulations are very small <120 nm (double arrows) and scattered in the cytoplasm. C, capillary. The pericordonal basal lamina (PBL) is still present.

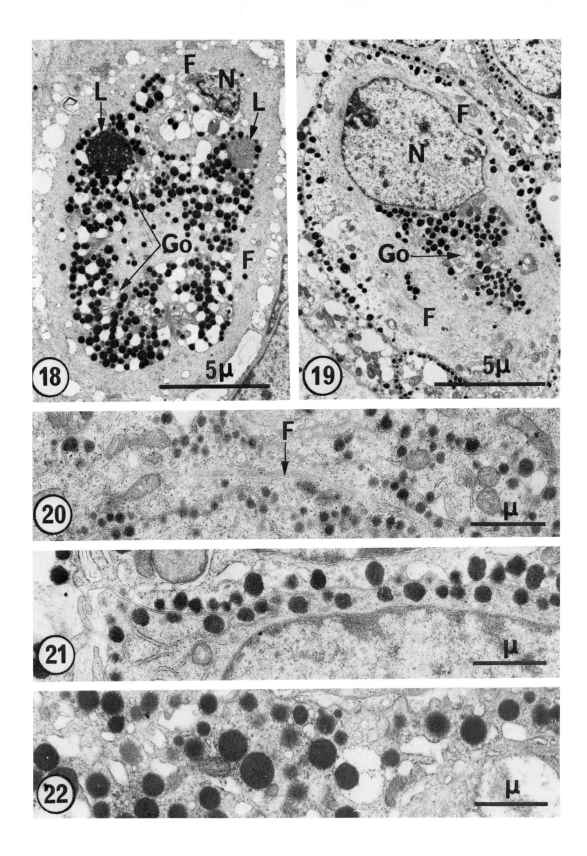

(b) It also is possible that granules are not released and are destroyed *in situ*. However lysosomal structures do not appear developed enough to support this hypothesis. On the other hand, exocytoses have been observed (Foncin, 1971).

(c) Finally, secretion may be restricted to the elaboration of a granule bearing a hormonally inactive secretory product. Immunocytochemistry alone might answer such a question.

Another category of nonfunctional adenomas are those which contain *follicular cells*. Adenomas with follicular formations have been known for a long time (Roussy and Oberling, 1933), although their analysis in the electron microscope is quite recent (Fukuda, 1973; Olivier *et al.*, 1973b). These follicular formations, either microscopic or submicroscopic, are scattered among chromophobe cells. They are bounded by more or less abnormal follicular cells (Fig. 23, to be compared to Fig. 1). Some of the characteristics of the normal human follicular cells (Fig. 1) described by Salazar (1968), Bergland and Torack (1969), and Vila-Porcile *et al.* (1971) are found: microvilli and junctional complexes at the apical (or follicular) pole, "foot" lying on the pericordonal basal lamina, stacking up of organelles from one pole to the other, elongated mitochondria with tightened cristae, and lysosomes. But on the other hand, the cells show abnormal features: vesiculation of the Golgi apparatus, mitochondria of abnormal sizes, and most important, submicroscopic granules accumulating at the basal (or vascular) pole. These granules have never been described up to now in normal follicular cells and this is therefore a new feature of tumor follicular cells.

Tumor follicular cells are an instance of a tumor cell able to retain a high degree of differentiation while acquiring a new character (i.e., appearance of secretion granules), which represents a deviation from its original type. This phenomenon, evident here for it involves morphological structures, might also occur in other cells at the biochemical level.

B. Description of Experimental Tumors

Some tumors offer the great advantage of being possibly followed chronologically throughout their development. This is the case of those experimental

---

FIGS. 18–22. Cushing's disease. Fig. 18. Crooke's cell in normal residual pituitary. Perinuclear ring of filaments (F), enclosing secretory granules, large lysosomes (L), Golgi elements (Go, double arrow) and vacuoles. Along the cell periphery, a row of vacuoles. N, nucleus. Fig. 19. Crooke's cell in the tumoral pituitary of the same patient. Large ring of filaments (F) around the nucleus (N) and a perinuclear concentration of granules and organelles. Small marginal granules along the cell periphery. Figs. 20–22. Three different aspects of the secretory granules in Cushing's disease. Fig. 20. Very small granules of variable opacity, with a dark core separated from the granule membrane by a light narrow space. Bundle of filaments (F) within the cytoplasm. Fig. 21. Slightly polyhedral granules, lying along the plasma membrane. Their average diameter is larger than in the above picture (Fig. 20). Cytoplasm is filled with ribosomes. Fig. 22. Opaque granules, with variable diameters. The cytoplasm is vacuolated.

tumors related to a "disturbance in the inter- and intracellular communications" (Furth, 1968) caused by an endocrine homeostatic perturbation. Their study should make it possible therefore to solve cytogenesis problems.

Little is known about experimental gonadotroph tumors: gonadotroph cell proliferation followed by adenoma formation has been observed (Griesbach and Purves, 1960); however, their functional activity has not been studied.

The three fundamental varieties of functional experimental tumors are corticotroph, thyrotroph, and mammosomatotroph (Furth and Clifton, 1966). These three categories in fact include numerous variants, especially among the transplanted tumors. Corticotroph tumors have been rarely studied from the point of view of cytology (Bahn *et al.*, 1957; Farquhar and Furth, 1959); it is why we are only summarizing cytological data on thyrotroph and mammosomatotroph tumors.

1. *Thyrotroph Tumors*

In mice, they can be induced through different processes, all of them causing a hypothyroidism, which leads to pituitary stimulation and then to thyrotroph hormone hypersecretion: the most frequently used method is radiothyroidectomy. Three stages can be considered in their evolution: simple hyperplasia, hyperplasia with microscopic adenomas, and finally gross tumors (Halmi and Gude, 1954; Dent *et al.*, 1955), these last tumors being transplantable. In the first stage, pituitary hypertrophy is due to capillary dilatation on the one hand, to the presence of numerous "thyroidectomy cells" on the other, these last cells being quite familiar from experimental pituitary cytophysiology of different species (Farquhar and Rinehart, 1954; Barnes, 1963; Shiino *et al.*, 1973).

They are hypertrophied cells with indented nuclei and a ballooning of the ergastoplasm cisternae. Secretory granules of thyrotroph type are still found, although their genesis appears modified as they are condensed within the reticulum as well as in the Golgi cavities (Barnes, 1963). The granules are less numerous than in a normal thyrotroph cell and their number varies from one cell to the other, so much so that many cells appear chromophobe. After thyroxine administration, these cells are able to accumulate granules, probably as the thyroxine blocks granule release and thus allows the formation of a new granular stock (Messier, 1966). The hyperplastic thyroidectomy cells are seen through the whole gland, among the other cellular types, which, however, are somewhat modified and perhaps involved in tumorigenesis (see p. 258).

Microadenomas are observed a few months later. They consist of chromophobe cells with a large nucleus and little cytoplasm, quite often without granules, even in the electron microscope. Under the light microscope, these cells resemble thyroidectomy cells because of their vacuolar aspect (Halmi and Gude, 1954). At this stage, necrosis is quite frequent.

At the gross tumor stage (approximately 1 year later), the whole pituitary organization has disappeared with only a few remnants of the original parenchyma. Hemorrhagic and necrotic foci are numerous. All tumor cells are of the chromophobe type, frequently in mitosis. Some cells still present a few characteristics of the thyroidectomy cell, with a vacuolar aspect due to the dilatation of their ergastoplasm.

Transplanted tumors have a similar cytology. Their granules are very small (diameter: 30 to 80 nm according to Kamat *et al.*, 1960, and 150 nm in cells observed by Ueda, 1971), and less numerous in autonomous varieties (developing in euthyroid hosts) as compared to dependent varieties (developing in thyroidectomized hosts).

Because of the aspect of the ergastoplasm, the tumor cells have been considered as still deriving from thyroidectomy elements (Farquhar and Furth, 1959; Lundin and Schelin, 1964).

Therefore, at every stage, hyperplastic, adenomatous, or tumorous, the common point resides in the presence of some cells which can be considered as thyroidectomy cells, according to their cytological characteristics. The problem is that of their origin. At first view, thyroidectomy cells seem to be modified thyrotroph cells, derived from division of thyrotroph cells, as demonstrated by Stratmann *et al.* (1972) in the rat. However, the origin of thyroidectomy cells has been widely discussed in the case of the mouse pituitary, in which tumors can be induced.

In the mouse pituitary, at the early hyperplastic stage, thyroidectomy cells are more numerous than the initially observable thyrotroph cells. But, according to Messier (1966), no mitoses are found to account for the increased number of these cells, which probably arose from a "mobilization" of existing pituitary cells, possibly chromophobe "of thyrotropic nature" (Messier, 1969).

According to Dingemans (1969) this mobilization might be due to transformation of gonadotroph cells, somatotroph cells, or of cells with "haloed secretory granules" into thyrotropic cells, because he considers that true undifferentiated cells do not exist (Dingemans, 1971; Dingemans and Feltkamp, 1972).

Later, the increased number of thyroidectomy cells evidently implies a proliferation process. What cells are dividing? Kwa (1961), on the basis of tumor growth, suggested "that the percentage of cells of the normal pituitary that are induced to proliferate by radiothyroidectomy . . . may be estimated to be between 10% and 50% of the cell population of the normal pituitary: it therefore precludes the concept of radiothyroidectomy induced tumours being exclusively derived from the TSH producing, PAS-positive, basophile cell type of the pituitary." Messier (1969), using tritiated thymidine and light microscopy, demonstrated "that increased cell proliferation was initiated at some time between 30 days and 4 months after radiothyroidectomy." One year later, labeled cells become 16 times more numerous than in an intact animal. This increased value is brought back to normal level after several days of thyroxine treatment. This responsiveness to the homeostatic action of thyroxine "may be taken as indirect evidence of the thyrotropic nature of the dividing cells" (Messier, 1969).

On the other hand, the problem of pituitary cells stimulation triggered by radiothyroidectomy is probably more complex than that of the "thyroidectomy" cell origin.

Clifton (1966) established a model which "assumes persistence of most of the original cell population at a constant level throughout tumor induction, an exponential increase in the size of a cell population that initially constitutes a small percentage of the total population, and the appearance of a 3rd more rapidly increasing cell population some time after radiothyroidectomy." The population endowed with a rapid growth might correspond to the selection of a clone of cellular variants holding "a reproductive advantage." According to the results from radioautography, such cells seem to be located in the adenomatous formations appearing before the stage of gross tumor.

Tumors obtained through radiothyroidectomy can exhibit functional activities other than thyrotroph. Thyrotroph tumors show a gonadotroph functional activity which in some mutant autonomous variants can even be prevalent (Messier and Furth, 1962). The gonadotroph elaboration could be explained "on the basis of a subunit common to gonadotropic hormones and thyrotropic hormone and/or by derepression of the genetic code associated with neoplastic transformation of normal thyrotropes" (Furth *et al.,* 1973b). This double hormonal elaboration could be demonstrated by immunocytochemistry, but it has not been determined whether one or two cellular stemlines can be involved.

In fact, some authors consider that thyrotroph and gonadotroph cells are both involved in thyroidectomy (Théret and Renault, 1964, 1965; Dingemans, 1969). Ueda and Mori (1967) could even observe a transplantable thyrotroph tumor with some cells of a gonadotroph nature.

Thyrotroph tumors can have a somatotroph activity particularly important in autonomous stemlines (Furth and Clifton, 1966). This was shown by immunocytochemistry in the light microscope (Furth *et al.,* 1973a) and cannot be explained by the presence of common subunits as in the case of the gonadotroph activity. The actual mechanism through which the tumor reaches a somatotroph potentiality remains completely unknown.

Ueda (1971) in a thyrotroph autonomous tumor which became mammosomatotroph observed cells appearing like thyrotroph as well as cells with 200 to 600 nm granules similar to these in mammosomatotroph tumors (cf. below), perhaps involving two stemlines?

Potvliège (1968) and Schelin (1969) observed striking changes in the morphology of thyrotroph and prolactin cells, when stimulating the pituitary of goitrogen-treated rats by estrogen. This indicates that two cell types can proliferate simultaneously under the influence of an hormonal derangement. However, it is not known whether these two cell types originate from one or two stemlines.

2. Mammotroph and Mammosomatotroph Tumors

These tumors are induced by a massive and prolongated administration of natural or synthetic estrogens (for instance, stilbestrol). During their induction, they go through several stages while animals become cachectic because of the toxicity of the treatment. First, hyperplasia is obtained, afterward hyperplasia with adenomas and, finally, tumors (Wolfe and Wright, 1938; Clifton and Meyer, 1956). After serial transplantations the tumors become autonomous and very proliferative (Furth et al., 1956). A mammotroph activity has been noted in tumors "in situ" (Lacour, 1950; Bielchowsky, 1954), and a mammosomatotroph activity in dependent and autonomous grafted tumors (Furth et al., 1956). Some autonomous grafted tumors show associated corticotroph activity such as the MtT/F4 strain (Takemoto et al., 1962).

Cytological pictures of hyperplasia and tumors before transplantation have been widely discussed. The abundance of chromophobe cells was noticed by Cramer and Horning (1936) and by Wolfe and Wright (1938). But, as early as 1938, Erdheim (in Zondek, 1938) claimed that the involved cells were of chromophil nature, and Lacour (1950) and Clifton and Meyer (1956) pointed out that these cells were of an acidophil type.

According to Lundin and Schelin (1962) and Schelin and Lundin (1971), the two mammotroph and somatotroph types could not be individualized in these acidophil cells as is the case in the normal animal. However, most authors, after Lacour's suggestions (1950), consider hyperplastic cells as mammotroph (or prolactin) cells, which in the electron microscope show stereotyped cytological modifications (Hymer et al., 1961; Shimazaki et al., 1962; Watari and Tsukagoshi, 1969; Waelbroeck-Van Gaver and Potvliège, 1969). These modifications are a hypertrophy of the ergastoplasm with a lamellar and whorled aspect of the cisternae leading to "Nebenkerne" pictures (Fig. 24), together with a hypertrophy of the Golgi apparatus (Haguenau and Bernhard, 1955; Haguenau and Lacour, 1955). Exocytoses are frequent. The nucleolus increases in size (Fig. 24). All these characteristics are interpreted as pictures of mammotroph secretory hyperactivity of the cell. At this stage, the cells might develop from preexisting prolactin cells, in which mitoses may be seen. However, it cannot be demonstrated that other cells are involved in the proliferation, for mitotic figures are found in nongranular cells with the light microscope (Clifton and Meyer, 1956).

At the end of several months of permanent estrogenization, hyperplasia becomes tumorous with frequent hemorrhagic areas. Cells from other categories are still found and some of them, thyrotroph or corticotroph, seem to be activated (Waelbroeck-Van Gaver and Potvliège, 1969). There also are unidentifiable cells which account for the chromophobe aspect of the tumor in the light microscope (Shimazaki et al., 1962). The prolactin cells keep their characteristics especially the "Nebenkerne."

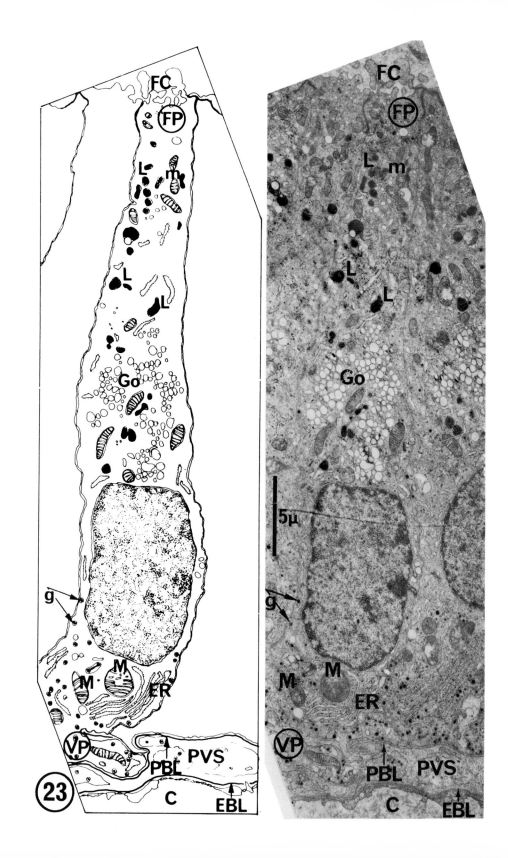

Interruption of the estrogenization at this stage causes a decrease in the pituitary volume. Mammotroph cells remain and seem to survive either "active" or "quiescent." The latter form could correspond to cells "permanently altered by the estrogenic treatment" (Waelbroeck-Van Gaver and Potvliège, 1969).

After transplantation, Waelbroeck-Van Gaver and Potvliège (1969) have studied with the electron microscope a tumor that had become autonomous and functionally mammosomatotroph. It was impossible to identify most of the cells as they completely lacked granules (chromophobe cells). But there were still some which could be related to three cellular types: mammotroph, somatotroph, and corticotroph. Compared to normal morphofunctional cells, however, they showed prominent abnormalities of their nucleus, ergastoplasm, and Golgi apparatus. In the mammotroph cell, the "Nebenkerne" had completely disappeared.

At this stage, the cytogenetical modifications already observed before transplantation have increased, probably with isolation of aberrant cellular stemlines (Waelbroeck-Van Gaver, 1969).

The aspect of indefinitely transplantable autonomous strains is still different from that of all the above described tumors. For instance the MtT/F4 strain is functionally mammosomatocorticotroph and cytologically monomorphous (Furth and Clifton, 1966). It is a rapidly growing malignant tumor completely chromophobe in the light microscope. Under the electron microscope (Schelin et al., 1964; Pantic et al., 1971; Pelletier et al., 1971), cells do not show any pituitary cell characteristic (Fig. 25): ergastoplasm and Golgi apparatus are reduced, free ribosomes are numerous and secretory granules are scarce or lacking. In the sample studied by Schelin et al. (1964), the granules were 350 nm and in that studied by Pelletier et al. (1971), the diameter varied between 150 and 220 nm. According to these data, it is impossible to ascertain whether, in spite of their uniform aspect, the cells are equivalent or not, or if the three hormones are secreted by one or several cells of the MtT/F4 tumor.

But in another form of transplantable tumor, Tashjian et al. (1968) could find an answer to the problem raised by these multisecretory tumoral strains. They established a clonal strain from the MtT/W5, a tumor known to secrete both somatotroph and mammotroph hormones (Takemoto et al., 1962). In the electron microscope, these cultures consisted of very poorly granulated cells. The granules vary from 150 to 250 nm. The organelles involved in the synthesis processes are well developed (Gourdji et al., 1972). In spite of the autonomy of the tumor, the

---

FIG. 23. Adenoma with follicular cells. The cell is organized between a follicular "pole" (FP) at the follicular cavity level (FC) and a vascular pole (VP) in contact with the pericordonal basal lamina (PBL), perivascular space (PVS), and blood capillary keeping its fenestrations (C). EBL, endothelial basal lamina. Note the vesiculation of the Golgi apparatus (Go), the well-developed ergastoplasm (ER), the small secretory granules (g, double arrow) at the cell basal pole. Note also the increasing size of mitochondria (m,M) from the follicular pole down to the vascular pole. L, lysosomes.

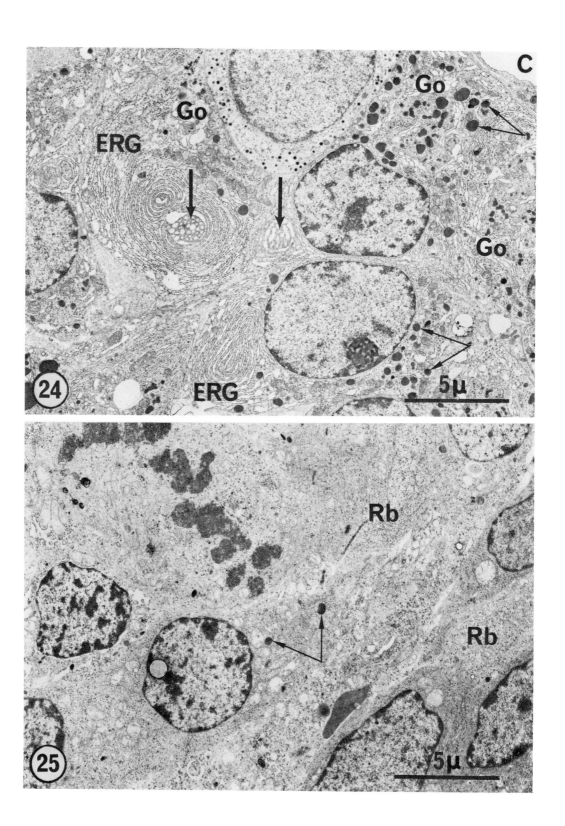

secretion of these cells can be modulated by exogenous factors, hydrocortisone or TRF (Tashjian *et al.,* 1970, 1971). With TRF the cytology is modified: the Golgi area is hypertrophied and the number of granules increases (Gourdji *et al.,* 1972) (see chapter by Tixier-Vidal).

V. Summary

The features of spontaneous and experimental tumors are described in relation to features of normal pituitary cells as reviewed in chapters by Herlant and by Farquhar *et al.* They are also related to the present knowledge on normal pituitary cell cytogenesis in fetal and adult pituitary.

In the fetal pituitary, chromophobe cells are considered as "undifferentiated" cells. However, such a notion remains only a morphological one, because it has not been ascertained whether they are multipotent stem cells or already determined as unipotent stemlines to be later morphologically expressed at the time of granule elaboration. It has not been ascertained whether or not such cells are able to carry out hormone secretion.

In the adult pituitary, similar problems arise with poorly granulated cells. Do uni- or multipotent reserve cells exist, arising either from persistent embryonic stem elements or from secretory cells in a particular functional state?

Finally, although the electron microscope has brought new criteria for cellular identification, some cells are still difficult to classify. Their significance remains debatable in the normal pituitary, thus explaining the difficulties found in the interpretation of tumor cells.

The tumors developed from the pituitary parenchyma can be classified on either morphological or endocrinological bases. The morphological analysis in many cases does not allow a precise cellular identification. The functional endocrinological analysis demonstrates secretory activity of the tumor, but may fail either because of the inadequacy of the method or because of the difficulty in determining whether the secretion arises from the tumor or from the residual parenchyma. On the other hand, analysis of functional activity indicates only the plasma hormone release and gives no indication as regards the intratumoral hormone storage phenomena.

Figs. 24 and 25. Experimental tumors. Fig. 24. Rat pituitary after estrogenization. Vascularized parenchyma still keeping its organization, although made of prolactin cells. Some of these are granular (double arrow) with large Golgi areas (Go). The others are invaded by an extremely developed ergastoplasm (ERG), appearing as whorls: Nebenkerne (vertical arrows). C, blood capillary. Regular nuclei with large nucleolus. Fig. 25. MtT/F4 642nd passage. Compact parenchyma, without any organization. Cytoplasm is filled with free ribosomes (Rb), nuclei are irregular and without nucleoli. One cell is in mitosis.

Pituitary tumors show various biological and cytological aspects related to their "spontaneous" or "induced" origin.

Spontaneous tumors, mainly studied in man, are only disclosed at a well-developed stage. Their causes are completely unknown. They appear relatively stable during their evolution, at least in the absence of any endocrine disturbance (such as adrenalectomy in Cushing's disease). Human spontaneous tumors have been divided into functional and nonfunctional tumors, depending on whether or not they are secreting.

Three main categories of functional tumors are described: tumors in acromegaly involving the somatotroph function, tumors in Cushing's disease involving the corticomelanotroph function, and tumors in amenorrhea–galactorrhea involving the mammotroph function. In each of these categories of tumors with a specific secretion, three cellular aspects can be observed: (1) granular cells of the classical morphofunctional type corresponding to the related secretion, (2) more or less abnormal cells, because of their low granule content and the modifications of their organelles, but still linked to the granular type involved by some of their morphological characteristics, and (3) finally, cells that are no longer identifiable, as they are almost without granules ("chromophobe cells"), or show abnormal granules.

In each of the tumor functional categories, several tumor patterns can be observed from tumors where the classic morphofunctional type prevails up to tumors only made of chromophobe cells. The latter generally correspond to rapidly evolving forms.

In addition to the three basic functional categories above, cases where one tumor seems to release several hormones have been described. However, no clear link between the histological and endocrinological aspects has been ascertained.

The nonfunctional human adenomas consist of cells which are chromophobe under the light microscope, but show under the electron microscope a low amount of granules, generally small sized. The origin and significance of such cells, when compared to normal embryonic or normal adult types, are still unknown, except for the follicular cells sometimes present.

Experimental tumors differ from human tumors since they undergo successive biological and cytological transformations during their evolution. They start with an endocrine disturbance which induces the stimulation of one secretory function. Their growth remains for a long time dependent on this very stimulation. Later, following successive transplantations, they develop autonomous and malignant characteristics.

Cytologically, the most studied tumors are the thyrotroph tumors of the mouse, and the mammosomatotroph tumors of rat and mouse.

The stimulated morphofunctional type can be identified at the very beginning (hyperplasia stage). Cellular multiplication and tumor development cause tumor cells to lose the cytological characteristics of the initially stimulated morphofunctional type. When autonomous, the cells have almost completely lost these characteristics. In addition, the functional capacities are modified and sometimes new hormonal secretory capacities, in addition to the initial potencies, may arise. In some cases, contrary to the normal pituitary cell, one tumor cell may secrete several hormones.

Such changes are associated with chromosome modifications, stressing the "transformed" nature of the tumor cells. They could result from the selection of variant cell clones which possess a "reproductive advantage," explaining the trend "for progression from bad to worse" (Furth, 1959).

Note added in proof: Since this chapter was written, an extensive study on the cytogenesis in the adenohypophysis has been published: Svalander, C. (1974). Ultrastructure of the fetal rat adenohypophysis. *Acta Endocrinol. (Copenhagen)* **76**, *Suppl. 188,* 1–113.

ACKNOWLEDGMENTS

We would like to thank Professors Aboulker, David, Guiot, Pertuiset, and Professors Hurst, Bricaire, Gilbert-Dreyfus, and Klotz for the authorization to use their cases.

We are endebted to Dr. Haguenau and Dr. Tixier-Vidal for helpful discussions concerning the manuscript.

We gratefully acknowledge the technical assistance of Mrs. Brandi, Mrs. Drouet, and Miss Marie, and the help of Mrs. de Venoge and Mrs. Combrier in the preparation of the manuscript.

REFERENCES

Abe, K., Island, D. P., Liddle, G. W., Fleischer, N., and Nicholson, W. E. (1967). Radioimmunology evidence for a MSH (melanocyte stimulating hormone) in human pituitary and tumor tissues. *J. Clin. Endocrinol. Metab.* **27**, 46–52.

Andersen, H., von Bülow, F. A., and Møllgård, K. (1971). The early development of the pars distalis of human foetal pituitary gland. *Z. Anat. Entwicklungsgesch.* **135**, 117–138.

Bahn, R., Furth, J., Anderson, E., and Gadsden, E. (1957). Morphologic and functional changes associated with transplantable ACTH-producing pituitary tumors of mice. *Amer. J. Pathol.* **33**, 1075–1085.

Bahn, R., Ross, G. T., and MacCarty, C. S. (1960). Melanocyte stimulating hormone and ACTH activities of pituitary tumors with Cushing's syndrome. *Proc. Staff Meet. Mayo Clin.* **35**, 623–652.

Bancroft, F. C., and Tashjian, A. H. (1971). Growth in suspension culture of rat pituitary cells which produce growth hormones and prolactin. *Exp. Cell Res.* **64**, 125–128.

Bancroft, F. C., Levine, L., and Tashjian, A. H. (1969). Control of growth hormone production by a clonal strain of rat pituitary cells. Stimulation by hydrocortisone. *J. Cell Biol.* **43**, 432–441.

Barnes, B. G. (1963). The fine structure of mouse adenohypophysis in various physiological states. *In* "Cytologie de l'adénohypophyse" (J. Benoit and C. Da Lage, eds.), pp. 73–90. CNRS, Paris.

Bates, R. W., Anderson, E., and Furth, J. (1957). Thyrotrophin potency of the transplantable tumors of mice through four transfers. *Endocrinology* **61**, 549–554.

Bates, R. W., Milkovic, S., and Garrison, M. M. (1962). Concentration of prolactin, growth hormone and ACTH in blood and tumor of rats with transplantable mammotropic pituitary tumors. *Endocrinology* **71**, 943–948.

Bayreuther, K. (1960). Chromosome in primary neoplastic growth. *Nature (London)* **186**, 6–9.

Bergland, R. M., and Torack, R. M. (1969). An ultrastructural study of follicular cells in the human anterior pituitary. *Amer. J. Pathol.* **57**, 273–297.

Bernhard, W. (1969). Ultrastructure of the cancer cell. *In* "Handbook of Molecular Cytology" (A. Lima-de-Faria, ed.), pp. 687–715. North-Holland Publ., Amsterdam.

Bielchowsky, F. (1954). Functional acidophilic tumours of the pituitary of the rat. *Brit. J. Cancer* **8**, 154–160.

Biggart, J. H., and Dott, N. M. (1936). Pituitary tumours. Their classification and treatment. *Brit. Med. J.* **2**, 1153–1155.

Breustedt, H. J. (1971). Methodological aspects in the immunohistological detection of pituitary hormones. *Horm. Metab. Res.* **3**, Suppl., 18–21.

Bricaire, H., Luton, J. P., and Turpin, G. (1971). Tumeurs de l'hypophyse au cours des syndromes de Cushing. *In* "Les adénomes hypophysaires sécrétants. Endocrinopathies et immunologie," pp. 73–117. Masson, Paris.

Brilmayer, H., Marguth, R., and Müller, W. (1957). Das Mischtypadenom und seine Abgrenzung gegen den chromophoben Hypophysentumor. *Acta Neuroveg.* **15**, 352–373.

Brozman, M. (1967). Histochemical localization of ACTH and TSH in the human hypophysis. *Acta Histochem.* **26**, 261–270.

Buonassisi, V., Sato, G., and Cohen, A. I. (1962). Hormone-producing cultures of adrenal and pituitary tumor origin. *Proc. Nat. Acad. Sci. U. S.* **48**, 1184–1190.

Capen, C. C., and Koestner, A. (1967). Functional chromophobe adenomas of the canine adenohypophysis. An ultrastructural evaluation of a neoplasm of pituitary corticotrophs. *Pathol. Vet.* **4**, 326–347.

Capen, C. C., Martin, S. L., and Koestner, A. (1967a). Neoplasms in the adenohypophysis of dogs. A clinical and pathologic study. *Pathol. Vet.* **4**, 301–325.

Capen, C. C., Martin, S. L., and Koestner, A. (1967b). The ultrastructure and histopathology of an acidophil adenoma of the canine adenohypophysis. *Pathol. Vet.* **4**, 348–365.

Cardell, R. P., and Knighton, R. S. (1966). The cytology of a human pituitary tumor: An electron microscopic study. *Trans. Amer. Microsc. Soc.* **85**, 58–78.

Clifton, K. H. (1966). Cell population kinetics during the induction of thyrotropic pituitary tumors. *Cancer Res.* **26**, 374–381.

Clifton, K. H., and Meyer, R. K. (1956). Mechanism of anterior pituitary tumor induction by estrogen. *Anat. Rec.* **125**, 65–81.

Cohen, A. I., and Furth, J. (1959). Corticotropin assay with transplantable adrenocortical tumor slices: Application to the assay of adrenotropic pituitary tumors. *Cancer Res.* **19**, 72–78.

Conklin, J. L. (1968). The development of human foetal adenohypophysis. *Anat. Rec.* **160**, 79–92.

Costello, R. T. (1935). Subclinical adenomas of the pituitary body. *Proc. Staff Meet. Mayo Clin.* **10**, 449–453.

Costero, I., and Berdet, H. (1939). Estudio anatómico de 135 tumores de la hipófisis y del tracto hipofisario. *Monogr. Soc. Med. Hosp. Gen. Mex.* No. 1, pp. 1–116.

Cramer, W., and Horning, E. S. (1936). Experimental production by estrin of pituitary tumors with hypopituitarism and of mammary cancer. *Lancet* **1**, 247–248.

Cushing, H. (1932). The basophil adenomas of the pituitary body and their clinical manifestation (pituitary basophilism). *Bull. Johns Hopkins Hosp.* **50**, 137–195.

Cushing, H. (1933). "Dyspituitarism": Twenty years later with special consideration of the pituitary adenomas. *Arch. Int. Med.* **51**, 487–557.

D'Abrera, V. St. E., Burke, W. J., Bleasel, K. F., and Bader, L. (1973). Carcinoma of the pituitary gland. *J. Pathol.* **109**, 335–343.

Daikoku, S., Kinutani, M., and Watanabe, Y. G. (1973). Role of hypothalamus on development of adenohypophysis: An electron microscopic study. *Neuroendocrinology* **11**, 284–305.

Deaton, P. C., and Dugger, G. S. (1972). The ultrastructure of the nonadenomatous anterior lobe of the pituitary gland in man. *Surg., Gynecol. Obstet.* **135**, 901–907.

De Cicco, F. A., Dekker, A., and Yunis, E. J. (1972). Fine structure of Crooke's hyaline change in the human pituitary gland. *Arch. Pathol.* **94**, 65–70.

Deminatti, M., and Vendrely, C. (1959). Etude de la teneur en acide desoxyribonucléique des cellules préhypophysaires chez le cobaye. *C. R. Soc. Biol.* **153**, 840–843.

Dent, J. N., Gadsden, E. L., and Furth, J. (1955). On the relation between thyroid depression and pituitary tumor induction in mice. *Cancer Res.* **15**, 70–75.

Dingemans, K. P. (1969). On the origin of thyroidectomy cells. *J. Ultrastruct. Res.* **26**, 480–500.

Dingemans, K. P. (1971). Undifferentiated cells in the mouse adenohypophysis. *Electron Microsc., Proc. Int. Congr., 7th, 1970* Vol. 3, pp. 563–564.

Dingemans, K. P., and Feltkamp, C. A. (1972). Nongranulated cells in the mouse adenohypophysis. *Z. Zellforsch. Mikrosk. Anat.* **124**, 387–405.

Di Paolo, J. A., Messier, B., and Hauschka, T. S. (1964). Chromosome constitution of thyrotropic pituitary tumors in relation to autonomy. *Acta Unio Int. Contra Cancrum* **20**, 1343–1346.

Dott, N. M., and Bailey, P. (1925). A consideration of the hypophysial adenomata. *Brit. J. Surg.* **13**, 314–366.

Dourov, N., and Herlant, M. (1970). Apparition de cellules de Crooke dans l'hypophyse du nourrisson à la suite d'un traitement par l'ACTH. *Ann. Endocrinol.* **31**, 735–741.

Dubois, M. P. (1971). Les cellules à hormones glycoprotéidiques du lobe antérieur de l'hypophyse: Séparation par immunofluorescence des cellules thyréotropes et des cellules gonadotropes dans l'hypophyse des bovins, ovins et porcins. *Ann. Rech. Vet.* **2**, 197–222.

Dubois, M. P. (1973). Recherche par immunofluorescence des cellules adenohypophysaires élaborant les hormones polypeptidiques ACTH, αMSH, βMSH. *Bull. Ass. Anat. (Nancy)* **156**, 63–66.

Dubois, P. (1967). Etude au microscope électronique de la pars distalis de l'hypophyse de l'embryon humain. *Bull. Ass. Anat.* **138**, 434–441.

Dubois, P. (1968). Données ultrastructurales sur l'antéhypophyse d'un embryon humain à la huitième semaine de son développement. *C. R. Soc. Biol.* **162**, 689–692.

Dubois, P., and Dumont, L. (1966). Nouvelles observations au microscope électronique sur l'antéhypophyse humaine du troisième au cinquième mois du développement embryonnaire. *C. R. Soc. Biol.* **160**, 2105–2107.

Dubois, P., Vargues-Regairaz, H., and Dubois, M. P. (1973). Human foetal anterior pituitary immunofluorescent evidence for corticotropin and melanotropin activities. *Z. Zellforsch. Mikrosk. Anat.* **145**, 131–143.

Dupouy, J. P., and Magre, S. (1973). Ultrastructure des cellules granulées de l'hypophyse foetale du rat. Identification des cellules corticotropes et thyréotropes. *Arch. Anat. Microsc. Morphol. Exp.* **62**, 185–206.

Ezrin, C., and Murray, S. (1963). The cells of the human adenohypophysis in pregnancy, thyroid disease and adrenal cortical disorders. *In* "Cytologie de l'adénohypophyse" (J. Benoit and C. Da Lage), pp. 183–200. CNRS, Paris.

Falin, L. I. (1961). The development of human hypophysis and differentiation of cells of its anterior lobe during embryonic life. *Acta Anat.* **44**, 188–205.

Fand, S. B. (1973). Polyploidy in human pituitary: Coexistence of hyperplasia and hypertrophy. *Amer. J. Pathol.* **70,** 61a–62a (abstr.).

Farquhar, M. G., and Furth, J. (1959). Electron microscopy of experimental pituitary tumors. *Amer. J. Pathol.* **35,** 698 (abstr.).

Farquhar, M. G., and Rinehart, J. F. (1954). Cytologic alterations in the anterior pituitary gland following thyroidectomy: An electron microscope study. *Endocrinology* **55,** 857–876.

Feltkamp, C. A., and Kwa, H. G. (1965). Transformation of pituitary tumours from prolactin into TSH-cell types. I. Electron microscopical observations. *Acta Endocrinol. (Copenhagen), Suppl.* **100,** 161.

Fink, G., and Smith, G. C. (1971). Ultrastructural features of the developing hypothalamo-hypophysial axis in the rat. A correlative study. *Z. Zellforsch. Mikrosk. Anat.* **119,** 208–226.

Foncin, J. F. (1966). Etude sur l'hypophyse humaine au microscope électronique. *Pathol. Biol.* **14,** 893–902.

Foncin, J. F. (1971). Morphologie ultrastructurale de l'hypophyse humaine. *Neurochir.* **17,** Suppl. 1, 10–24.

Foncin, J. F., and Le Beau, J. (1963). Etude en microscopie optique et électronique d'une tumeur hypophysaire à fonction adrénocorticotrope. *C. R. Soc. Biol.* **157,** 249–252.

Foncin, J. F., and Le Beau, J. (1964). Identification au microscope électronique des cellules adrenocorticotropes de l'hypophyse humaine. *C. R. Soc. Biol.* **158,** 2276–2279.

Foster, C. L. (1956). Some observations upon the cytology of the pars distalis of surgically removed human pituitary. *Quart. J. Microsc. Sci.* **97,** 379–391.

Fukuda, T. (1973). Agranular stellate cells (so-called follicular cells) in human fetal and adult adenohypophysis and in pituitary adenoma. *Virchows Arch., A* **359,** 19–30.

Furth, J. (1959). A meeting of ways in cancer research: Thoughts on the evolution and nature of neoplasms. *Cancer Res.* **19,** 241–259.

Furth, J. (1968). Hormones and neoplasia. *In* "Thule International Symposia on Cancer and Aging" (A. Engel and T. Larsson, eds.), pp. 131–151. Nordiska Bokhandelns Förlag, Stockholm.

Furth, J. (1969). Pituitary cybernetics and neoplasia. *Harvey Lect.* **63,** 47–71.

Furth, J., and Clifton, K. H. (1958). Experimental pituitary tumors. *Ciba Found. Colloq. Endocrinol. [Proc.]* **12,** 3–17.

Furth, J., and Clifton, K. H. (1966). Experimental pituitary tumors. *In* "The Pituitary Gland" (G. W. Harris and B. T. Donovan, eds.), Vol. II, pp. 460–497. Univ. of California Press, Berkeley.

Furth, J., Clifton, K. H., Gadsden, E. L., and Buffett, R. F. (1956). Dependant and autonomous mammotropic pituitary tumors in rats; their somatotropic features. *Cancer Res.* **16,** 608–616.

Furth, J., Martin, J. M., Moy, P., and Ueda, G. (1973a). Growth hormone activity of thyrotropic pituitary tumors. *Proc. Soc. Exp. Biol. Med.* **142,** 511–515.

Furth, J., Moy, P., Schalch, D. S., and Ueda, G. (1973b). Gonadotropic activities of thyrotropic tumors: Demonstration by immunohistochemical staining. *Proc. Soc. Exp. Biol. Med.* **142,** 1180–1184.

Gardner, W. U. (1941). The effect of estrogen on the incidence of mammary and pituitary tumors of hybrid mice. *Cancer Res.* **1,** 345–358.

Gardner, W. U., Pfeiffer, C. A., Trentin, J. J., and Wolstenholme, J. T. (1953). Hormonal factors in experimental carcinogenesis. *In* "The Physiopathology of Cancer" (F. Homburger and W. H. Fishman, eds.), pp. 225–297. Harper (Hoeber), New York.

Goluboff, L. C., MacRae, M. E., Ezrin, C., and Sellers, A. E. (1970). Autoradiography of tritiated thymidine labeled anterior pituitary cells in propylthiouracil treated rats. *Endocrinology* **87**, 1113–1118.

Gourdji, D., Kerdelhué, B., and Tixier-Vidal, A. (1972). Ultrastructure d'un clone de cellules hypophysaires secretant de la prolactine (clone GH 3). Modifications induites par l'hormone hypothalamique de libération de l'hormone thyréotrope (TRF). *C. R. Acad. Sci., Ser. D* **274**, 437–440.

Griesbach, W. E., and Purves, H. D. (1960). Basophil adenomata in the rat hypophysis after gonadectomy. *Brit. J. Cancer* **14**, 49–59.

Gusek, W. (1962). Vergleichende licht und elektronenmikroskopische Untersuchungen menschlicher Hypophyseadenome bei Akromegalie. *Endokrinologie* **42**, 257–283.

Guyda, H., Robert, F., Colle, E., and Hardy, J. (1973). Histologic, ultrastructural and hormonal characterization of a pituitary tumor secreting both hGH and prolactin. *J. Clin. Endocrinol. Metab.* **36**, 531–547.

Hagen, T. C., Lawrence, A. M., and Kirsteins, L. (1971). *In vitro* release of monkey pituitary growth hormone by acromegalic plasma. *J. Clin. Endocrinol. Metab.* **33**, 448–451.

Haguenau, F., and Bernhard, W. (1955). L'appareil de Golgi dans les cellules normales et cancéreuses de Vertébrés. Rappel historique et étude au microscope électronique. *Arch. Anat. Microsc. Morphol. Exp.* **44**, 27–55.

Haguenau, F., and Lacour, F. (1955). Cytologie électronique de tumeurs hypophysaires expérimentales; leur appareil de Golgi. *In* "Symposium on the Fine Structure of Cells" (P. Noordhoff, ed.), pp. 316–322. Wiley (Interscience), New York.

Halmi, N. S., and Gude, W. D. (1954). The morphogenesis of pituitary tumors induced by radiothyroidectomy in the mouse and the effects of their transplantation on the pituitary body of the host. *Amer. J. Pathol.* **30**, 403–419.

Herlant, M., and Pasteels, J. L. (1967). Histophysiology of human anterior pituitary. *Methods Achiev. Exp. Pathol.* **3**, 250–305.

Herlant, M., and Pasteels, J. L. (1971). Les cellules productrices de prolactine, la régulation de leur sécrétion et leur rôle dans les adénomes sécrétant la prolactine. *In* "Les adénomes hypophysaires sécrétants. Endocrinopathies et Immunologie," pp. 21–41. Masson, Paris.

Herlant, M., Laine, E., Fossati, P., and Linquette, M. (1965). Syndrome aménorrhée-galactorrhée par adénome hypophysaire à cellules à prolactine. *Ann. Endocrinol.* **26**, 65–71.

Herlant, M., Linquette, M., Laine, E., Fossati, P., May, J. P., and Lefebvre, J. (1966). Adénome hypophysaire à cellules thyréotropes, avec syndrome aménorrhée-galactorrhée, chez une malade porteuse d'un myxoedème congénital par ectopie thyroïdienne. *Ann. Endocrinol.* **27**, 181–198.

Hirano, A., Tomiyasu, U., and Zimmerman, H. M. (1972). The fine structure of blood vessels in chromophobe adenomas. *Acta Neuropathol.* **22**, 200–207.

Hunt, T. E. (1942). Mitotic activity in the anterior hypophysis of female rats. *Anat. Rec.* **82**, 263–276.

Hunt, T. E., and Hunt, E. A. (1966). A radioautographic study of the proliferative activity of adrenocortical and hypophyseal cells of the rat at different periods of the estrous cycle. *Anat. Rec.* **156**, 361–367.

Hymer, W. C., McShan, W. H., and Christiansen, R. G. (1961). Electron microscopic studies of anterior pituitary glands from lactating and estrogen treated rats. *Endocrinology* **69**, 81–90.

Ishikawa, H., Watanabe, T., and Yoshimura, F. (1971). ACTH, GH, Prolactin and α-MSH activities of six kinds of cells isolated from the rat adenohypophysis. *Endocrinol. Jap.* **18**, 223–226.

Jackson, I. (1965). Hyperthyroidism in a patient with a pituitary chromophobe adenoma. *J. Clin. Endocrinol. Metab.* **25**, 491–494.

Jacobs, L. S., and Daughaday, W. H. (1973). Pathophysiology and control of prolactin secretion in patients with pituitary and hypothalamic disease. *In* "Human Prolactin" (J. L. Pasteels and C. Robyn, eds.), pp. 84–97. Excerpta Med. Found., Amsterdam.

Kamat, V. B., Wallach, D. F. H., Crigler, J. F., and Ladman, A. J. (1960). The intracellular localization of hormonal activity in transplantable thyrotropin-secreting pituitary tumors in mice. *J. Biophys. Biochem. Cytol.* **7**, 219–225.

Kernohan, J. W., and Sayre, G. P. (1956). Tumors of the pituitary gland and infundibulum. *In* "Atlas of Tumor Pathology," Vol. X, Part 36, pp. 1–81. Armed Forces Inst. Pathol., Washington, D. C.

Klotz, H. P., Racadot, J., and Olivier, L. (1964). Apport des nouvelles données morphologiques à la connaissance des adénomes hypophysaires dans l'acromégalie. *Sem. Hop.* **55**, 3025–3031.

Kohler, P. O., Bridson, W. E., Rayford, P. L., and Kohler, S. E. (1969). Adenomas in culture. *Metab., Clin. Exp.* **18**, 782–788.

Kovacs, K., and Horvath, E. (1973). Pituitary "chromophobe" adenoma composed of oncocytes. A light and electron microscopic study. *Arch. Pathol.* **95**, 235–239.

Kracht, J., and Hachmeister, U. (1967). Immunohistological studies on Crooke cells. *Endokrinologie* **51**, 164–169.

Kraus, E. J. (1926). Die Hypophyse. *In* "Handbuch der speziellen pathologischen Anatomie und Histologie" (F. Henke and O. Lubarsch, eds.) Vol. 8, pp. 810–950. Springer-Verlag, Berlin and New York.

Kuromatsu, C. (1968). The fine structure of the human pituitary chromophobe adenoma with special reference to the classification of this tumor. *Arch. Histol. Jap.* **29**, 41–61.

Kwa, H. G. (1961). "An Experimental Study of Pituitary Tumors. Genesis, Cytology and Hormone Content." Springer-Verlag, Berlin and New York.

Kwa, H. G., and Feltkamp, C. A. (1965). Transformation of pituitary tumours from prolactin into TSH cell types. II. Endocrine aspects. *Acta Endocrinol. (Copenhagen), Suppl.* **100**, 162.

Lacassagne, A. (1959). Sur les cancers glandulaires d'origine hormonale. *Biol. Med. (Paris)* **48**, 1–20.

Lacour, F. (1950). Recherches sur la relation entre les cellules hypophysaires à granulations orangées (cellules de Romeis) et les phénomènes de lactation. *C. R. Soc. Biol.* **144**, 248–249.

Lamberg, B. A., Ripatti, J., Gordin, A., Juustila, H., Sivula, A., and Björkesten, G. (1969). Chromophobe pituitary adenoma with acromegaly and TSH induced hyperthyroidism associated with parathyroid adenoma. *Acta Endocrinol. (Copenhagen)* **60**, 157–172.

Landolt, A. M., and Oswald, U. W. (1973). Histology and ultrastructure of an oncocytic adenoma of the human pituitary. *Cancer* **31**, 1099–1105.

Le Beau, J., and Foncin, F. (1972). Contribution à l'étude ultrastructurale des adénomes à prolactine. *Ann. Endocrinol.* **33**, 353–356.

Leblond, C. P., and Walker, B. E. (1956). Renewal of cell populations. *Physiol. Rev.* **36**, 255–276.

Leleux, P., and Robyn, C. (1971). Immunohistochemistry of individual adenohypophysial cells. *Acta Endocrinol. (Copenhagen) Suppl. 153,* 168–189.

Levan, A. (1969). Chromosome abnormalities and carcinogenesis. *In* "Handbook of Molecular Cytology" (A. Lima-de-Faria, ed.), pp. 717–731. North-Holland Publ., Amsterdam.

Lewis, P. D., and Van Noorden, S. (1972). Pituitary abnormalities in acromegaly. *Arch. Pathol.* **94**, 119–126.

Leznoff, A., Fishman, J., Talbot, M., McGarry, E. E., Beck, J. C., and Rose, B. (1962). The cytological localization of ACTH in the human pituitary. *J. Clin. Invest.* **41**, 1720–1724.

Linquette, M., Herlant, M., Laine, E., Fossati, P., and Dupont-Lecompte, J. (1967). Adénome à prolactine chez une jeune fille dont la mère était porteuse d'un adénome hypophysaire avec aménorrhée-galactorrhée. *Ann. Endocrinol.* **28**, 773–780.

Linquette, M., Herlant, M., Fossati, P., May, J. P., Decoulx, M., and Fourlinnie, J. C. (1969). Adénome hypophysaire à cellules thyréotropes avec hyperthyroïdie. *Ann. Endocr.* **30**, 731–740.

Linquette, M., Fossati, P., Decoulx, M., Lefebvre, J., and Laine, E. (1971). Adénomes thyréotropes. *In* "Les adénomes hypophysaires secrétants. Endocrinopathies et immunologie," pp. 155–175. Masson, Paris.

Lundin, P. M., and Schelin, U. (1962). Light and electron microscopical studies on the pituitary in stilbol-treated rats. *Acta Pathol. Microbiol. Scand.* **54**, 66–74.

Lundin, P. M., and Schelin, U. (1964). Light and electron microscopic studies on thyrotrophic pituitary adenomas in the mouse. *Lab. Invest.* **13**, 62–68.

Luse, S. A. (1962). Electron microscopy of brain tumors. *In* "Biology and Treatment of Intracranial Tumors" (W. S. Field and P. C. Sharkey, eds.), pp. 73–103. Thomas, Springfield, Illinois.

McCormick, R. V., Reed, C., Murray, R., and Ray, B. (1951). Coexisting acromegaly and Cushing's syndrome; discussion of hormone production by pituitary acidophilic cell. *Amer. J. Med.* **10**, 662–670.

McCormick, W. F., and Halmi, N. S. (1971). Absence of chromophobe adenomas from a large serie of pituitary tumors. *Arch. Pathol.* **92**, 231–238.

McEuen, C. S., Selye, H., and Collip, J. B. (1936). Some effects of prolonged administration of oestrin in rats. *Lancet* **1**, 775–776.

Mahesh, V. B., Dalla Pria, S., and Greenblatt, R. B. (1969). Abnormal lactation with Cushing's syndrome. A case report. *J. Clin. Endocrinol. Metab.* **29**, 978–981.

Mark, J. (1971). Chromosomal characterization of human pituitary adenomas. *Acta Neuropathol.* **19**, 99–109.

Mastro, A., Hymer, W. C., and Therrien, C. D. (1969a). DNA synthesis in adult rat anterior pituitary glands in organ culture. *Expl. Cell Res.* **54**, 407–414.

Mastro, A., Shelton, E., and Hymer, W. C. (1969b). DNA synthesis in the rat anterior pituitary. An electron microscope radioautographic study. *J. Cell Biol.* **43**, 626–629.

Mautalen, C. A., and Mellinger, R. C. (1965). Non suppressible adrenocortical function in a patient with untreated acromegaly. *J. Clin. Endocrinol. Metab.* **25**, 1423–1428.

Meneghelli, V., Benedetti, A., and Mazzocchi, G. (1964). Studio al microscopio elettronico sull' adenoma cromofobo ipofisario dell'uomo. *Arch. de Vecchi* **44**, 893–914.

Messier, B. (1966). Changes in the number of mouse thyrotropes following radiothyroidectomy. *Acta Endocrinol. (Copenhagen)* **52**, 391–398.

Messier, B. (1969). Effect of exogenous thyroxine on ^{3}H thymidine uptake in mouse pituitary gland. *Acta Endocrinol. (Copenhagen)* **61**, 133–136.

Messier, B., and Furth, J. (1962). A reversely responsive variant of a thyrotropic tumor with gonadotropic activity. *Cancer Res.* **22**, 804–808.

Montgomery, D. A. D. (1963). Pituitary tumors in Cushing's syndrome. *Quart. J. Med.* [N.S.] **32**, 365–366 (abstr.).

Müller, W. (1954). Zur Frage der hypophysären Tumoren vom Mischtyp. *Acta Neuroveg.* **8**, 451–465.

Murakami, M., Yoshida, T., Nakayama, Y., Hashimoto, J., and Hirata, S. (1968). The fine structure of the pars intermedia of the pituitary in the human foetus. *Arch. Histol. Jap.* **30**, 61–73.

Nelson, D. H., and Sprunt, J. G. (1965). Pituitary tumors post-adrenalectomy for Cushing's syndrome. *Proc. Int. Congr. Endocrinol., 2nd 1964* Int. Congr. Ser. No. 83, pp. 1053–1057.

Nouët, J. C., and Kujas, M. (1973). Influence de la méthode de fixation sur l'observation des divisions cellulaires dans l'adénohypophyse du rat mâle. *Z. Zellforsch. Mikrosk. Anat.* **143**, 535–547.

Nyhan, W. L., and Green, M. (1964). Hyperthyroidism in a patient with a pituitary adenoma. *J. Pediat.* **65**, 583–589.

Ohtsuka, Y., Ishikawa, H., Omoto, T., Takasaki, Y., and Yoshimura, F. (1971). Effect of CRF on the morphological and functional differenciation of the cultured chromophobes isolated from rat anterior pituitaries. *Endocrinol. Jap.* **18**, 133–153.

Oliva, H., Navarro, V., and Obrador, S. (1966). Microscopia electronica de los adenomas cromofobos de la hipofisis. *Acta Neurochir.* **14**, 141–153.

Olivier, L., Porcile, E., de Brye, C., and Racadot, J. (1965). Etude de quelques adénomes hypophysaires chez l'homme en microscopie électronique. *Bull. Ass. Anat. (Nancy)* **127**, 1258–1265.

Olivier, L., Vila-Porcile, E., Peillon, F., and Racadot, J. (1972). Etude en microscopie électronique des grains de sécrétion "basophiles" dans les cellules hypophysaires tumorales de la maladie de Cushing. *C. R. Soc. Biol.* **166**, 1591–1595.

Olivier, L., Vila-Porcile, E., Peillon, F., and Racadot, J. (1973a). Crinophagie dans les cellules de Crooke de l'hypophyse humaine. *J. Microsc. (Paris)* **17**, 84a.

Olivier, L., Vila-Porcile, E., Racadot, O., Peillon, F., and Racadot, J. (1973b). Personal observations.

Paiz, C., and Hennigar, G. R. (1970). Electron microscopy and histochemical correlation of human anterior pituitary cells. *Amer. J. Pathol.* **59**, 43–52.

Pantic, V., Ozegovic, B., Genbacev, O., and Milkovic, S. (1971). Ultrastructure of transplantable pituitary tumor cells producing luteotropic and adrenocorticotropic hormones. *J. Microsc. (Paris)* **12**, 225–232.

Peake, G. T., McKeel, D. W., Jarett, L., and Daughaday, W. H. (1969). Ultrastructural, histologic and hormonal characterization of a prolactin-rich human pituitary tumor. *J. Clin. Endocrinol. Metab.* **29**, 1383–1393.

Pearse, A. G. E. (1952). Observations on the localization, nature and chemical constitution of some components of the anterior hypophysis. *J. Pathol. Bacteriol.* **64**, 791–809.

Pearse, A. G. E. (1962). Cytology and cytochemistry of adenomas of the human hypophysis. *Acta Unio Int. Contra Cancrum* **18**, 302–304.

Pearse, A. G. E., and Van Noorden, S. (1963). The functional cytology of the human adenohypophysis. *Can. Med. Ass. J.* **88**, 462–471.

Peillon, F., Vila-Porcile, E., Olivier, L., and Racadot, J. (1970). L'action des oestrogènes sur les adénomes hypophysaires chez l'homme. Documents histopathologiques en microscopie optique et électronique et apport de l'expérimentation. *Ann. Endocrinol.* **31**, 259–270.

Peillon, F., Gourmelen, M., Brandi, A. M., and Donnadieu, M. (1972). Adénomes somatotropes humains en culture organotypique, ultrastructure et sécrétion étudiée à l'aide de leucine tritiée. *C. R. Acad. Sci.* **275**, 2251–2254.

Pelletier, G., Peillon, F., Pham Hun Trung, M. T., and Racadot, J. (1971). Etude de la morphologie et de la sécrétion d'une tumeur corticotrope expérimentale du rat. *Rev. Eur. Etud. Clin. Biol.* **16**, 79–83.

Phifer, R. F., Spicer, S. S., and Orth, D. N. (1970). Specific demonstrations of the human hypophyseal cells which produce adrenocorticotrophic hormone. *J. Clin. Endocrinol. Metab.* **31**, 347–361.

Phifer, R. F., Spicer, S. S., and Hennigar, G. R. (1973). Histochemical reactivity and staining properties of functionally defined cell types in the human adenohypophysis. *Amer. J. Pathol.* **73**, 569–587.

Porcile, E., and Racadot, J. (1966). Ultrastructure des cellules de Crooke observées dans l'hypophyse humaine au cours de la maladie de Cushing. *C. R. Acad. Sci.* **263**, 948–951.

Porcile, E., Olivier, L., and Racadot, J. (1968). Identification des types de grains de sécrétion de l'adénohypophyse au microscope électronique par la méthode acide chromique-phosphotungstique. *J. Microsc. (Paris)* **7**, 51a–52a (abstr.).

Potvliège, P. R. (1968). Effects of estrogen on pituitary morphology in goitrogen treated rats. An electron microscopic study. *Anat. Rec.* **160**, 595–606.

Purves, H. D. (1966). Cytology of the adenohypophysis. *In* "The Pituitary Gland" (G. W. Harris and B. T. Donovan, eds.), Vol. I, pp. 147–232. Univ. of California Press, Berkeley.

Racadot, J. (1949). La différenciation des cellules pituitaires au cours du développement. *Arch. Anat. Microsc. Morphol. Exp.* **38**, 318–352.

Racadot, J., Olivier, L., Porcile, E., de Brye, C., and Klotz, H. P. (1964). Adénome hypophysaire de type "mixte" avec symptomatologie acromégalique. II. Etude au microscope optique et au microscope électronique. *Ann. Endocrinol.* **25**, 503–507.

Racadot, J., Peillon, F., Decourt, J., and Gilbert-Dreyfus (1966a). Les problèmes de la nature des adénomes hypophysaires dans la maladie de Cushing. A propos de deux observations. *Sem. Hop.* **42**, 469–487.

Racadot, J., Peillon, F., Decourt, J., and Gilbert-Dreyfus (1966b). Particularités histologiques de l'adénohypophyse (présence d'adénomes à cellules à MSH) dans deux cas de maladie de Cushing. *Ann. Endocrinol.* **27**, 59–64.

Racadot, J., Girard, F., Peillon, F., and Binoux, M. (1967). Contribution à la physiopathologie de la maladie de Cushing. Histologie de l'hypophyse. Etude de la régulation hypophyso-surrénalienne. *In* "Le syndrome de Cushing," pp. 1–59. Masson, Paris.

Racadot, J., Vila-Porcile, E., Peillon, F., and Olivier, L. (1971). Adénomes hypophysaires à cellules à prolactine: Etude structurale et ultrastructurale, corrélations anatomocliniques. *Ann. Endocrinol.* **32**, 298–305.

Racadot, J., Vila-Porcile, E., Olivier, L., and Peillon, F. (1974). Ultrastructure of pituitary tumors. *In* "Progress in Neurological Surgery" (H. Krayenbühl, ed.), Vol. VI. Karger, Basel (in press).

Rambourg, A. (1967). Détection des glycoprotéines en microscopie électronique. Coloration de la surface cellulaire et de l'appareil de Golgi par un mélange acide-chromique-phosphotungstique. *C. R. Acad. Sci.* **265**, 1426–1428.

Rambourg, A. (1969). Localisation ultrastructurale et nature du matériel coloré au niveau de la surface cellulaire par le mélange chromique-phosphotungstique. *J. Microsc. (Paris)* **8**, 325–342.

Robyn, C., Leleux, P., Vanhaelst, L., Golstein, J., Herlant, M., and Pasteels, J. L. (1973). Immunohistochemical study of the human pituitary with anti-luteinizing hormone, anti-follicle stimulating hormone and anti-thyrotrophin sera. *Acta. Endocrinol. (Copenhagen)* **72**, 625–642.

Romeis, B. (1940). Hypophyse. *In* "Handbuch der mikroskopischen Anatomie des Menschen" (W. von Möllendorff, ed.), Vol. 6, Part 3. Springer-Verlag, Berlin and New York.

Roussy, G., and Clunet, J. (1911). Les tumeurs du lobe anterieur de l'hypophyse. Essai de classification histologique. *Rev. Neurol.* **17**, 313–320.

Roussy, G., and Oberling, C. (1933). Contribution à l'étude des tumeurs hypophysaires. *Presse Med.* **92**, 1799–1804.

Russfield, A. B. (1958). Hypophyseal changes in hypothyroidism induced by radioactive iodine in man. *Arch. Pathol.* **66**, 79–88.

Russfield, A. B. (1966). Tumors of endocrine glands and secondary sex organs. *U. S. Pub. Health Serv., Publ.* **1332**, 1–24.

Russfield, A. B. (1968). Adenohypophysis. *In* "Endocrine Pathology" (M. B. Bloodworth, ed.), pp. 75–116. Williams & Wilkins, Baltimore, Maryland.

Salazar, H. (1968). Ultrastructural evidence for the existence of a nonsecretory, sustentacular cell in the human adenohypophysis. *Anat. Rec.* **160**, 419–420 (abstr.).

Sano, M., and Sasaki, F. (1969). Embryonic development of the mouse anterior pituitary studied by light and electron microscopy. *Z. Anat. Entwicklungsgesch.* **129**, 195–222.

Schechter, J. (1972). Ultrastructural changes in the capillary bed of human pituitary tumors. *Amer. J. Pathol.* **67**, 109–120.

Schechter, J. (1973). Electron microscopic studies of human pituitary tumors. I. Chromophobic adenomas. *Amer. J. Anat.* **138**, 371–386.

Schelin, U. (1962). Chromophobe and acidophil adenomas of the human pituitary gland. A light and electron microscopic study. *Acta Pathol. Microbiol. Scand., Suppl.* **158**, 1–80.

Schelin, U. (1969). Effects of simultaneous thyroidectomy and oestrone treatment on the pituitary cytology in the mouse. A light and electron microscopic study. *Acta Pathol. Microbiol. Scand.* **75**, 537–544.

Schelin, U., and Lundin, P. M. (1971). An electron microscopic study of normal and neoplasic acidophil cells of the rat pituitary. *Acta Endocrinol. (Copenhagen)* **67**, 29–39.

Schelin, U., Lundin, P. M., and Bartholdson, L. (1964). Light and electron microscopic studies on an autonomous stilbestrol-induced pituitary tumor in rats. *Endocrinology* **75**, 893–900.

Schochet, S. S., McCormick, W. F., and Halmi, N. S. (1972). Acidophil adenomas with intracytoplasmic filamentous aggregates. A light and electron microscopic study. *Arch. Pathol.* **94**, 17–22.

Setaló, G., and Nakane, P. K. (1972). Studies on the functional differenciation of cells in foetal anterior pituitary glands of rats with peroxidase-labeled antibody method. *Anat. Rec.* **172**, 403–404.

Severinghaus, A. E. (1933). A cytological study of the anterior pituitary of the rat, with special reference to the Golgi apparatus and to the cell relationship. *Anat. Rec.* **57**, 149–175.

Severinghaus, A. E. (1937). Cellular changes in the anterior hypophysis with special reference to its secretory activity. *Physiol. Rev.* **17**, 556–588.

Shiino, M., Williams, M. G., and Rennels, E. G. (1973). Thyroidectomy cells and their response to thyrotrophin releasing hormone (TRH) in the rat. *Z. Zellforsch. Mikrosk. Anat.* **138**, 327–332.

Shimazaki, M., Ueda, G., Ito, M., Mukobayashi, H., and Shirakawa, J. (1962). Electron microscopic studies of the estrogen-induced pituitary tumors. *Wakayama Med. Rep.* **7**, 1–6.

Siler, T. M., Morgenstern, L. L., and Greenwood, F. C. (1972). The release of prolactin and other peptide hormones from anterior pituitary tissue cultures. *Lactogenic Horm., Ciba Found. Symp.* pp. 207–222.

Sinha, D. K., and Meites, J. (1967). Direct effects of a hypothalamic extract on hormone secretion by a pituitary mammo-somatotropic tumor. *Endocrinology* **80**, 131–134.

Steelman, S. L., Kelly, T. L., Norgello, H., and Weber, G. F. (1956). Occurrence of melanocyte stimulating hormone (MSH) in a transplantable pituitary tumor. *Proc. Soc. Exp. Biol. Med.* **92**, 392–394.

Stokes, H., and Boda, J. M. (1968). Immunofluorescent localization of growth hormone and prolactin in the adenohypophysis of fetal sheep. *Endocrinology* 83, 1362–1366.

Stratmann, I. E., Ezrin, C., Sellers, E. A., and Simon, G. T. (1972). The origin of thyroidectomy cells as revealed by high resolution radioautography. *Endocrinology* 90, 728–734.

Takemoto, H., Yokoro, K., Furth, J., and Cohen, A. I. (1962). Adrenotropic activity of mammo-somatotropic tumors in rats and mice. I. Biologic aspects. *Cancer Res.* 22, 917–924.

Tashjian, A. H., Yasumura, Y., Levine, L., Sato, G. H., and Parker, M. L. (1968). Establishment of clonal strains of rat pituitary tumor cells that secrete growth hormone. *Endocrinology* 82, 342–352.

Tashjian, A. H., Bancroft, F. C., and Levine, L. (1970). Production of both prolactin and growth hormone by clonal strains of rat pituitary tumor cells. Differential effects of hydrocortisone and tissue extracts. *J. Cell Biol.* 47, 61–70.

Tashjian, A. H., Barowsky, N. J., and Jensen, D. K. (1971). Thyrotropin releasing hormone: Direct evidence for stimulation of prolactin production by pituitary cells in culture. *Biochem. Biophys. Res. Commun.* 43, 516–523.

Théret, C., and Renault, H. (1964). L'ultrastructure de tumeurs thyréotropes adénohypophysaires expérimentales après radiothyroïdectomie. *Bull. Cancer* 51, 505–534.

Théret, C., and Renault, H. (1965). Orientation gonadotrope de tumeurs adénohypophysaires-après radiothyroïdectomie par inhibition chlorpromazinique de l'hypothalamus. *Bull. Cancer* 52, 279–302.

Tomiyasu, U., Hirano, A., and Zimmerman, H. M. (1973). Fine structure of human pituitary adenomas. *Arch. Pathol.* 95, 287–292.

Ueda, G. (1971). An electron microscopic study of a transplantable thyrotropic pituitary tumor in mice. *Endocrinol. Jap.* 18, 27–35.

Ueda, G., and Mori, T. (1967). Astatine-211-induced transplantable pituitary tumor in the rat with a brief analytic review of cell types of pituitary tumors. *Amer. J. Pathol.* 51, 601–619.

Vila-Porcile, E., Olivier, L., and Racadot, O. (1971). Cellules folliculaires du lobe antérieur de l'hypophyse humaine. *Bull. Ass. Anat.* (Nancy) 152, 813 (abstr.).

von Lawzewitsch, I., Dickmann, G. H., Amezùa, L., and Pardal, C. (1972). Cytological and ultrastructural characterization of the human pituitary. *Acta Anat.* 81, 286–316.

von Wechsler, W., and Hossmann, K. A. (1965). Elektronenmikroskopische untersuchungen chromophober Hypophysen-Adenome des Menschen. *Zentralbl. Neurochir.* 26, 105–122.

Waelbroeck-Van Gaver, C. (1969). Tumeurs hypophysaires induites par les oestrogènes chez le rat. II. Etude cytogénétique. *Eur. J. Cancer* 5, 119–127.

Waelbroeck-Van Gaver, C., and Potvliège, P. (1969). Tumeurs hypophysaires induites par les oestrogènes chez le rat. I. Activité fonctionnelle, histologie et ultrastructure. *Eur. J. Cancer* 5, 99–117.

Wagermark, J., and Wersall, J. (1968). Ultrastructural features of Crooke's changes in pituitary basophil cells. *Acta Pathol. Microbiol. Scand.* 72, 367–373.

Watari, N., and Tsukagoshi, N. (1969). Electron microscopic observations on the estrogen-induced pituitary tumors. *Gunma Symp. Endocrinol.* 6, 297–314.

Werner, S. C., and Stewart, W. B. (1958). Hyperthyroidism in a patient with a pituitary chromophobe adenoma and a fragment of normal pituitary. *J. Clin. Endocrinol. Metab.* 18, 266–270.

Wolfe, J. M., and Wright, A. W. (1938). Histological effects induced in the anterior pituitary of rats by prolonged injection of estrin with particular reference to the production of pituitary adenomata. *Endocrinology* 23, 200–210.

Yoshida, Y. (1966). Electron microscopy of the anterior pituitary gland under normal and different experimental conditions. *Methods Achiev. Exp. Pathol.* **1**, 439–454.

Yoshimura, F., and Harumiya, K. (1965). Electron microscopy of anterior lobe of pituitary in normal and castrated rats. *Endocrinol. Jap.* **12**, 119–152.

Yoshimura, F., Harumiya, K., Ishikawa, H., and Ohtsuka, Y. (1969). Differenciation of isolated chromophobes into acidophils or basophils when transplanted into the hypophysiotrophic area of hypothalamus. *Endocrinol. Jap.* **16**, 531–540.

Yoshimura, F., Harumiya, K., and Kiyama, H. (1970). Light and electron microscopic studies of the cytogenesis of anterior pituitary cells in perinatal rats in reference to the development of target organs. *Arch. Histol. Jap.* **31**, 333–369.

Young, D. G., Bahn, R. C., and Randall, R. V. (1965). Pituitary tumors associated with acromegaly. *J. Clin. Endocrinol. Metab.* **25**, 249–259.

Young, R. L., Bradley, E. M., Goldzieher, J. W., Myers, P. W., and Lecocq, F. R. (1967). Spectrum of non puerperal galactorrhea: Report of two cases evolving through the various syndromes. *J. Clin. Endocrinol. Metab.* **27**, 461–466.

Zambrano, D., Amezua, L., Dickmann, G., and Franke, E. (1968). Ultrastructure of human pituitary adenomata. *Acta Neurochir.* **18**, 78–94.

Zondek, B. (1936). Tumours of the pituitary induced with follicular hormone. *Lancet* **1**, 776–778.

Zondek, B. (1938). Hypophyseal tumors induced by estrogenic hormone. *Amer. J. Cancer* **33**, 555–559.

SUBJECT INDEX

A 5
B 6
C 7
D 8
E 9
F 0
G 1
H 2
I 3
J 4